S0-CID-549

MOLECULAR ASPECTS OF DEVELOPMENT AND AGING OF THE NERVOUS SYSTEM

ADVANCES IN EXPERIMENTAL MEDICINE AND BIOLOGY

Recent Volumes in this Series

Volume 257
THE IMMUNE RESPONSE TO VIRAL INFECTIONS
Edited by B. A. Askonas, B. Moss, G. Torrigiani, and S. Gorini

Volume 258
COPPER BIOAVAILABILITY AND METABOLISM
Edited by Constance Kies

Volume 259
RENAL EICOSANOIDS
Edited by Michael J. Dunn, Carlo Patrono, and Giulio A. Cinotti

Volume 260
NEW PERSPECTIVES IN HEMODIALYSIS, PERITONEAL DIALYSIS, ARTERIOVENOUS HEMOFILTRATION, AND PLASMAPHERESIS
Edited by W. H. Hörl and P. J. Schollmeyer

Volume 261
CONTROL OF THE THYROID GLAND: Regulation of Its Normal Function and Growth
Edited by Ragnar Ekholm, Leonard D. Kohn, and Seymour H. Wollman

Volume 262
ANTIOXIDANT NUTRIENTS AND IMMUNE FUNCTIONS
Edited by Adrianne Bendich, Marshall Phillips, and Robert P. Tengerdy

Volume 263
RAPID METHODS IN CLINICAL MICROBIOLOGY: Present Status and Future Trends
Edited by Bruce Kleger, Donald Jungkind, Eileen Hinks, and Linda A. Miller

Volume 264
ANTIOXIDANTS IN THERAPY AND PREVENTIVE MEDICINE
Edited by Ingrid Emerit, Lester Packer, and Christian Auclair

Volume 265
MOLECULAR ASPECTS OF DEVELOPMENT AND AGING OF THE NERVOUS SYSTEM
Edited by Jean M. Lauder, Alain Privat, Ezio Giacobini, Paola S. Timiras, and Antonia Vernadakis

---

A Continuation Order Plan is available for this series. A continuation order will bring delivery of each new volume immediately upon publication. Volumes are billed only upon actual shipment. For further information please contact the publisher.

MOLECULAR ASPECTS OF DEVELOPMENT AND AGING OF THE NERVOUS SYSTEM

Edited by

Jean M. Lauder
The University of North Carolina School of Medicine
Chapel Hill, North Carolina

Alain Privat
Institute of Biology, INSERM
Montpellier, France

Ezio Giacobini
Southern Illinois University School of Medicine
Springfield, Illinois

Paola S. Timiras
University of California
Berkeley, California

and

Antonia Vernadakis
University of Colorado School of Medicine
Denver, Colorado

PLENUM PRESS • NEW YORK AND LONDON

Library of Congress Cataloging in Publication Data

Molecular aspects of development and aging of the nervous system / edited by Jean M. Lauder . . . [et al.].
p. cm. – (Advances in experimental medicine and biology; v. 265)
Proceedings of a conference held by the Institute of Developmental Neuroscience and Aging, June 15–18, 1988, in Athens, Greece.
Includes bibliographical references.
ISBN 0-306-43408-3
1. Nervous system – Aging – Molecular aspects – Congresses. 2. Nervous system – Growth – Molecular aspects – Congresses. I. Lauder, Jean M. II. Institute of Developmental Neuroscience & Aging. III. Series.
[DNLM: 1. Aging – congresses. 2. Nervous System – growth & development – congresses. 3. Nervous System – physiology – congresses. WD1 AD559 v. 265 / WL 102 M718 1988]
QP356.2.M63 1990
612.8 – dc20
DNLM/DLC 89-70995
for Library of Congress CIP

Proceedings of the Second Conference of the Institute of Developmental Neuroscience and Aging, on Molecular Aspects of Development and Aging of the Nervous System, held June 15–18, 1988, in Athens, Greece

Printed in the United States of America

INTRODUCTION

The rapidly expanding fields of molecular and cellular neurobiology are the newest frontiers of neuroscience. This book represents the continuing efforts of the Institute of Developmental Neuroscience and Aging (IDNA) to disseminate the most recent advances on the developing and aging nervous system at the molecular and cellular levels. A group of neuroscientists presented and discussed their findings at a recent IDNA conference held in Athens, Greece, June 15-18, 1988. This meeting was sponsored by the National Hellenic Research Foundation, FIDIA, the Ministry of Research and Technology, the Tourism Organization of Greece, and the National Institute of Child Health and Human Development, NIH. The Directors of the IDNA are grateful to the local committee, Drs. Eleni Fleischer, Costas Sekeris, Michael Alexis, Theony Valcana, and Elias Kouvelas, for their efforts in organizing this meeting and for their successful integration of science and culture for the participants.

This volume provides a comprehensive overview of the information presented at this conference, including in-depth discussions of each topic by the participants. The chapters are grouped into five general categories which correspond to the subject areas covered during the meeting. These include: Gene and Phenotypic Expression, Growth Factors and Oncogenes, Cytoskeletal and Extracellular Molecules, Neurotransmitters and Hormones, and Molecular Aspects of Aging and Alzheimer's Disease.

The section on **Gene and Phenotypic Expression** includes discussions of transient gene expression in the nervous system (Herschman), developmental regulation of myelin-associated genes (Gordon et al.), molecular analyses of glial scarring (Rataboul et al.) and glial fibrillary protein (GFAP) synthesis (Tardy and Nunez) using GFAP cDNA probes, and cell lineage studies of neural crest development (Baroffio et al.).

Growth Factors and Oncogenes includes information concerning trophic factors stored and secreted by neurons (Unsicker et al.), neuropeptides as neuronal growth regulatory factors (Davila-Garcia et al.), effects of basic fibroblast growth factor on cultured mesencephalic neurons (Ferrari et al.), role of trophic factors in neuronal aging (Perez-Polo), localization of the SRC oncogene in neuronal growth cones (Maness et al.), the role of RAS oncogenes in carcinogenesis and differentiation (Spandidos and Anderson), neuronal expression of nerve growth factor precursor-like immunoreactivity (Senut et al.), and development of the NILE glycoprotein (Batistatou and Kouvelas) in the CNS.

Chapters included in **Cytoskeletal and Extracellular Molecules** discuss the influences of the cytoskeleton (Adler and Madreperla), extracellular matrix (Rome et al.), proteases (Seeds et al.), gangliosides (Skaper et al.) and neural cell adhesion molecules (Nybroe and Bock) on myelinogenesis, neuronal differentiation, migration and plasticity.

Neurotansmitters and Hormones contains chapters on the developmental expression and roles played by neurotransmitters and hormones during the ontogeny of neural and non-neural tissues. The first chapters discuss neurotransmitter control of morphogenesis (Lindemann Shuey et al.) and neuronal growth cone activity (Kater and Mills), followed by a discussion of how cyclic AMP can substitute for neurotransmitters in controlling glial differentiation (Hertz). The remaining chapters describe the ontogeny of adrenergic receptors in the spinal cord (Bernstein-Goral and Bohn), early nutritional influences on the developing serotonergic system (Hernandez), and ontogeny of the glucocorticoid receptor in the CNS (Alexis et al.).

The book concludes with a section on **Molecular Aspects of Aging and Alzheimer's Disease** including molecular genetic approaches to therapy (Giacobini) and research on Alzheimer's disease (Neve, Saitoh et al., Fisher and Oster-Granite), and studies of the role of cytoskeletal proteins in aging of the nervous system (Sternberg et al.).

The editors are grateful to the scientific contributors to this volume, and to the local organizing committee and contributors of financial support for making both the conference and this book a success.

The Editors

CONTENTS

GENE AND PHENOTYPIC EXPRESSION

GROWTH FACTORS AND ONCOGENES

TRANSIENT INDUCTION OF GENE EXPRESSION IN THE NERVOUS SYSTEM IN RESPONSE TO EXTRACELLULAR SIGNALS

Harvey R. Herschman

Department of Biological Chemistry, and
Laboratory of Biomedical and Environmental Sciences
UCLA Center for the Health Sciences; Los Angeles, CA 90024

My laboratory has, for a number of years, been interested in the molecular mechanisms by which extracellular effectors such as growth factors and hormones mediate biological responses in target cells. We have utilized both genetic approaches to this problem and, more recently, molecular approaches. Our genetic studies have been concerned primarily with induction of cell division in quiescent, non-dividing cells. In those studies we selected mutant cell lines unable to divide in response to the polypeptide growth factor epidermal growth factor (EGF) (Pruss and Herschman, 1977; Terwilliger and Herschman, 1984) or the mitogen/tumor promoter tetradecanoyl phorbol acetate (TPA) (Butler-Gralla and Herschman, 1981). We isolated a number of independently derived 3T3 murine fibroblast variant cell lines unable to mount a mitogenic response either to EGF or to TPA and characterized the biochemical and physiological deficits in the EGF nonproliferative or the TPA nonproliferative derivatives. At the same time that we were isolating and characterizing mitogen nonproliferative variant lines, we also initiated a molecular program designed to identify genes induced in response to the mitogen/tumor promoter TPA. Our studies with the cDNAs cloned as a result of our search for TPA-induced gene expression have led us to a consideration of these genes, in response to a number of biological response modifiers, in cells of nervous system origin. In this chapter I will describe our isolation of these cDNAs, and their expression in several *in vitro* preparations derived from the nervous system.

ISOLATION OF THE TIS cDNAs

We reasoned that, at some point in the mitogenic response to TPA, new gene expression would occur; gene expression distinct from that of the quiescent, non-dividing cell. We also decided to look first for the induction of new gene expression occurring as close to the time of mitogen addition as possible. We decided, therefore, to isolate cDNAs homologous to genes expressed as "primary responses" (Yamamoto and Alberts, 1976) to mitogen addition; i.e., genes whose induction in response to mitogen could occur as a result of already existing intracellular factors, without the requirement that any new proteins be synthesized. We selected as the mitogen to be used the potent biological response modifier TPA. There were several reasons for choosing TPA: (1) Our TPA non-responsive mutants of 3T3 cells contained protein kinase C, and were thus blocked in proliferation at a step(s) distal to activation of the enzyme (Bishop et al., 1985). (2) TPA stimulates a variety of biological responses in addition to mitogenesis. We anticipated that some of the cDNAs we might isolate

Molecular Aspects of Development and Aging of the Nervous System
Edited by J.M. Lauder
Plenum Press, New York, 1990

would perhaps play important roles in other biological processes. (3) TPA potently activates a single intracellular second messenger pathway. We thought, therefore, that we might isolate cDNAs that, in some cases, might be uniquely induced by TPA and not by other cellular effectors. (4) Finally, TPA is perhaps the most potent tumor promoter known. We reasoned that we might also isolate cDNAs for genes whose expression is important in the process of tumor progression.

To isolate TPA Induced Sequences, which we refer to as TIS genes, we utilized the following strategy: Confluent, non-dividing 3T3 cells were treated with a maximally stimulating level of TPA, along with the protein synthesis inhibitor cycloheximide (CHX). CHX was included to restrict the expression of TPA-induced genes to primary responses; any gene whose induction by TPA required the presence of an intervening protein induced by TPA would not be activated under these inducing conditions. RNA was harvested from the 3T3 cells treated with TPA and CHX after a three hour exposure period. PolyA^+ mRNA was prepared and the population of sequences from these TPA+CHX treated cells was copied into double-stranded DNA and cloned into the λgt10 cloning vector. These phage were then plated out onto bacterial lawns at a density of 500-1000 bacteriophage per plate. The phage DNAs were transferred to nitrocellulose filters; each plate was used to make duplicate filter lifts. Thus, two identical filter lifts from each plate were available. Two ^{32}P-labeled single-stranded cDNA probes were then prepared. The first, or "[+] probe" was made from polyA^+ mRNA prepared from 3T3 cells that had been exposed to TPA+CHX for three hours. The second, or "[-] probe", was prepared from 3T3 cells that had been exposed for three hours only to CHX. The [+] probe contained labeled sequences homologous to TPA-induced mRNAs, as well as all other cellular RNAs and those induced by CHX alone. The [-] probe contained labeled sequences homologous to constitutive cellular messages and CHX induced messages, but not to messages induced by TPA. Duplicate filter lifts were hybridized with either the [+] probe or the [-] probe, and autoradiographs were prepared from these filters. The duplicate filters were then carefully examined, looking for autoradiographic spots that were present on the filter probed with the [+] probe but absent from the duplicate filter labeled with the [-] probe. Such spots on the autoradiogram should represent bacteriophage containing a cDNA insert from a gene whose mRNA is present in cells treated with TPA, but absent from cells that were not exposed to TPA. Phage from such plaques were picked, purified, and rescreened by a second "+/-" screen. Approximately 50,000 bacteriophage were screened by this procedure; 50 plaques that consistently gave much more intense signals with the [+] probe than with the [-] probe were selected for further study. Cross-hybridization studies with the various TIS cDNAs eliminated some redundancies in which homologous sequences were cloned as separate phage plaques. Because it was possible that our cDNAs were non-overlapping segments of a single mRNA in some cases, we also restricted our subsequent analysis to cDNAs that, when used as probes in northern analysis, detected mRNAs of distinct sizes. By a combination of cross-hybridization, size-determination of the cDNA inserts, and mRNA sizing by northern analysis we selected seven TIS cDNAs, each representing a unique TPA-induced message, for further study (Lim et al., 1987).

TIS GENE EXPRESSION IN 3T3 CELLS

Time Courses of TIS Gene Expression

Exposure of 3T3 cells to TPA resulted in rapid, transient expression of the TIS genes. In many cases induction of mRNA accumulation could be observed as early as 10 minutes after the addition of TPA. Frequently, baseline values of TIS gene expression could not be observed on northern blots of RNAs prepared from 3T3 cells. The peak period of the expression of the various TIS genes following TPA addition occurred between 30-90 minutes, depending on the various TIS gene in question. In nearly all cases the level of TIS mRNA accumulation in the cells subsided to base-line values within three to four hours after the initial addition of TPA (Lim et al., 1987).

Superinduction of the TIS Genes in the Presence of Cycloheximide

In contrast to the transient induction of the TIS genes under normal circumstances, when induction of TIS mRNA accumulation was examined in the presence of both TPA and CHX, accumulation of these mRNAs continued well beyond normal, transient periods. As a consequence, levels of these RNAs in TPA plus CHX treated cells vastly exceeded those present in cells exposed only to TPA. This phenomenon is referred to as "superinduction" (Lim et al., 1987).

Induction of the TIS Genes in 3T3 Cells by Other Mitogens

We thought it possible, because we were activating a specific second messenger pathway with TPA, that induction of at least some of the TIS genes might be unique to TPA. To examine this question we treated 3T3 cells with the polypeptide mitogens epidermal growth factor (EGF) or fibroblast growth factor (FGF), and examined the cells, by northern analysis, for the expression of TIS gene messages. We found that all seven of the TIS gene messages are induced in 3T3 cells by both EGF as well as FGF; none of the genes was uniquely induced by TPA (Lim, Varnum, O'Brien, and Herschman; submitted for publication).

CHARACTERIZATION OF THE TIS GENE cDNAs

The rapid and transient induction of the TIS genes, and their superinduction by CHX, are characteristics that are familiar to cell biologists and molecular biologists interested in questions of cellular proliferation. The treatment of quiescent cells with mitogens is known to induce the rapid and transient induction of expression of the c-fos protooncogene. Moreover, concomitant addition of a mitogen and CHX also results in superinduced expression of c-fos message. We, in fact, anticipated that we should clone the c-fos message as a part of our TIS gene family; to do so would, in a sense, "validate" our search for TPA-induced primary response genes associated with the mitogenic response. We screened the seven TIS cDNA probes with a v-fos probe (as well as several other sequences known to be induced by TPA) and found that the TIS28 cDNA was, indeed, a partial clone of the c-fos message.

Several other laboratories have also isolated primary response genes from mitogen-stimulated 3T3 cells (Cochran et al., 1983; Lau and Nathans, 1987; Almendral et al., 1988). One such gene, known as "egr-1" (for early growth response; Sukhatme et al., 1987, 1988) had been described by Vikas Sukhatme and his colleagues. Sequencing of a partial cDNA of the TIS8 gene by Brian Varnum in my laboratory demonstrated that TIS8, isolated in our laboratory, was identical to egr-1.

We have now isolated nearly full-length cDNAs for TIS1, TIS7, TIS11, and TIS21. Sequence data is complete for TIS11, and shows that it is a gene product not identified by other research groups. A partial cDNA identical to the TIS7 cDNA was previously cloned and sequenced from cells exposed to virus infection (Skup et al., 1982). Thus, one TIS gene, TIS8/egr-1, is a protein that contains a zinc-finger binding region and is likely to be a transcription factor or a factor modifying transcription capability (Sukhatme et al., 1987, 1988). A second TIS gene, TIS28/c-fos, is a protein known to bind to and modulate the transcriptional activity of a transcription factor (Chiu et al., 1988). A third TIS gene, TIS7, appears to be a factor destined for secretion from the cell (Skup et al., 1982). The function of this factor is at present an open question. Although interesting and tantalizing sequence information is present for TIS11 (Varnum et al., 1988), TIS1, and TIS21, we do not as yet have information regarding their cellular localization or their potential functions.

DEVELOPMENTAL EXPRESSION OF THE TIS GENES

The TIS genes were isolated as a family of genes, whose co-ordinate expression was induced by TPA. The c-fos gene is by far the most well studied of the TIS genes. If, indeed, these genes are all induced by common mechanisms, than it might be redundant to study the structure of the various TIS genes and the regulatory regions governing their expression. To pursue this question we examined the expression of the various TIS gene mRNAs in organs of adult and developing mice (Tippetts et al., 1988). We found that the TIS gene mRNAs showed a combination of differential organ enrichment and developmental expression. When both organ distribution and developmental appearance of the various TIS gene messages is examined, it is clear that each of the TIS genes has a unique developmental tissue distribution pattern. Each TIS gene must have unique regulatory features governing its expression and/or unique aspects of its mRNA processing or stability that contribute to a unique pattern of expression in developing animals.

TIS GENE EXPRESSION IN PC12 PHEOCHROMOCYTOMA CELLS

Although c-fos induction is most often associated with cellular proliferation, it is well known that expression of this gene often also occurs in biological responses in which differentiation, rather than proliferation, is induced by an extracellular effector. Probably the best described of these experimental systems is the induction of c-fos expression in the PC12 pheochromocytoma cell line. This cell line, when exposed to nerve growth factor (NGF) stops dividing, and differentiates morphologically, electrically, and biochemically into a cell that more closely resembles a sympathetic neuron (for a review see Guroff, 1985). Treatment of PC12 cells with NGF causes a rapid and transient induction of the expression of c-fos mRNA (Greenberg et al., 1985). Exposure of PC12 cells to EGF, depolarizing agents, TPA and calcium channel agonists can also induce the transient accumulation of c-fos mRNA (Morgan and Curran, 1986).

Because c-fos is a member of the TIS gene family, we decided to study the specificity of the induction of TIS genes to the mitogenic response of 3T3 cells. We turned first to the induction of TIS genes in PC12 cells, since this system had been so well described for c-fos induction. We find that the majority of the TIS genes examined (TIS1, TIS7, TIS8, TIS11, TIS21, and TIS28/c-fos) can be induced in PC12 cells by TPA, NGF, EGF and depolarizing conditions (Kujubu et al., 1987).

Expression of all the NGF-inducible TIS genes can be superinduced by CHX in PC12 cells. Curran and Morgan (1985) showed that peripheral benzodiazepines did not stimulate induction of c-fos mRNA accumulation but could, in cooperation with NGF administration, also superinduce c-fos expression. Those TIS genes whose expression could be induced in PC12 cells were also superinducible by benzodiazepines (Kujubu et al., 1987).

Expression of one TIS gene, TIS10, could not be detected in PC12 cells, despite combined administration of agents (e.g. TPA plus CHX) leading to maximal superinduction of the other TIS genes. One trivial explanation for this observation might have been that our TIS10 probe is only a partial sequence from a murine gene, and PC12 cells are of rat origin. Simple lack of sequence similarity in the cross-hybridization study might, therefore, have accounted for the lack of a detectable TIS10 message in PC12 cells. However, rat-1 cells, a rat embryonic cell line, accumulate a message of the appropriate size that hybridizes well with the TIS10 cDNA probe when exposed to TPA (Kujubu and Herschman, unpublished observation). The lack of detectable TIS10 message in PC12 cells is not due to failure to detect by cross-hybridization an appropriate sequence. We conclude that the lack of TIS10 expression in PC12 cells must result from one of two interesting possibilities. On the one hand, PC12 cells may be the tumorigenic equivalent of a developmental lineage in which the expression of the TIS10 gene is silenced in response to all stimuli. Comparative chromatin structure experiments of the TIS10 gene in PC12 cells and cells capable of expressing this gene should be of great interest in this regard. On the other hand, PC12 cells may have a mutation in the TIS10 gene that prevents proper transcription, splicing, or messenger stabilization. Cloning of the TIS10 gene from PC12 and tissue/cells capable of expressing TIS10 will be required to clarify this possibility. In any event, it is clear from these studies that expression of detectable levels of TIS10 mRNA is not necessary for either survival or NGF-induced differentiation of PC12 cells.

TIS GENE EXPRESSION IN MYELOID CELLS

In addition to characterizing the expression of the TIS genes in the organs of developing animals and in PC12 cells, we extended our analysis of TIS gene expression to cells of the myeloid series. We pursued this line of investigation in order (1) to further characterize the spectrum of cell types that could express the TIS genes, (2) to examine other cell-type specific biological response modifiers (that induce either proliferation or differentiation) for their ability to induce TIS gene expression, and (3) to see if other cell types might, like PC12 cells, express a restricted subset of the TIS genes.

We first examined the induction of TIS genes in a myeloid cell proliferation response. 32Dclone3 cells are a murine cell line dependent on granulocyte-macrophage colony stimulating factor (GM-CSF) for proliferation (Greenberger et al., 1983). When either GM-CSF or interleukin-3 (IL-3) are provided to these cells, they will proliferate. If, however, neither of these factors is present, the 32D cells will withdraw from the cell cycle. If either GM-CSF or IL-3 is given to quiescent cells, the cells will reenter the cell cycle and divide. To examine the expression of the TIS genes we removed GM-CSF from the medium, and brought populations of 32D cells to quiescence. Either TPA or GM-CSF was then added to cells and, at short time intervals, RNA samples were prepared for northern analysis of TIS gene message accumulation. Both TPA and GM-CSF were able to induce the induction of several of the TIS genes; e.g. TIS7, TIS8, TIS10, and TIS11. However, neither TPA nor GM-CSF could induce the accumulation of detectable levels of TIS1 mRNA accumulation. Cycloheximide was able to superinduce the TPA and GM-CSF induction of TIS11 mRNA accumulation in these cells. In contrast, even under superinducing conditions no TIS1 message could be detected in 32D cells exposed to either TPA or GM-CSF (Varnum, Lim, Kaufman, Gasson, Greenberger, and Herschman, submitted for publication).

Neuterophils are post-mitotic cells of the myeloid lineage. When exposed to GM-CSF they do not differentiate, but respond with a characteristic set of biochemical and physiological changes. Sufficient sequence similarity exists between the murine TIS1, TIS8, and TIS11 sequences to allow detection of these messages in human cells (Varnum and Herschman, unpublished). When human neuterophils were exposed to either TPA or to GM-CSF, we could detect the induced accumulation of TIS8 and TIS11 messages. However, as with the myeloid murine cell line 32D, neither TPA nor GM-CSF were able to induce the accumulation of TIS1 mRNA in human neuterophils.

Myeloid cells are apparently restricted, at least in some instances, to the expression of a particular subset of the TIS genes. These data are similar to the restricted expression of a subset of TIS genes in PC12 cells. Note, however, that in the case of PC12 cells TIS10 could not be expressed, while in 32D cells and neuterophils it is the TIS1 gene that cannot be expressed. These data suggest that different classes of cells may be developmentally restricted in the various subclasses of TIS gene they may express. These data have obvious implications for the way in which the biological specificity of distinct cellular responses to a common ligand may be developmentally determined.

TIS GENE EXPRESSION IN CULTURED RAT ASTROCYTES

A number of mitogenic factors, including EGF, FGF, and ganglioside GM_1 stimulate cell division in cultures of rat astrocytes. In contrast, agents that cause elevation of cAMP induce a rapid, dramatic morphological change known as "stellation" in cultured astrocytes. Thus, like PC12 cells, astrocytes in culture are a population of cells that can be induced, in response to appropriate agents, to undergo either a differentiation response or a proliferative response. We wanted to examine TIS gene expression in a population of cells that could, in culture, both differentiate and proliferate and was a more "normal" cell population than the cell lines (such as PC12) that we had used previously. Secondary cultures of rat astrocytes are relatively simple to prepare (McCarthy and de Vellis, 1980), are quite consistent from one culture to the next, and can be induced -- under appropriate culture conditions -- to proliferate on the one hand, or undergo morphological and biochemical differentiation on the other.

Induction of TIS Genes in Astrocytes by TPA, EGF, and FGF

When secondary cultures of rat astrocytes were exposed to the mitogens TPA, EGF or FGF rapid, transient induction of all the TIS genes examined (TIS1, TIS7, TIS8, TIS10, TIS11, and TIS28/c-fos) occurred. Cultured rat astrocytes, in contrast to PC12 cells or to myeloid cells, were able to express all the TIS genes isolated from the library prepared from TPA-treated 3T3 cells (Arenander, Lim, Varnum, Cole, de Vellis, and Herschman, submitted for publication).

Careful dose response curves for the induction of the TIS genes in response to TPA, EGF and FGF were carried out for the astrocyte cultures. We found that distinctions in the inducibility of the various TIS genes occurred. For two of the TIS genes, TIS1 and TIS11, it was clear that TPA was a far better inducer of message accumulation than were either EGF or FGF. Although both polypeptide growth factors could induce accumulation of TIS1 and TIS11 message, maximal induction of these genes in response to EGF or FGF was substantially less than that observed for maximal TPA induction. In contrast, when maximally inducing levels of TPA, EGF and FGF were utilized, the levels of TIS8 and TIS28/c-fos messages were essentially the same for these three agents. These data suggested that TPA and the polypeptide mitogens might induce TIS genes by separate pathways. The EGF/FGF pathway appeared as potent in astrocytes as the TPA pathway for TIS8 and TIS28/c-fos. In contrast, the TPA pathway was much more effective than the EGF/FGF pathway for TIS1 and TIS11.

Induction of the TIS Genes in Astrocytes by Combinations of Mitogens

If independent pathways for TIS gene induction occur for TPA and the polypeptide mitogens, one might expect that these two pathways could interact. We exposed cultured astrocytes to combinations of EGF and FGF, EGF and TPA, or FGF and TPA. The combination of EGF and FGF (each at maximally inducing levels) was no more effective for TIS gene induction than either peptide mitogen alone. Thus, if separate pathways for TIS gene induction exist for EGF and FGF, they must saturate some common, limiting step. In contrast, when EGF (or FGF) was co-administered to astrocytes along with TPA, induction of the TIS genes was at least as great, for accumulation of all the TIS gene messages, as the sum of that observed for the individual inducers. These data suggest that TIS gene message accumulation can be induced by separate, independent and additive pathways in cultured astrocytes (Arenander, Lim, Varnum, Cole, de Vellis, and Herschman, submitted for publication). Experiments designed to test this question were also carried out in 3T3 cells. When protein kinase C was down-regulated by pretreatment with TPA subsequent TPA induction of the TIS genes was blocked. In contrast, induction of the TIS genes by either EGF or FGF was only slightly impaired (Lim, Varnum, O'Brien, and Herschman, submitted for publication). These data suggest that, in 3T3 cells, there also exist separate and independent pathways to induce TIS gene expression.

Induction of the TIS Genes in Astrocytes by Agents that Stimulation Stellation

When secondary cultures of rat astrocytes were exposed to either dibutyryl cyclic AMP or to forskolin, induction of the TIS genes was rapidly and transiently induced. The level of TIS gene expression and the time courses of expression were essentially the same for TIS gene induction by mitogen and by differentiation-inducing agents (Arenander, Lim, Varnum, Cole, de Vellis, and Herschman, submitted for publication).

Induction of the TIS Genes in Astrocytes by Ganglioside GM_1

GM_1 is both a potent astrocyte mitogen and an inhibitor of the stellation response. When secondary rat astrocyte cultures were exposed to GM_1 TIS gene induction occurred (Arenander, Lim, Varnum, Cole, de Vellis, and Herschman, submitted for publication). PC12 cells also demonstrate morphological and biochemical responses to GM_1. However, GM_1 was unable to induce accumulation of TIS gene messages in PC12 cells, in contrast to its activity as an inducer of TIS gene expression in cultured astrocyte cultures.

Conclusions from the Study of TIS Gene Induction in Cultured Astrocytes

Induction studies using combinations of mitogens suggest that at least two pathways for TIS gene induction may exist, a protein kinase C mediated pathway and a protein kinase C independent pathway utilized by polypeptide mitogens such as EGF and FGF. Blackshear and his colleagues (1987), using human astrocytomas, have reached a similar conclusion for c-fos induction. They have also postulated that a third pathway for c-fos induction may exist, by which carbachol can induce expression of this gene. Forskolin is thought to work through cAMP-dependent pathways, suggesting yet another independent route to TIS gene expression. The mitogenic pathways of GM_1 and the polypeptide mitogen EGF are known to respond differently to inhibitors of ion transport (Skaper and Varon, 1987), suggesting that their intracellular pathways signalling proliferation are separable. If their mitogenic and TIS gene inducing pathways are similar, these data suggest that GM_1 may induce TIS gene accumulation by still another pathway. Finally, at least some cellular specificity for ligand induction of TIS gene expression appears to exist, since GM_1 can induce TIS gene expression in astrocytes but not in PC12 cells, despite clear biological effects of GM_1 on these cells.

ACKNOWLEDGEMENTS

This work was supported by NIH award GM24797 and DOE contract DE FC03 87ER 60615.

REFERENCES

Almendral, J. M., Sommer, D., MacDonald-Bravo, H., Burckhardt, J., Perera, J., and Bravo, R., 1988, Complexity of the early genetic response to growth factors in mouse fibroblasts, Mol. Cell. Biol., 8:2140.

Bishop, R., Martinez, R., Weber, M. J., Blackshear, P. J., Beatty, S., Lim, R., and Herschman, H. R., 1985, Protein phosphorylation in a TPA non-proliferative variant of 3T3 cells, Mol. Cell. Biol., 5:2231.

Blackshear, P. J., Stumpo, D. J., Huang, J., Nemenoff, R. A., and Spach, D. H., 1987, Protein kinase C-dependent and -independent pathways of proto-oncogene induction in human astrocytoma cells, J. Biol. Chem. 262:7774.

Butler-Gralla, E., and Herschman, H. R., 1981, Variants of 3T3 cells lacking mitogenic response to the tumor promoter tetradecanoyl-phorbol-acetate, J. Cell. Physiol., 107:59.

Chiu, R., Boyle, W. J., Meek, J., Smeal, T., Hunter, T., and Karin, M., 1988, The c-fos protein interacts with c-jun/AP-1 to stimulate transcription of AP-1 responsive genes, Cell, 54:541.

Cochran, B. H., Reffel, A. C., and Stiles, C. D., 1983, Molecular cloning of gene sequences regulated by platelet-derived growth factor, Cell, 33:939.

Curran, T., and Morgan, J., 1985, Superinduction of c-fos by nerve growth factor in the presence of peripherally-active benzodiazepine, Science, 229:1265.

Greenberg, M., Green, L., and Ziff, E., 1985, Nerve growth factor and epidermal growth factor induce rapid transient changes in proto-oncogene transcription in PC-12 cells, J. Biol. Chem., 260:14101.

Greenberger, J. S., Sakakeeny, M. A., Humphries, R. K., Eaves, C. J., and Eckner, R. J., 1983, Demonstration of permanent factor-dependent multipotential (erythroid/neutrophil/basophil) hematopoietic progenitor cell lines, Proc. Natl. Acad. Sci. USA, 80:2931.

Guroff, G., 1985, PC12 cells as a model of neuronal differentiation, in: "Cell Culture in the Neurosciences," J. Bottenstein and G. Sato, eds., pp. 245-272, Plenum Press, New York.

Kujubu, D. A., Lim, R. W., Varnum, B. C., and Herschman, H. R., 1987, Induction of transiently expressed genes in PC-12 pheochromocytoma cells, Oncogene, 1:257.

Lau, L. F., and Nathans, D., 1987, Expression of a set of growth-related immediate early genes in BALB/c 3T3 cells: coordinate regulation with c-for or c-myc, Proc. Natl. Acad. Sci. USA, 84:1182.

Lim, R. W., Varnum, B. C., and Herschman, H. R., 1987, Cloning of tetradecanoyl phorbol ester induced "primary response" sequences and their expression in density-arrested Swiss 3T3 cells and a TPA nonproliferative variant, Oncogene, 1:263.

McCarthy, K. D., and de Vellis, J., 1980, Preparation of separate astroglial and oligodendroglial cell cultures from rat cerebral tissue, J. Cell Biol., 85:890.

Morgan, J., and Curran, T., 1986, Role of ion flux in the control of c-fos expression, Nature, 322:552.

Pruss, R. M., and Herschman, H. R., 1977, Variants of 3T3 cells lacking mitogenic response to epidermal growth factor, Proc. Natl. Acad. Sci. USA, 74:3918.

Skaper, S. D., and Varon, S., 1987, Ionic responses and growth stimulation in rat astroglial cells: differential mechanisms of gangliosides and serum, J. Cell Physiol., 130:453.

Skup, D., Windass, J. D., Sor, F., George, H., Williams, B. R. G., Fukuhara, H., De Maeyer-Guignard, J., and De Maeyer, E., 1982, Molecular cloning of partial cDNA copies of two distinct mouse IFN-β mRNAs, Nucleic Acids Res., 10:3069.

Sukhatme, V. P., Kartha, S., Toback, F. G., Taub, R., Hoover, R. G., and Tsai-Morris, C., 1987, A novel early growth response gene rapidly induced by fibroblast, epithelial cell and lymphocyte mitogens, Oncogene Res., 1:343.

Sukhatme, V. P., Cao, X., Chang, L. C., Tsai-Morris, C., Stamenkovich, D., Ferre, P. C., Cohen, D. R., Edwards, S. A., Shows, T. B., Curran, T., LeBeau, M. M., and Adamson, E. D., 1988, A zinc finger-encoding gene coregulated with c-fos during growth and differentiation, and after cellular depolarization, Cell, 53:37.

Terwilliger, E., and Herschman, H. R., 1984, 3T3 variants unable to bind epidermal growth factor cannot complement in co-culture, Biochem. Biophys. Res. Commun., 118:60.
Tippetts, M. T., Varnum, B. C., Lim, R. W., and Herschman, H. R., 1988, Tumor promoter inducible genes are differentially expressed in the developing mouse, Mol. Cell. Biol., 8:4570.
Varnum, B. C., Lim, R. W., Sukhatme, V. P., and Herschman, H. R., 1988, Nucleotide sequence of a cDNA encoding TIS11, a message induced in Swiss 3T3 cells by the tumor promoter tetradecanoyl phorbol acetate, Oncogene, in press.
Yamamoto, K. R., and Alberts, B. M., 1976, Steroid receptors: elements for modulation of eukaryotic transcription, Annu. Rev. Biochem., 45:721.

DEVELOPMENTAL REGULATION OF MYELIN-ASSOCIATED GENES IN THE NORMAL AND THE MYELIN DEFICIENT MUTANT RAT

M. N. Gordon, S. Kumar, A. Espinosa de los Monteros, S. Scully
M.-S. Zhang, J. Huber, R. A. Cole and J. de Vellis

Departments of Anatomy and Psychiatry, UCLA School of Medicine
Laboratory of Biomedical and Environmental Sciences, Mental Retardation Research Center, and Brain Research Institute, University of California, Los Angeles, CA, USA 90024-1759

ABSTRACT

Oligodendrocyte development and myelinogenesis, both in vivo and in vitro, are characterized by the sequential and coordinate expression of markers which participate in the differentiation of oligodendrocytes as a prerequisite for myelination. The myelin deficient (*md*) rat shows greatly reduced mRNA expression for several oligodendrocyte markers: glycerol phosphate dehydrogenase (GPDH), myelin basic protein (MBP) and proteolipid protein (PLP). Brain GPDH mRNA levels are initially equivalent in *md* and unaffected littermates, but the mutant rats fail to display the normal developmental increase in gene expression. Immunostaining of brain tissue sections also reveals decreased expression of these oligodendrocyte markers. The number of oligodendrocytes containing GPDH-like immunoreactivity is reduced in mutant rats, and in general these cells appear morphologically less complex with shorter processes. However, the intensity of staining in many oligodendrocytes appears equivalent to that observed in unaffected rats. Expression of the neuronal marker, glutamic acid decarboxylase, and the astrocyte markers, glutamine synthetase and glial fibrillary acidic protein, are largely unaffected at either the mRNA or protein level. Mixed glial cultures prepared from the brains of neonatal male *md* rats possess fewer oligodendrocytes compared to cultures derived from unaffected littermates, and the temporal sequence of marker development is delayed. Although an abnormality in the PLP gene is suspected in the *md* rat, these findings document profound deficits in many oligodendrocyte gene products.

INTRODUCTION

The myelin deficient (*md*) rat mutant fails to elaborate central nervous system (CNS) myelin during postnatal development, although peripheral nerves are fully myelinated on the expected ontogenic schedule (Csiza and DeLahunta, 1979; Dentinger et al., 1982; 1985). This condition is carried in the Wistar rat lineage, and is inherited as an X-linked, recessive trait. Symptoms are first apparent in affected males as a mild tremor beginning near the end of the second week of life. Tremors become progressively worse over the ensuing ten days, affecting primarily the hindquarters and tail. During this time, rats are slow to initiate locomotor activity, and total open field locomotion becomes increasingly reduced (Fig. 1). The affected rats also display an unusual, slow, circling retropulsion. By postnatal day 22-23 (P22-23), symptoms progress to include generalized tonic seizures, initially induced by handling, loud sounds or light exposure, but later occurring spontaneously. Skeletal musculature appears completely contracted during seizures, with arching of the back and full

Molecular Aspects of Development and Aging of the Nervous System
Edited by J.M. Lauder
Plenum Press, New York, 1990

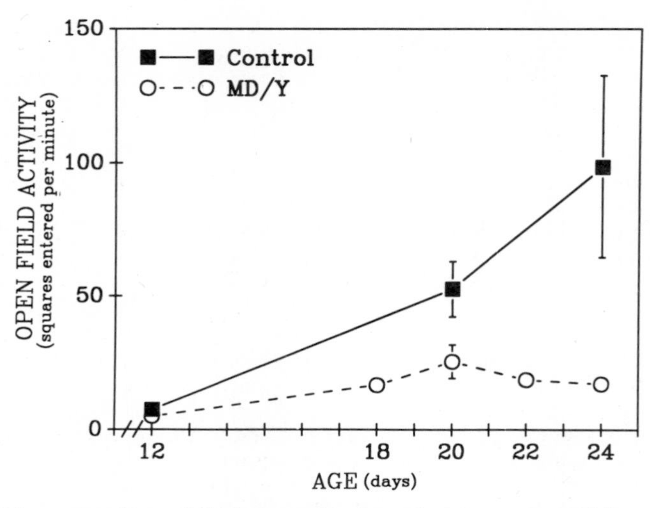

Figure 1. Open-field locomotion in *md* mutant rats and in unaffected littermates. Developing rats were placed in the center of a small open field arena measuring 120 x 60 cm, delineated by a grid work of squares measuring 6 cm per side. The time required for each rat to enter 60 squares was recorded; five such trials were averaged for each rat. Data are presented as the mean number of squares entered per minute ± SEM (n = 2-5 rats per group). The symbols encompass some SEM values. c, control; *md*, mutant. Analysis of variance demonstrated significant main effects of age and genotype, and a significant interaction term.

extension of all limbs and tail. Each seizure episode lasts 30-60 sec., and is followed by a hypoactive period. Death results by one month of age (median age of death in our colony = P26), probably due to apneic spells during seizures, but can be delayed by anticonvulsant medication (Dentinger et al., 1982).

These behavioral manifestations of the *md* trait occur during the ontogenic window during which myelination is the most active in non-mutant rats. Myelin deposition in the forebrain of non-mutant rats begins about postnatal day 10-14, with rapid increases in myelin-associated lipids, proteins and their respective biosynthetic enzymes over the ensuing month; spinal cord and more caudal brain areas are myelinated slightly earlier. In contrast, histological and ultrastructural examinations have failed to detect normal myelin in the *md* CNS (Csiza and DeLahunta, 1979; Dentinger et al., 1982; 1985; Duncan et al., 1987b; Rosenbluth, 1987). A few myelin lamellae are sometimes formed, but they fail to compact properly, and no intraperiod line can be visualized (Duncan et al., 1987b). Ultrastructurally identified, immature oligodendrocytes appear to be present in normal numbers in the *md* CNS at P3-7 (Dentinger et al., 1982). Some mature oligodendrocytes appear morphologically normal after P10, but such cells are reduced in number compared with unaffected littermates, and they fail to myelinate axons. Most oligodendrocytes fail to mature, resulting in increased ratios of immature : mature oligodendrocytes (Dentinger et al., 1982; Jackson et al., 1987). These immature oligodendrocytes display a protracted period of proliferation (Jackson et al., 1987). Abnormal oligodendrocytes have also been described. Duncan et al. (1987a;b) described cells with a unique distention of the rough endoplasmic reticulum, filled with a flocculent electron-dense substance, while Dentinger et al. (1982)

reported the presence of abnormal oligodendrocytes containing lipid inclusions. It is not clear whether these latter cells are true oligodendrocytes, or derive from the macrophage/microglia cell lineage. Finally, an increased incidence of pyknotic oligodendrocytes has also been reported (Dentinger et al., 1982; Jackson et al., 1987). Thus, the morphological differentiation of oligodendroglia appears to be impaired in the *md* mutant rat. We have used RNA blot hybridization, enzyme activity measurements and immunohistochemistry to explore the expression of oligodendrocyte-specific genes in order to assess the degree of oligodendrocyte maturation at a biochemical level. As markers of myelination, we have examined the expression of the two major protein components of myelin, myelin basic protein (MBP) and proteolipid protein (PLP). We have also examined the expression of glycerol phosphate dehydrogenase (GPDH). This cytoplasmic enzyme is localized exclusively to oligodendrocytes in the CNS (Leveille et al., 1980) where it plays an important role in phosphatide biosynthesis (Laatsch, 1962). GPDH is expressed in the pre-myelinating oligodendrocyte at a younger developmental age than MBP or PLP both in vivo and in vitro (Kumar et al., 1988b), and therefore provides a useful marker to assess the status of *md* oligodendrocytes prior to myelination. In addition, GPDH is expressed in peripheral tissues; comparisons of gene expression in the CNS and periphery allow assessment of the specificity of any observed oligodendrocyte impairments.

Although oligodendrocyte abnormalities are probably directly responsible for the myelination failure, *md* mutants also show astrogliosis in the CNS (Dentinger et al., 1982; 1985; Connor et al., 1987; Duncan et al., 1987a; Koeppen et al., 1988). Astrocyte processes interdigitate between the oligodendrocyte cell processes and the axolemma, or within the few lamellae of rudimentary myelin that sometimes form, suggesting that impairments in cell-cell recognition contribute to the myelination failure in the *md* mutants (Rosenbluth, 1987). Consequently, we have also examined the expression of two astrocyte-specific genes during development of the *md* mutant, glial fibrillary acidic protein (GFAP) and glutamine synthetase (GS). The development of the neuron-specific marker, glutamic acid decarboxylase, was also monitored.

GENE EXPRESSION IN OLIGODENDROCYTES

Northern blot analysis revealed the expression of all oligodendrocyte-specific RNA molecules in the *md* brain (Kumar et al., 1988a). GPDH and MBP cDNAs each hybridized to a single RNA species of the expected size (4 kb and 2.1 kb, respectively). In control rats, the PLP probe recognized two RNA species of 3.2 and 1.6 kb, as expected, which are generated by alternative splicing of the single gene (Milner et al., 1985). The functional significance of these alternative transcripts is not known. In the *md* brain, only the larger transcript was detectable, and it appeared to be slightly smaller than the wild-type transcript.

Although transcripts of all markers were present in the *md* mutant brain, slot blot hybridization demonstrated that each was reduced in level (Fig. 2; Kumar et al., 1988a). At the youngest age examined (P15), brain GPDH RNA levels were comparable in control and *md* rats. However, the 5-fold increase in RNA level occurring between P15 and P25 in unaffected rats failed to materialize in the *md* brain, resulting in an 80% reduction in GPDH RNA levels at P25 (Fig. 2; Table I). At the protein level, immunohistochemistry demonstrated a reduction in the number of oligodendrocytes containing GPDH-like immunoreactivity in the major white matter tracts of the forebrain (Fig. 3). In general, these cells appeared morphologically less complex, with reduced process elaboration. Most oligodendrocytes containing GPDH-like immunoreactivity in the *md* brain appeared to be less intensely stained than those in age-matched controls, but some cells contained very intense immunoreactivity. The reduction in cell number was approximately 40-50% in the anterior corpus callosum (Table I), and is therefore less severe than the 80% reduction in RNA level. Thus, the failure to produce GPDH RNA in the *md* brain does not result solely from death of the oligodendrocytes, but rather reflects their failure to differentiate on schedule. Unexpectedly, the specific activity of GPDH in whole brain homogenates was not reduced in the *md* rats (Table I). Experiments are currently in progress to examine this discordance between the immunological data and the enzyme activity measurements.

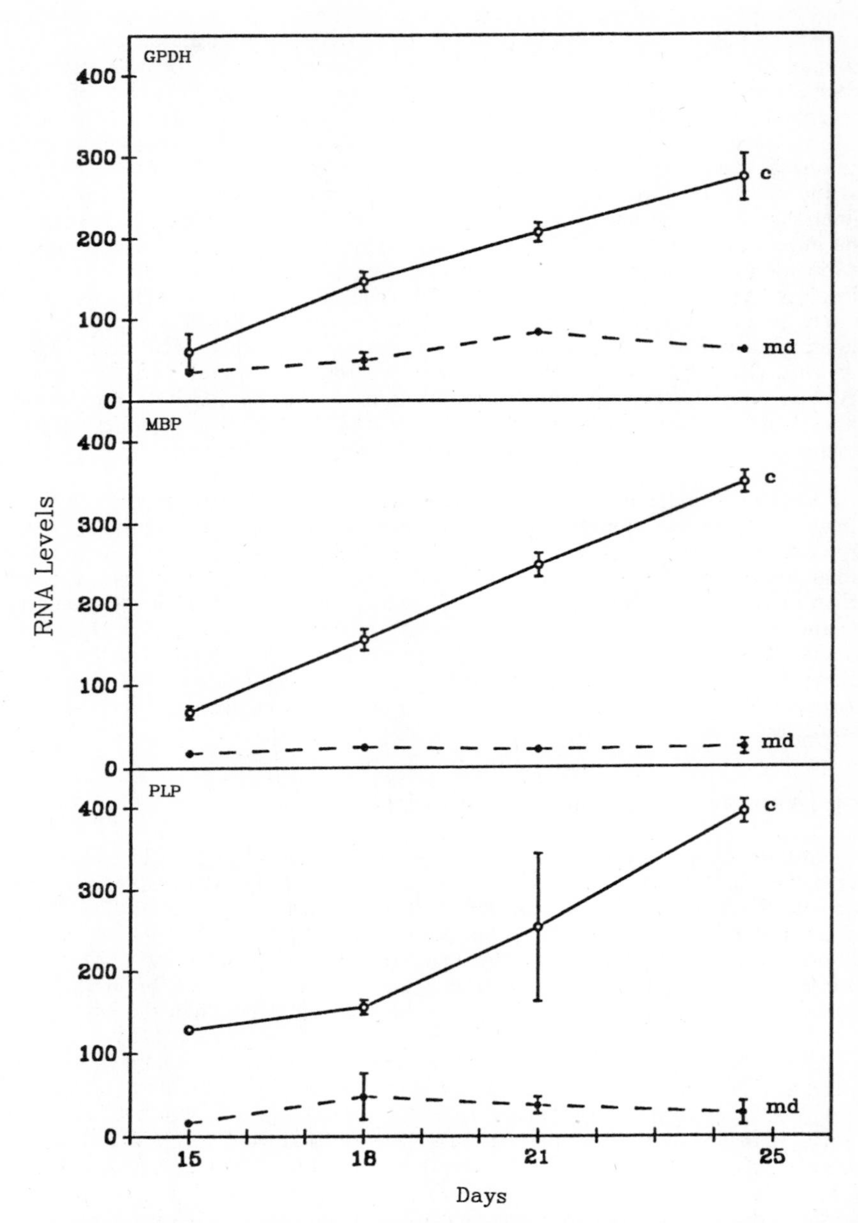

Figure 2. Oligodendrocyte marker RNA expression during development. Poly (A+) RNA was isolated from the brains of *md* and unaffected littermates on postnatal days 15, 18, 21 and 25. RNA was spotted on a nitrocellulose membrane, hybridized to [^{32}P]-labeled GPDH cDNA, autoradiographed, and densitometrically scanned. The same blot was sequentially probed with cDNAs for MBP and PLP, and finally with a cDNA for pCHOB, a non-developmentally regulated gene, to normalize for minor variations in the amount of RNA loaded in each slot. Data are presented as normalized densitometer readings in arbitrary units. Mean $\pm$ SEM (n = 3) are displayed. Some SEM values are encompassed by the symbol. c: control; *md*: mutant. The *md* rats failed to display the age-related developmental increase in RNA levels observed in control rats. For each marker, analysis of variance demonstrated significant main effects of age and genotype, and significant interaction terms. (From Kumar et al., J. Neurosci. Res., in press, 1988).

Table I. Alterations in Gene Expression Observed in Affected *md* Rat Brain and in Cell Cultures Derived From *md* Brain.

MARKER	RNA LEVEL	ENZYME SPECIFIC ACTIVITY (Units per mg protein)			IMMUNOCHEMISTRY (Cell Number)	
					IN VIVO	IN VITRO
	% change	+/Y	*md*/Y	% change	% change	% change
GPDH	-80%	29.9 ± 1.3	30.3 ± 1.2	-0%	-40%	-65%
MBP	-93%				> -40%	-80%
PLP	-93%				> -40%	-85%
GAD	- 0%	1.76 ± 0.22	1.60 ± 0.04	-0%	- 0%	
GFAP	- 0%				increased	
GS	- 0%	14.4 ± 0.6	13.7 ± 0.6	-0%		

RNA levels and enzyme activities were measured using standard techniques in whole brain homogenates from affected *md* rats and from unaffected littermates, aged P20-25. The values for percent change in RNA levels were calculated from the data presented in Fig. 2. Oligodendrocytes containing GPDH-like immunoreactivity were counted in camera lucida drawings of the anterior corpus callosum in one hemisphere at a level corresponding to A8620 *u*m of the Konig and Klippel atlas. Representative sections from which these counts were generated are displayed in Fig. 3. Although cells containing MBP- or PLP-like immunoreactivity were not counted, they were far fewer in number than GPDH-positive cells. Immunoreactive oligodendrocytes in cell culture were counted after 15 days in culture in 3 representative areas of cultures derived from 3 *md* rats. For all dependent variables, n = 3-4.

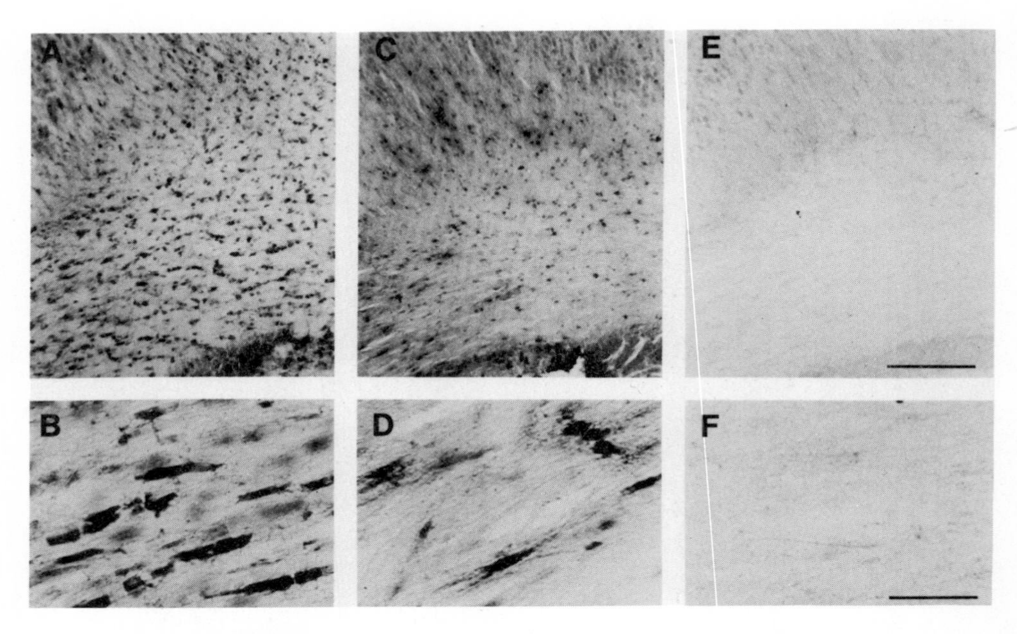

Figure 3. GPDH-like immunoreactive oligodendrocytes in the anterior corpus callosum of P25 control and *md* rats. After perfusion of anesthetized rats with phosphate-buffered 4% para-formaldehyde, brains were sectioned at 100 *u*m using a Vibratome. Immunochemistry was performed according to standard procedures using a polyclonal antiserum directed against GPDH, Vectastain ABC kits (Vector Labs, Burlingame, CA), and the chromogen diaminobenzidine. Sections from control and *md* rats were processed simultaneously; photographs were exposed and printed for equivalent times. A, B: control rat; C, D: mutant rat; E, F: immunochemical control in which GPDH antiserum was replaced by normal rabbit serum. The number of immunoreactive cells appears less in the corpus callosum of *md* rats compared to age-matched unaffected littermates. Scale bar represents 200 *u*m in A, C and E and 50 *u*m in B, D, and F.

In the peripheral tissues, skeletal muscle of the hindlimb and liver, GPDH RNA levels (Kumar et al., 1988a) and GPDH specific activity (data not shown) were not reduced in the affected P21 *md* rat compared with control littermates. The GPDH gene thus appears intact in the *md* mutant, and the reduction in gene expression in the oligodendrocytes is secondary to the differentiation failure.

MBP and PLP RNAs were profoundly reduced throughout development, and were affected to a greater degree than was GPDH (Fig. 2). At the youngest age examined (P15), MBP and PLP RNAs were reduced 4- and 8-fold, respectively, in the affected *md* individuals compared with unaffected littermates. The levels of both messages increased 3-5-fold between P15 and P25 in control rats, but failed to increase in the *md* brain. Consequently, by P25, the *md* mutants expressed only 7% each of control MBP and PLP RNA levels (Table I). Similar results were reported by Zeller et al. (1985), while Duncan et al. (1987b) demonstrated reduced expression of MBP and PLP RNAs by in situ hybridization in the P23 *md* spinal cord.

At the protein level, the reduction in MBP and PLP expression also appears larger than the reduction in GPDH expression by immunohistochemistry (Fig. 5 vs. Fig. 3). Only occasional oligodendrocytes contain MBP- or PLP-like immunoreactivity in white matter tracts of the *md* forebrain. Such cells are very faintly labeled, and appear smaller and

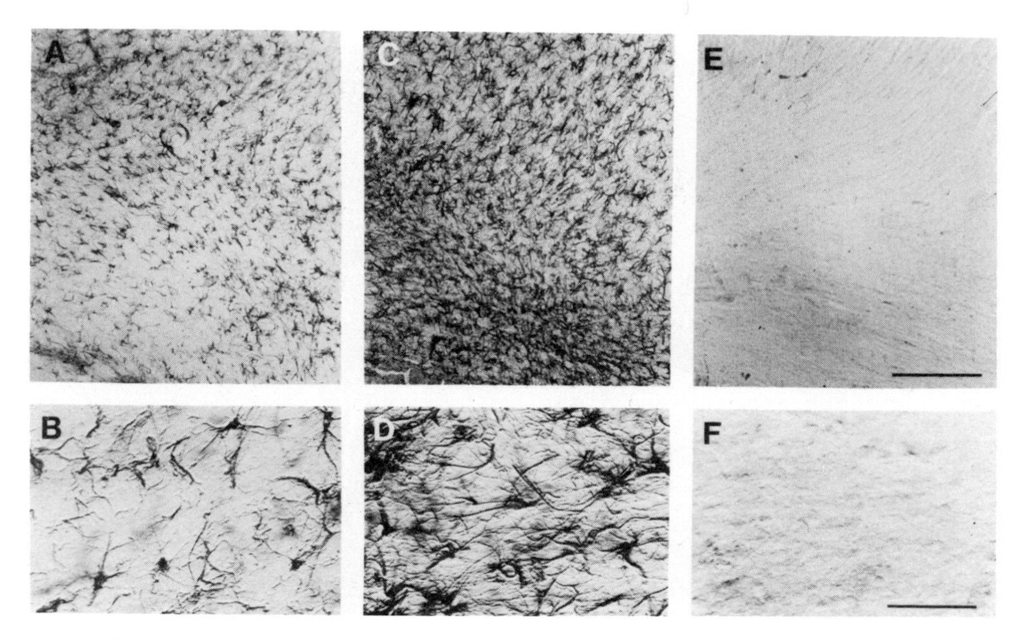

Figure 4. GFAP-like immunoreactivity in the corpus callosum of control and *md* rats. Adjacent sections to those displayed in Fig. 3 were processed as described above using a monoclonal antibody directed against GFAP (Boehringer Mannheim Biochemicals, Indiananopolis, IN). A, B: control rat; C, D: mutant rat; E, F: immunochemical control in which the GFAP antibody was replaced by a heterologous monoclonal antibody (anti-217C) which does not recognize any CNS antigen. GFAP-like immunoreactivity is increased in the mutants. Scale bar represents 200 *u*m in A, C, and E, and 50 *u*m in B, D, and F.

morphologically less complex than cells from control brains. Duncan et al. (1987b) has also demonstrated less MBP-like immunoreactivity in myelin in the P23 *md* spinal cord, and was unable to detect PLP-like immunoreactivity in myelin.

Yanagisawa et al. (1986) quantified MBP and PLP levels in the *md* CNS by RIA and Western blotting techniques. MBP was reduced by 99% in the *md* brain, and 75-90% in the spinal cord, at all developmental ages. Western blot analysis demonstrated the presence of all molecular weight variants of MBP in the expected ratios in the *md* spinal cord; a similar analysis in brain tissue was precluded by the extremely low levels of MBP there. A preferential expression of 21 kDa MBP was observed in the X-linked dysmyelination disorders of mice (*jimpy*) and dogs (shaking pup) [Yanagisawa et al., 1987]. Proteolipid protein (PLP) was undetectable by immunoblotting techniques, suggesting a reduction of at least 99.8-99.9% of control PLP levels in the *md* CNS (Yanagisawa et al., 1986).

GENE EXPRESSION OF NON-OLIGODENDROGLIAL MARKERS

RNA levels of the three non-oligodendroglial markers each increased 2-3-fold during development, although each marker displayed a characteristic developmental pattern (Kumar et al., 1988a). Thus, GAD RNA increased 3-fold between P15 and P18, then plateaued until P25. GS RNA increased 2-fold between P15 and P18, then declined 30% between P18 and P25. GFAP RNA levels increased progressively from P15 to P25, with a maximal increase of 2-fold. However, for all markers, mutant and control RNA levels were not statistically distinguishable at any developmental age (Table I).

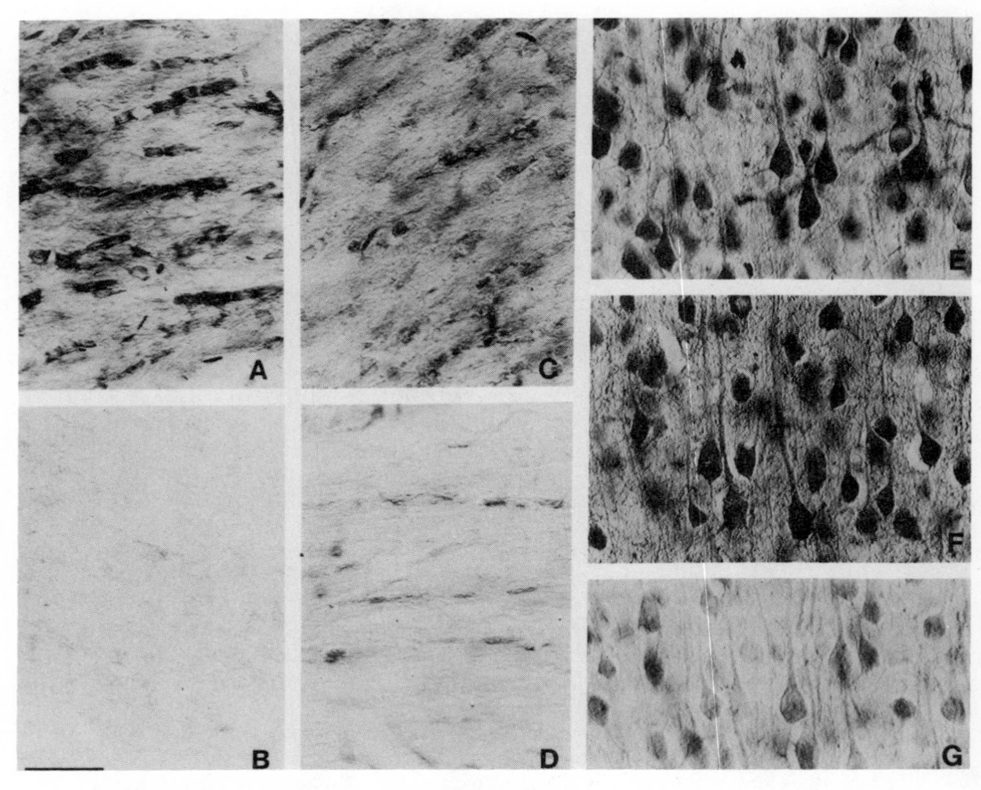

Figure 5. Immunoreactive cells in the corpus callosum (A-D) or anterior cingulate cortex (E-G). A: MBP-like immunoreactivity, control rat. B: MBP-like immunoreactivity, *md* rat. C: PLP-like immunoreactivity, control rat. D: PLP-like immunoreactivity, *md* rat. E: GAD-like immunoreactivity, control rat. F: GAD-like immunoreactivity, *md* rat. G: immunochemical control employing pre-immune sheep serum in place of the GAD antiserum. The appropriate immunochemical control for MBP and PLP is displayed in Fig. 3F. The number of oligodendrocytes possessing MBP- or PLP-like immunoreactivity appears reduced in the *md* brain, but there is no striking abnormality in the GAD-like immunoreactivity. Scale bar represents 50 *u*m in all panels.

At the protein level, GAD and GS specific activities were comparable in whole brain homogenates from P20-25 *md* and unaffected littermates (Table I). Similarly, GAD-like immunoreactive neurons in the *md* cerebral cortex (Fig. 5) and hippocampus were not grossly abnormal in number or morphology. GFAP-like immunoreactivity does appear to be increased in the major white matter tracts in the *md* forebrain (Fig. 4). This phenomenon becomes increasingly obvious at the later developmental ages. The increased immunoreactivity may reflect increased protein levels via post-transcriptional regulation, since GFAP RNA levels are not increased. On the other hand, difficulty in penetration of immunological reagents into myelin is a well-known phenomenon. It may be that the failure to generate myelin in the major white matter tracts of the *md* brain allows better GFAP antibody penetration, and enhanced immunoreactivity of the intertwined astrocytes. Resolution of this issue awaits quantitative measurements of GFAP.

OLIGODENDROCYTE DIFFERENTIATION IN VITRO

Cultivation of neural cells in vitro allows rigorous control over the extracellular chemical milieu, and, through the use of enriched cell populations, can simplify cellular interactions modifying differentiation programs. We have employed cell cultures derived from *md* brain to discriminate intrinsic regulation of differentiation from extrinsically-regulated, malleable patterns of development.

Mixed glial cultures are prepared from individual neonatal *md* and unaffected male rat brains as described previously (Espinosa de los Monteros et al., 1986). In general, the pattern of differentiation of *md* oligodendrocytes in vitro mirrors that observed in vivo. Using phase-contrast microscopy, we observed that cells possessing the morphological criteria characteristic of oligodendrocytes do differentiate in cultures from *md* rats, but are reduced in number (Fig. 6). The number of oligodendroglia containing GPDH-like immunoreactivity is reduced about 50% and 65% after 9 and 15 days in vitro, respectively, compared to cultures derived from unaffected littermates. As observed in vivo, the expression of MBP and PLP is more affected, and an even smaller percentage of the oligodendrocytes present display MBP- or PLP-like immunoreactivity. In control cultures, MBP- and PLP-like immunoreactivity is first seen in the cell soma, but by 15 days in vitro, the protein is localized primarily in the membranous expanses of cell processes (Fig. 6). Oligodendrocytes in *md* cultures, on the other hand, fail to elaborate these processes, and immunoreactivity is primarily confined to the cell soma (Fig. 6). In addition, *md* oligodendrocytes continue to express early oligodendrocyte markers, such as galactocerebroside and transferrin, for longer times in culture than do control oligodendroglia. Taken together, these results indicate that *md* oligodendrocytes are reduced in number and are less mature than those from their unaffected counterparts. In conclusion, the study of differentiation of *md* oligodendrocytes in cell culture represents a good model of in vivo differentiation and will allow exploration of intervention therapies designed to ameliorate the *md* condition.

CONCLUSIONS AND FUTURE DIRECTIONS

Although the genetic lesion responsible for the *md* trait has not yet been confirmed, the sex-linked mode of inheritance, the array of behavioral symptomology and the biochemical deficits are strikingly similar to the *jimpy* mouse. The only known genetic abnormality in this mutant mouse is a single point transition in the gene encoding PLP, which alters a splice consensus sequence at the border of intron 4 and exon 5, and results in a truncated mRNA (loss of 74 bases) and in a truncated and frameshifted protein (Macklin et al., 1987; Nave et al., 1987). PLP mRNA is expressed in the *jimpy* mouse, albeit at low levels (Nave et al., 1986), but the protein is undetectable by Western blot (Sorg et al., 1986). Experiments in progress will determine the genetic lesion in *md* rats, and will examine regulation of PLP gene expression in *md* rats. Post-transcriptional regulatory factors must exist because PLP RNA is detectable, but the protein is not.

The only known function of the PLP protein concerns it's role as a structural component of myelin, where it is thought to be required for compaction of myelin and in generating the proper intraperiod line within the myelin lamellae (Duncan et al., 1987b). The molecular mechanisms by which the loss of PLP could generate the pleiotropic effects characteristic of the *jimpy* mouse and the *md* rat remain enigmatic. Using cell culture systems and brain grafting techniques, we hope to explore this potential regulatory role of PLP in oligodendrocyte differentiation.

ACKNOWLEDGEMENTS

The authors appreciate the generosity and advice of Dr. Robin S. Fisher. Antiserum to GAD was provided through the Laboratory of Clinical Science, NIMH where it was developed under the supervision of Dr. Irwin J. Kopin with Drs. W. Oertel, D.E. Schmechel and M. Tappaz. Effective use in immunocytochemistry was greatly aided through the laboratory of E. Mugnaini, Univ. Of Connecticut, Storrs. Probes used in these experiments

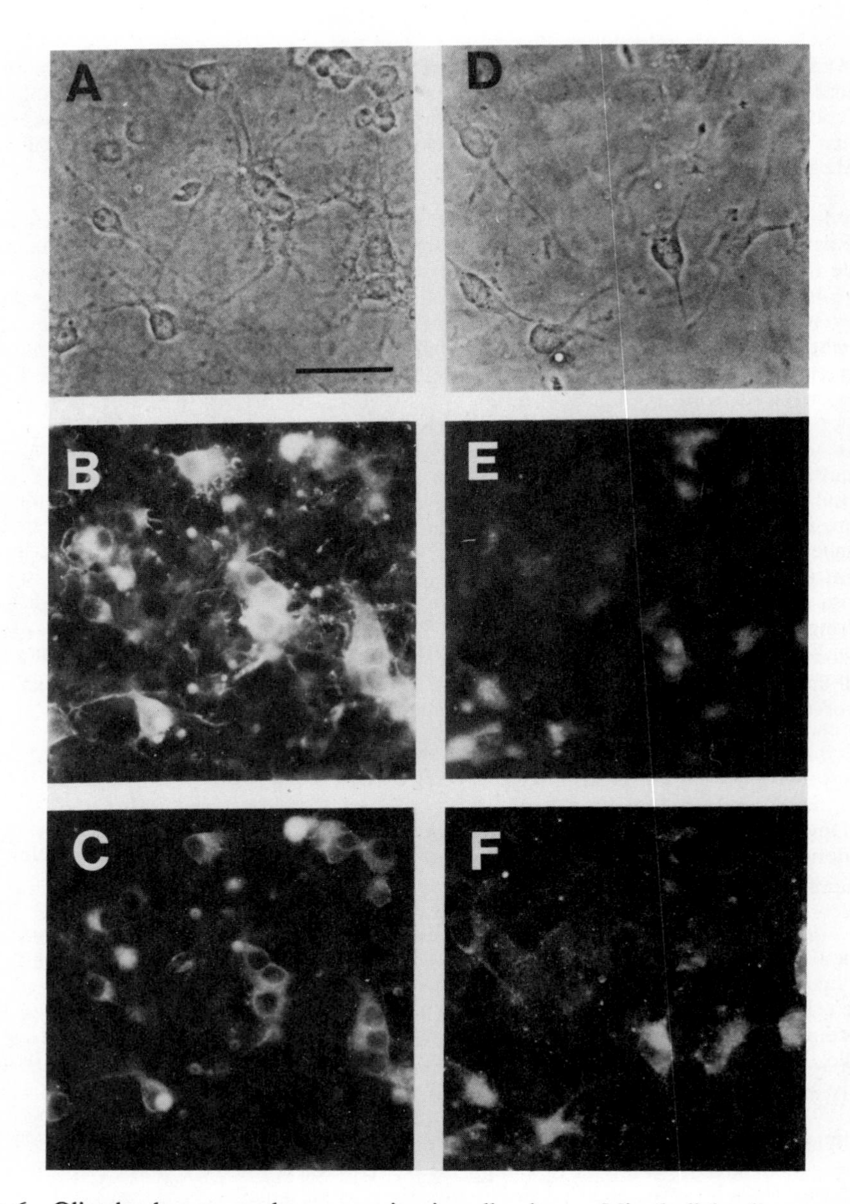

Figure 6. Oligodendrocyte marker expression in cell culture. Mixed glial cultures were prepared from individual, neonatal control and *md* rat brains. After 15 days in vitro, cultures were immunostained using fluorescent techniques for the simultaneous presence of MBP- and PLP-like immunoreactivities. MBP-like immunoreactivity is visualized using a mouse monoclonal antibody (Boehringer Mannheim Biochemicals, Indianapolis, IN) and secondary antibodies conjugated to fluorescein, while PLP-like is visualized using a rabbit antiserum and secondary antibodies conjugated to rhodamine. Panel A: control culture, phase contrast microscopy; Panel B: MBP-like immunoreactivity in the same microscope field; Panel C: PLP-like immunoreactivity in the same microscope field; Panel D: *md* culture, phase contrast microscopy; Panel E: MBP-like immunoreactivity in the same microscopic field; Panel F: PLP-like immunoreactivity in the same microscopic field. For all panels, scale bar = 50 um.

were provided previously by Drs. J. Riordan, R. A. Lazzarini, S. Lewis, R. H. Wilson and A. Tobin; their gifts are gratefully acknowledged. Supported by NICHHD 06576 and DOE Contract DE-FC03-87-ER60615 to Dr. Jean de Vellis. Marcia N. Gordon was supported by Training Grants HD07228 and HD07032.

REFERENCES

Connor, J.R., Phillips, T.M., Lashman, M.R., Barron, K.D., Fine, R.E., and Csiza, C.K., 1987, Regional variation in the levels of transferrin in the CNS of normal and myelin deficient rats, J. Neurochem. 49:1523.

Csiza, C.K., and de Lahunta, A., 1979, Myelin deficiency (md). A neurologic mutant in the Wistar rat, Am. J. Pathol. 95:215.

Dentinger, M.P., Barron, K.D., and Csiza, C.K., 1982, Ultrastructure of the central nervous system in a myelin deficient rat, J. Neurocytol. 11:671.

Dentinger, M.P., Barron, K.D., and Csiza, C.K., 1985, Glial and axonal development in optic nerve of myelin deficient rat mutant, Brain Res. 344:255.

Duncan, I.D., Hammang, J.P., and Jackson, K.F., 1987a, Myelin mosaicism in female heterozygotes of the canine shaking pup and myelin deficient rat mutants, Brain Res. 402:168.

Duncan, I.D., Hammang, J.P., and Trapp, B.D., 1987b, Abnormal compact myelin in the myelin deficient rat: Absence of proteolipid protein correlates with a defect in the intraperiod line, Proc. Natl. Acad. Sci. USA 84:6287.

Espinosa de los Monteros, M.A., Roussel, G., and Nussbaum, J.L., 1986, A procedure for long-term culture of oligodendrocytes, Develop. Brain Res. 24:117.

Jackson,K.F., Hammang, J.P., and Duncan, I.D., 1987, Further observations on the glial cell population of the optic nerve of the myelin deficient rat, Soc. Neurosci. Abstracts 13:888.

Koeppen, A.H., Barron, K.D., Csiza, and Greenfield, E.A., 1988, Comparative immunocytochemistry of Pelizaeus-Merzbacher disease, the jimpy mouse, and the myelin deficient rat, J. Neurol. Sci. 84:315.

Kumar, S., Gordon, M.N., Espinosa de los Monteros, M.A., and de Vellis, J., 1988, Developmental expression of neural cell type specific mRNA markers in the myelin deficient mutant rat brain: Inhibition of oligodendrocyte differentiation, J. Neurosci. Res., in press.

Kumar, S., Chiappelli, F., Cole, R., and de Vellis, J., 1988, Glucocorticoids induce oligodendrocyte differentiation by post-transcriptional regulation of myelin basic protein and proteolipid protein, and by transcriptional regulation of glycerol phosphate dehydrogenase, submitted.

Laatsch, R.H., 1962, Glycerol phosphate dehydrogenase activity of developing rat in the central nervous system, J. Neurochem. 9:487.

Leveille, P.J., McGinnis, J.F., Maxwell, D.S., and de Vellis, J., 1980, Immunocytochemical localization of glycerol-3-phosphate dehydrogenase in rat oligodendrocytes, Brain Res. 196:287.

Macklin, W.B., Gardinier, M.V., King, K.D., and Kampf, K., 1987, An AG --> GG transition at a splice site in the myelin proteolipid protein gene in jimpy mice results in the removal of an exon, FEBS Lett. 223:417.

Milner, R.J., Nave, K.A., Lenoir, D., Ogata, J., and Sutcliffe, J.G., 1985, Nucleotide sequences of two mRNAs for rat brain myelin proteolipid protein, Cell 42:931.

Nave, K.A., Bloom, F.E., and Milner, R.J., 1987, A single nucleotide difference in the gene for myelin proteolipid protein defines the jimpy mutation in mouse, J. Neurochem. 49:1873.

Nave, K.A., Lai, C., Bloom, F.E., and Milner, R.J., 1986, Jimpy mutant mouse: A 74-base deletion in the mRNA for myelin proteolipid protein and evidence for a primary defect in RNA splicing, Proc. Natl. Acad. Sci. USA 83:9264.

Rosenbluth, J., 1987, Abnormal axoglial junctions in the myelin-deficient rat mutant, J. Neurocytol. 16:497.

Sorg, B.J.A., Agrawal, D., Agrawal, H.C., and Campagnoni, A.T., 1986, Expression of myelin proteolipid protein and basic protein in normal and dysmyelinating mutant mice, J. Neurochem. 46:379.

Yanagisawa, K., Duncan, I.D., Hammang, J.P., and Quarles, R.H., 1986, Myelin deficient rat: Analysis of myelin proteins, J. Neurochem. 47:1901.
Yanagisawa, K., Moller, J.R., Duncan, I. D., and Quarles, R.H., 1987, Disproportional expression of proteolipid protein and DM-20 in the X-linked, dysmyelinating shaking pup mutant, J. Neurochem. 49:1912.
Zeller, N.K., Hudson, L.D., Lazzarini, R.A, and Dubois-Dalcq, M., 1985, The developmental expression of MBP and PLP in myelin deficient rat, J. Cell Biol. 101:434 (abstract).

ANALYSIS OF GLIAL SCARRING IN THE MAMMALIAN CNS WITH A GFAP CDNA PROBE

Pierre Rataboul[1], Philippe Vernier[2] and Alain Privat[3]

[1] B I Chimie, 131 bd Carnot, 78110 Le Vesinet, France
[2] Lab. Neurobiologie cellulaire et moléculaire, Dept Génétique Moléculaire, CNRS, Gif-sur-Yvette, France
[3] Lab. Neurobiologie du Développement, EPHE - INSERM U.249 - CNRS UPR 41, Institut de Biologie, 34060 Montpellier, France

INTRODUCTION

Reactive gliosis, leading to the formation of glial scar, is the response of astrocytes to CNS injury. It is an ubiquitous reaction observed in a large number of pathological conditions, such as mechanical and chemical lesions, as well as degenerative processes (Fulcrand and Privat, 1977 ; Eng and De Armond, 1982). This reaction is twofold, being characterized by an astrocyte multiplication (hyperplasia) as well as an hypertrophy of the perikarya and processes (Hain et al., 1960 ; Nathaniel and Nathaniel, 1981). The main intracellular event is an increase of the number of gliafilaments, paralleled by a raise in GFAP immunoreactivity.

GFAP, since its first isolation from multiple sclerosis plaques, (Eng et al., 1970 ; Eng et al., 1971) has been demonstrated to be a major constituent of astrocytic intermediate filaments. It is chemically and immunologically distinct from the other classes of intermediate filament proteins, (Lazarides, 1980) and has been used as a marker for astrocyte development and reactivity (Latov et al., 1979). During CNS development in rodents, GFAP is detected in late gestation (Dupouey et al., 1985 ; Malloch et al., 1987) and its emergence is probably related to the reorganization of the cytoskeleton accompanying changes of cell shape (Fedoroff et al., 1984). In cultured cells, the appearance of immunoreactive GFAP is correlated with the transition from a motile cell to a fixed form (Fedoroff, 1986).

Molecular Aspects of Development and Aging of the Nervous System
Edited by J.M. Lauder
Plenum Press, New York, 1990

Very little is known about the significance of the increase of GFAP immunoreactivity during reactive astrocytosis. One element is the increase in volume, and the outgrowth of numerous processes which occurs during astrocytic scarring, which must be sustained by an increased synthesis of cytoskeletal proteins. This outgrowth of processes fulfills at least two goals : reconstruction of the blood-brain barrier, and then of the ionic homeostasy of neurons and removal of neuronal and myelinic debris (Fulcrand and Privat, 1977).

However, reactive gliosis can also have negative effects by impeding axonal regeneration and remyelination, especially in lesions involving long distance projections.

The elevation of GFAP immunoreactivity can then be considered as a reflection of the scarring phenomenon in the CNS. However, the sole immunocytochemical detection of the protein does not provide information on either glial filament synthesis and turnover, or on their modulation in pathological conditions. These questions can now be approached at the molecular level by the use of cDNA probes. To investigate these phenomena, we have identified cDNA clones encoding human GFAP, using as starting material an astrocytoma tumor overexpressing the protein. We have found that the nucleotide sequence of human GFAP exhibits an extensive homology with that previously reported for mouse (Lewis et al., 1984), thus allowing us to use it in rodent studies. This cDNA probe was used to analyze at molecular level the early events underlying the increase in the GFAP immunoreactivity in two models of gliosis, secondary to neurotoxic lesions of the rat CNS.

First, we lesioned unilateraly the dopaminergic neurons of the substantia nigra with 6-OHDA, leading to the degeneration of the corresponding terminals in the ipsilateral striatum. These experimental conditions, which mimic part of the pathological features of human Parkinson's disease, allow an accurate analysis of the chronic effect of the neuronal degeneration on the striatal projections. Then, in a second series of experiments, we performed excitotoxic lesions of the striatum with ibotenic acid (a condition known to represent a model of human Huntington's disease) in order to study the reactivity of astrocytes at the lesion site, as well as in the substantia nigra where striatal neurons partly project. Preliminary results on the short term effect of the lesion on both regions are presented here. The cDNA probe allowed the quantification of an early increase in GFAP mRNA levels in these two animal models.

TECHNIQUES

Hybridization Probes

The GFAP probe was obtained from human astrocytoma (Grade III). Total RNA was extracted following the technique of Lomedico and Saunders (1976). Poly $(A)^+$ RNA was selected by oligo (dt) cellulose chromatography and was fractionated by centrifugation in a 5 - 20 % sucrose gradient. An aliquot of each RNA fraction was translated in a rabbit reticulolyte lysate, according to Pelham and Jackson (1976). The translation products were immunoprecipitated with anti-GFAP antibodies and loaded on SDS polyacrylamide gels.

The selected mRNAs were transcribed into cDNA, and the double stranded cDNA was inserted in a pBR 322 vector at the Pst I site. Recombinant clones were displayed on nitrocellulose filters, and selected by differential screening as described in Hanahan and Meselson (1980). Purified plasmids from selected clones were bound to nitrocellulose filters, and used to select complementary mRNA from human poly $(A)^+$ RNA. After hybridization, the selected RNAs were translated *in vitro*, and the translation product immunoprecipitated with GFAP antibodies. The GFAP probes consisted in PST I fragments cleaved from the selected pHGFA1 and pHGFA2 clones.

The preproenkephalin (PPE) probe was kindly provided by S.L. Sabol ; a 0.84 kb Sma I - Sac I fragment was obtained from the plasmid pRPE-2 (Yoshikawa et al., 1984).

The α-tubulin probe consisted of a 1.3 kb cDNA encoding chicken α-tubulin (unpublished result, F. Blanot and P. Rataboul). This clone corresponds to a highly conserved sequence hybridizing with the characteristic 1.8 kb long mRNA in rat.

The cDNA inserts were purified by electroelution after agarose electrophoresis. Double-stranded cDNA were labelled by nick translation (Rigby et al., 1977) either with [α-^{32}P] dCTP (Amersham, 400 Ci/mmole) to a specific activity of 15-20x10^7 cpm/µg or with biotinyl-7 d ATP (BRL) using the BRL nick translation system.

Northern Blot Analysis

Estimation of total RNA as well as quantification of specific mRNA were performed by Northern blotting according to Faucon-Biguet et al. (1986).

Toxic Lesions of Substantia Nigra and Striatum in the Rat

Male Wistar rats (200 - 250 g) were anaesthetized with pentobarbital. Lesions of the left substantia nigra were performed by infusion of 8 µg of 6-OHDA at the stereotaxic coordinates of A: 3.7, L: 2, V: 8,6 according to Paxinos and Watson (1982). Lesioned animals as well as sham operated rats were killed by decapitation 10, 25 and 120 days after the lesion (5 to 10 individuals per group). Left and right striata were dissected out and immediately frozen at -70°C until use.

In another series of experiments, striata were extensively lesioned by two injections of 20 µg of ibotenic acid, one in the head (A: 8.7, L: 2.5, V: 5) and one in the body (A: 10.7, L:3.5, V: 5) of the corpus striatum according to Paxinos and Watson (1982). Controls were naive animals of matching age. Ibotenic acid injected and control rats were killed 2 and 5 days after the lesion (7 animals in each lesioned group and 3 animals in each control group). The brains were rapidly dissected out and processed as described above. RNA from individual frozen striata and substantia nigra was prepared and quantified as previously reported (Faucon-Biguet et al., 1986). The assays of GFAP, PPE and α-tubulin mRNA were performed by a quantitative Northern blot analysis and filter hybridization of the corresponding probes as reported by Vernier et al. (1988).

GFAP immunocytochemistry was performed on animals perfused intracardially with a phosphate buffer rinse followed by a fixative made of 4% paraformaldehyde in cacodylate buffer. The brain was dissected out and sectioned in the frontal plane at 50 µm with a Vibratome. The sections were incubated in a solution of primary antiserum made in rabbit against GFAP (DAKO) for 18 h at 4°C and then processed by the DAB technique according to Sternberger et al. (1979).

RESULTS

Identification of cDNA clones encoding human GFAP

To identify cDNA clones encoding human GFAP we have constructed a cDNA library from a human astrocytoma in the pBR 322 vector. After fractionation of the poly(A)$^+$ RNA onto a 5 - 20 % sucrose gradient, aliquots of each fraction were translated *in vitro* and immunoprecipitated with a specific anti-GFAP serum. Fractions 1 to 3 were identified to be enriched in GFAP mRNA (Fig. 1), pooled and used for construction of the cDNA library.

Approximately 3,000 colonies with an average insert size of about 1 kb were submitted to a preliminary selection by differential screening. Clones were first hybridized with a ^{32}P cDNA transcribed from the same enriched mRNA fraction as the one used for the cloning procedure. A negative screening step was then performed with cDNA synthesized from human liver RNA. Less than 100 candidate colonies were retained and 25 randomly selected clones further investigated by the positive hybrid-selection technique. Individual mRNA isolated by this procedure were translated *in vitro.* Newly synthesized proteins were immunoprecipitated by anti-GFAP antibodies and analyzed on SDS-polyacrylamide gel.

Two clones coded for a protein with an apparent molecular weight of 49,000 daltons (Fig. 2) in agreement with that reported for human GFAP (Goldman et al., 1978 ; Newcombe et al., 1982 ; Bigbee et al., 1983). This argument and the sequence data described below established that these two clones, designated pHGFAP1 and pHGFAP2, encode human GFAP. The corresponding inserts were found to be approximately 3 kb and 550 bp long. Cleavage of pHGFAP1 by Pst I yields three 1.1, 1.45 and 0.45 kb fragments named HGFAP1a, HGFAP1b and HGFAP1c, respectively.

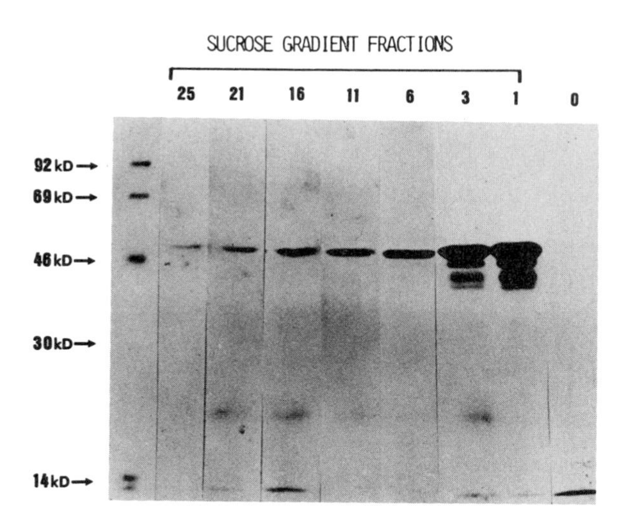

Fig. 1. Fractionation of poly(A)$^+$RNA from a human astrocytoma and identification of the GFAP mRNA translation product by immunoprecipitation. Poly(A)$^+$RNA (100 µg) was centrifuged on a 5-20% sucrose gradient. An aliquot of each fraction (1-25) was translated *in vitro* ; translation products were immunoprecipitated with a specific antiGFAP serum and analyzed by SDS-polyacrylamide gel electrophoresis. Lane 0 corresponds to *in vitro* translation control.

The cDNA inserts were partially sequenced and the data compared with those of Lewis et al. (1984) for mouse GFAP. The HGFAP1c fragment which exhibits a poly-A tail corresponds to the 3' end of the non-coding region. The adjacent HGFAP1b fragment encompasses the TGA stop codon. Altogether, the human GFAP mRNA possesses 1.8 kb of 3' untranslated sequences. This region is 0.4 kb longer than in mouse (Lewis et al., 1984 ; Balcarek and Cowan, 1985). The HGAP1c fragment is therefore characteristic of the human GFAP cDNA. Most of the coding region is contained in the HGFAP1a fragment which entirely includes the sequence of the HGFAP2 cDNA. Comparison of the nucleic acid sequence between human and mouse cDNAs has revealed nearly 90 % identity, overall. The identity is as high as 95 % in the C terminal domain between amino acids 310-403 (Fig. 3).

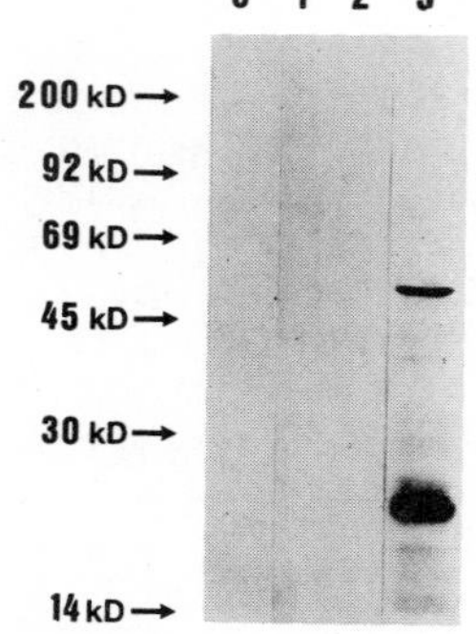

Fig. 2. Identification of a human GFAP cDNA clone by hybrid selection and *in vitro* translation. Recombinant plasmid DNAs immobilized on nitrocellulose were used to select mRNAs from human astrocytoma poly(A)$^+$RNA. Translation of the corresponding RNAs were analyzed by immunoprecipitation with specific anti-GFAP antibodies and SDS-polyacrylamide gel electrophoresis. **Lane 0** : pBR 322 ; **Lanes 1** and **2** : recombinant plasmids with undentified insert ; **Lane 3** : pHGFAP1.

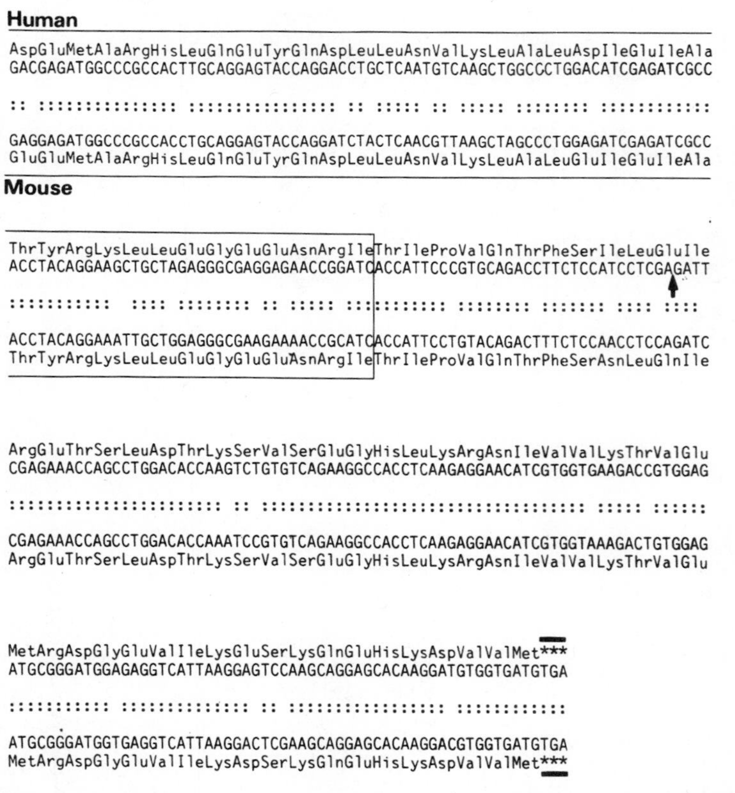

Fig.3. Alignment of the nucleotide and predicted C-terminal amino acid sequence of mouse and human GFAP cDNA. Helical region predicted by the intermediate filaments model (Geisler et al. 1982) is boxed. , Pst I site between HGFAP1a and HGFAP1b fragments ; --, stop codon.

Northern Blot Analysis

To determine the size of the human GFAP mRNA, total and poly(A)$^+$ RNA from various human tissues was analyzed on denaturing agarose gel ; corresponding blots, when hybridized with HGFAP 2 probe, reveal one single RNA species with an apparent size of 3.1 kb in human astrocytoma as well as in nontumoral brain tissues from an 80-year-old man (Fig. 4a and b). As expected, no hybridization signal was detected in liver RNA.

The extensive conservation of the human GFAP cDNA sequence among several mammalian species (Rataboul et al., 1988) enabled us to analyze the distribution of GFAP mRNA in various regions of the rat brain by gel blotting of poly (A)$^+$ RNA. In all tissues tested, the GFAP cDNA probe hybridized to a single RNA species which exhibits a motility of 2.8 kb (Fig. 5). As expected, no signal was detected in the pituitary gland even after long exposure time.

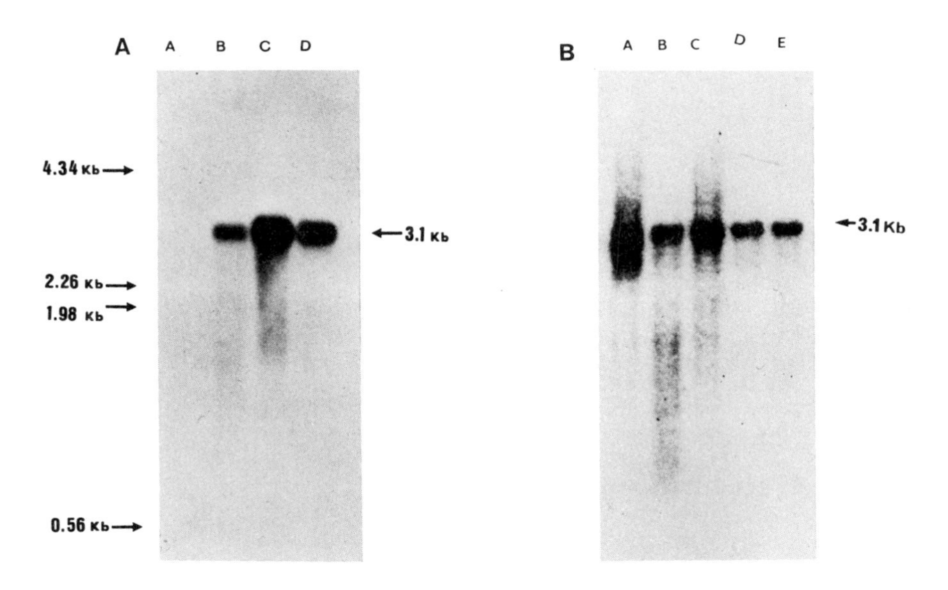

Fig. 4. Northern blot analysis of human GFAP mRNA from various tissues. Total or poly(A)$^+$RNAs were denatured and separated by electrophoresis in a 1% formaldehyde/agarose gel. After migration, RNAs were transferred to nitrocellulose and hybridized with the HGFAP1a probe. **a:** Human liver (2 µg poly(A)$^+$ RNA ; **Lane A**) ; human astrocytoma (500 ng poly(A)$^+$ RNA, **Lane B**) ; 500 ng poly(A)$^+$RNA from the sucrose gradient fractions 1 **(lane C)** and 3 **(lane D)**. **b:** Human cerebellum (3 µg total RNA, **lane A**), human substantia nigra (1 µg total RNA ; **lane B**), human cortex (2 µg total RNA, **lane C**), 350 ng poly(A)$^+$ RNA from two different human astrocytomas (**lanes D** and **E**).

Time Course of GFAP, PPE and α-Tubulin mRNA in the Striatum following 6-OHDA Lesion of substantia nigra

GFAP mRNA levels elevated only in the ipsilateral striatum after the dopaminergic deafferentiation elicited by 6-OHDA lesion of substantia nigra (Fig. 6). The highest amount of GFAP mRNA was observed 10 days after the lesion (1.4-fold), the shortest postlesion time we studied. This level then continuously decreased with time and returned to the control level 4 months post lesion.

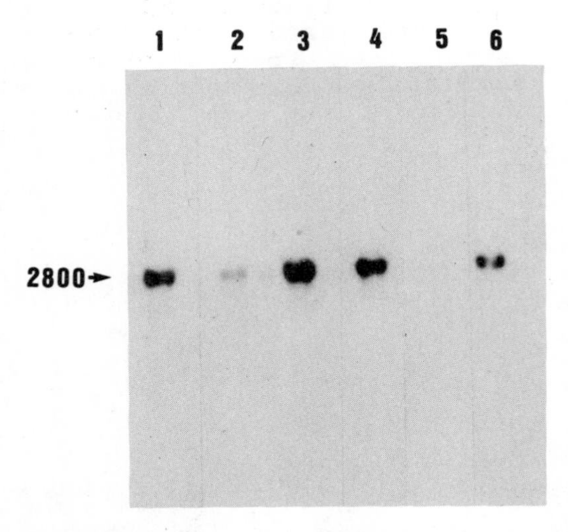

Fig. 5. Northern blot analysis of rat GFAP mRNA from various tissues. RNAs were extracted from: 1=total brain ; 2=striatum ; 3=cerebellum ; 4=cortex ; 5=pituitary gland ; 6=thalamus. Poly $(A)^+$ RNA (3 μg) from each tissue were electrophoresed in a 0.8 % formaldehyde/agarose gel, blotted and hybridized with the [^{32}P] labeled HGFAP 1a cDNA probe. Filters were exposed to autoradiography for 48 h with an intensification screen at -70°C.

The influence of denervation was also observed on the levels of two other mRNA species. PPE mRNA, a well-established marker of the neuronal target of the nigrostriatal pathway, increased very slowly to reach a maximum on day 25 and remained slightly elevated 3 months later ; α-tubulin mRNA level, a good marker of cellular outgrowth and synaptogenesis, showed no significant modification in the deafferented striatum as compared to controls. In addition, the lesion had no obvious effect on any of the mRNA levels studied in the contralateral striatum.

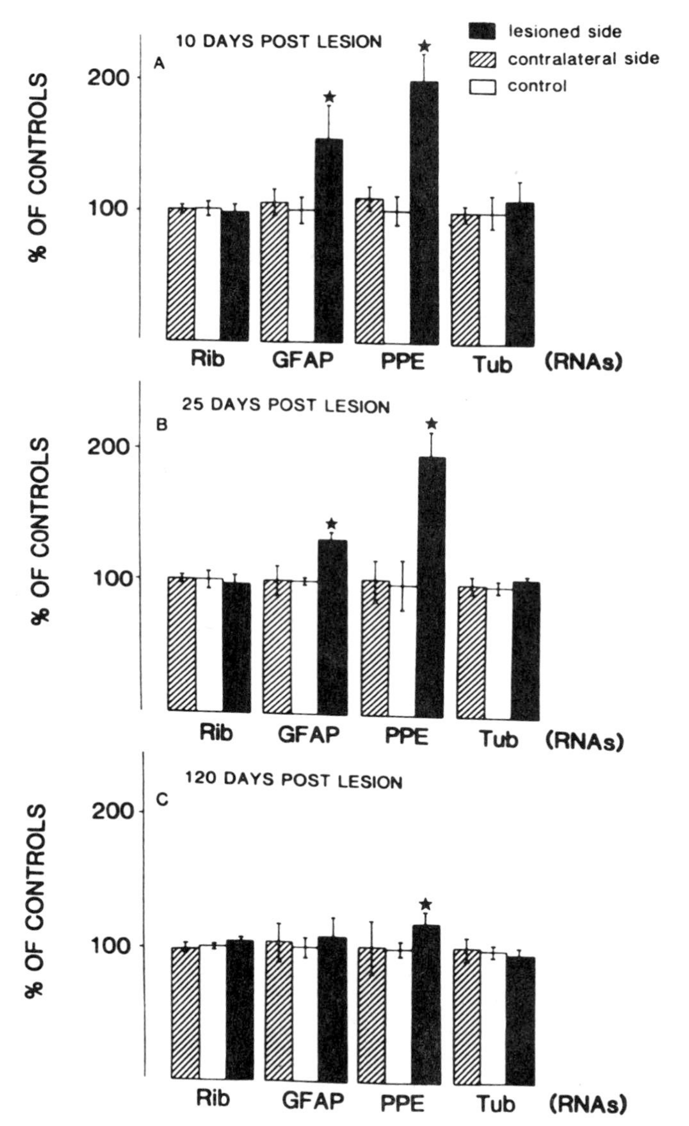

Fig. 6 . Time course change of the ribosomal (Rib), GFAP, PPE and α-tubulin mRNA levels in the deafferented and contralateral striatum after 6-OHDA lesion of substantia nigra. Ten days (A), 25 days (B) and 120 days (C) after the lesion, 8 μg of total RNA extracted from both striata were analyzed in the same gel from each postlesion time. Five to 10 lesioned and control animals were used per postlesion time. Amount of specific mRNA in deafferented and contralateral striatum is expressed as percent (± s.e.m.) of mRNA levels measured in the control striatum (* $p<0.005$, $n=5$ to 10).

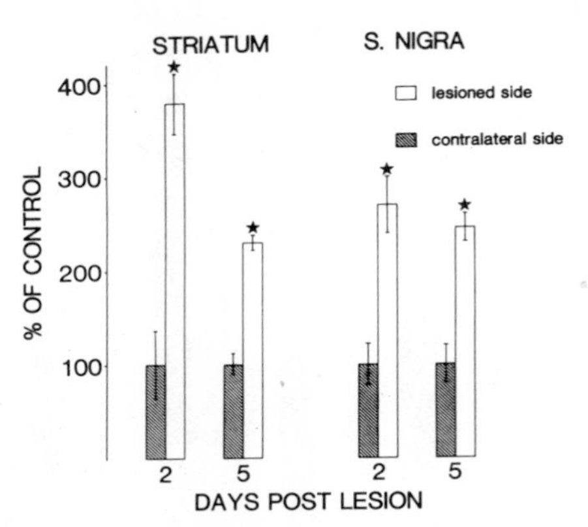

Fig. 7. Modulation of GFAP mRNA levels in the striatum and the substantia nigra following ibotenic acid lesion of the striatum. Two days (2) and 5 days (5) after unilateral ibotenate lesion of the striatum, 10 µg of total RNA extracted from both striata and substantia nigra were analyzed on the same gel. Amount of specific mRNA in deafferented tissues is expressed as percent (± s.e.m.) of mRNA levels measured in contralateral side (* $p<0.005$, $n=7$).

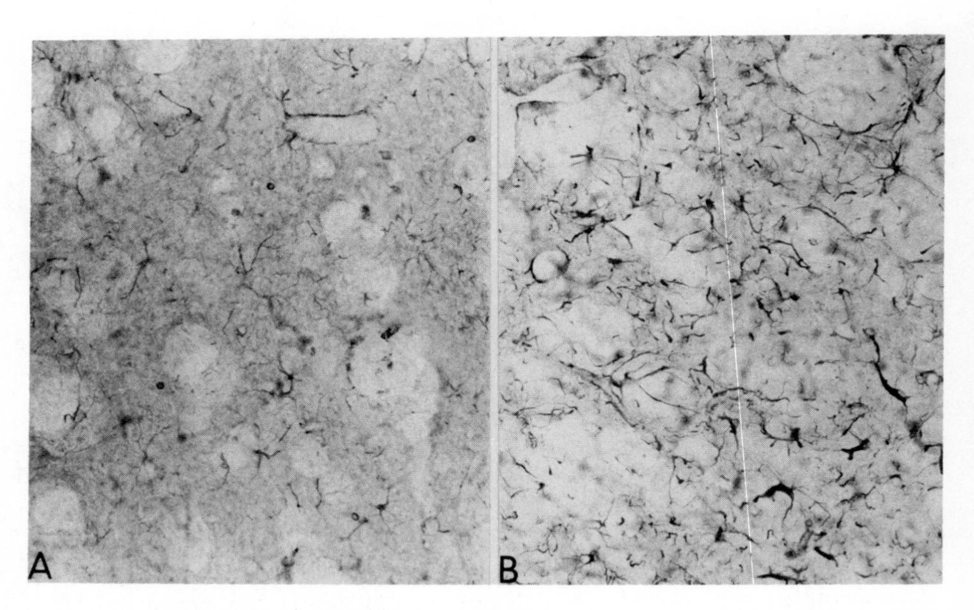

Fig. 8. Ibotenic acid-induced gliosis of striatum as seen 5 days after injection (A) as compared to the contralateral noninjected side (B). Note the increase of GFAP immunoreactivity in striatal astrocytes in the ipsilateral side.

Short Term Changes of GFAP mRNA Levels and Immunostaining After Excitotoxic Lesion of the Striatum

GFAP mRNA levels were markedly increased in the striatum lesioned by ibotenic acid, reaching 3.8-fold the level measured in the ipsilateral side 2 days after the lesion (Fig. 7). No change in GFAP mRNA levels was observed in the contralateral side of lesioned animals as compared to naive animals and were therefore used as controls. This level was already lower (2.2-fold the control level) 3 days later. These modifications contrasted with the more delayed increase of GFAP immunostaining in this structure. Indeed, GFAP immunoreactivity per se was also elevated within the cells and a greater number of astrocytes of larger size were labelled.

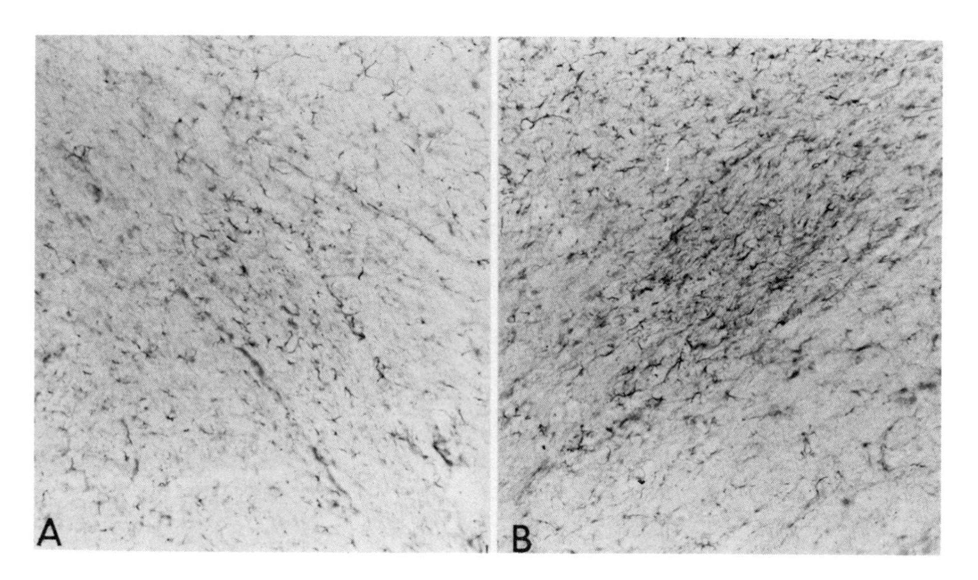

Fig. 9. Changes seen in the substantia nigra area 5 days after unilateral injection of ibotenic acid in the striatum. GFAP immunostaining is moderately increased in the ipsilateral substantia nigra (B) when compared to the corresponding striatum. No modification in GFAP immunoreactivity has been observed in the contralateral side (A).

These changes are much more pronounced on the 5th-postlesion day than on the second day (Fig. 88). GFAP mRNA levels were also elevated in the ipsilateral substantia nigra in response to the striatal lesion, although this increase was less pronounced than at the lesion site (2.5-fold above the control level, two days after the lesion). This level was maintained on day 5 postlesion (Fig. 7). Nigral GFAP immunoreactivity was still faint 2 days after ibotenate injection but was markedly enhanced at day 5 although it always remains less intense than in the striatum. Here again no modification of either GFAP mRNA levels (data not shown) or immunolabeling (Figs. 8 and 9) has been noticed in the contralateral side.

DISCUSSION

We have summarized here our recent work on the identification of cDNA clones encoding human GFA protein, and their use as a tool to study glial scarring in the central nervous system of mammals.

pHGFAP1 is an almost full length cDNA clone which encompasses the bulk of the 3' untranslated region of the human GFAP mRNA. In human, this 3' non coding sequence is 0.4 kb longer than that of the mouse (Lewis et al., 1984). The corresponding specific nucleotidic sequence is entirely included in the HGFAP1c fragment which permits the human GFAP sequence to be specifically recognized in hybrid cells or transgenic mice.

All intermediate filament subunits are built according to a common organization : each presents a central alpha-helical rod domain of 311-314 amino acids which has a highly conserved secondary structure (see Steinert et al., 1985 for review). Moreover desmin, vimentin and GFAP subunits have remarkably similar rod domain sequences (Geisler and Weber, 1981, 1982 ; Quax et al., 1983 ; Lewis et al., 1984). In contrast, the amino and carboxyl-terminal domains of intermediate filament subunits vary greatly in both size and amino acid sequence and may confer structural, functional and antigenic individuality. In this respect our data disclosed a very strong homology between human and mouse GFAP even in the end domains. For example, the 300 residues upstream of the stop codon, which comprise the codons translating the nonhelical COOH-terminal region, are 90 % homologous to the mouse cDNA, leading to 95 % homology for the amino acid sequence (Fig. 3). This high conservation of the protein structure in the two species substantiates the strong antigenic cross-reactivity of human antibodies which usually recognize most of the mammalian antigens. Our data support the evolutionary conservation of GFAP gene across species boundaries.

Southern blot experiments indicate that HGFAP mRNA is transcribed from a unique gene comprising two intrinsic Hind III recognition sites ; this is in good agreement with the results of Lewis et al. (1984), and Balcareck and Cowan (1985). A similar experiment, carried out with rat and chicken genomic DNA, again revealed a single band when hybridized with HGFAP1a probe. This finding is consistent with the interspecies cross reactivity of the probe which has also been demonstrated by Northern blot analysis. By this technique the probe characterized a 2.7 kb and 2.8 kb mRNA in mouse and rat respectively. In human, HGFAP1 or HGFAP2 cDNAs always revealed a unique mRNA species (3.1 kb) in the various tissues studied. *In vitro* translation of RNAs from the human astrocytoma gave a single band of about 49,000 daltons in good agreement with the molecular weight of GFAP described in the literature. However, there are consistent findings of lower molecular weight forms of GFAP in normal human brain (Goldman et al., 1978 ; Dahl and Bignami, 1973), in gliomas (Bigbee et al., 1983) or in demyelinating diseases (Newcombe et al., 1982). Thus, these proteins most probably correspond to breakdown products of the 49,000 daltons form as suggested by Bigbee et al. (1983), possibly due to the activity of a Calcium dependent protease (Schlaepfer and Zimmerman, 1981 ; De Armond et al., 1983).

Increased GFAP immunoreactivity has been described as one of the landmarks of reactive astrocytosis (Latov et al., 1979).

In this preliminary study we describe a very early increase of GFAP mRNA levels and the following long term decrease of mRNA amount promoted by two different kinds of toxic lesions in the rat. These experimental models involve anatomically and biochemically well defined systems allowing an accurate appreciation of the neuronal degeneration and glial reaction. GFAP mRNAs were quantified by a sensitive Northern blot analysis (Faucon-Biguet et al., 1986) using a human GFAP cDNA clone (Rataboul et al., 1988). The probe recognizes, in the various rat nervous tissues studied, a single 2.8 kb mRNA species whose tissue distribution nicely matches that of the protein (Patel et al., 1985 ; Weir et al., 1984). These data are different from the 2.0 kb long GFAP mRNA identified in the rat spinal cord by Kitamura et al. (1987), but fit well with the 2.7 kb mRNA species found in mouse brain (Lewis et al., 1984). These contradictory results might be due to inaccuracy in the molecular weight calibration.

In the 6-OHDA lesion model, the substantial increase in striatal GFAP mRNA levels that we observed 10 days after the toxic infusion in the substantia nigra, attenuates progressively throughout the postlesion time we have studied. Most probably, the maximal RNA level is reached earlier than 10 days. Indeed, in the ibotenate striatal lesion model, the strongest elevation of the GFAP transcripts (4-fold) is observed as soon as 2 days after the lesion, whereas it already decreases at 5 days postlesion. In this model, the amount of GFAP mRNA also rises dramatically in the substantia nigra, one of the projection areas of striatal neurons, although it follows a rather different time course. A similar level of GFAP mRNA is observed both at day 2 and 5 after the lesion and may result from the delayed degeneration of striatal terminals as compared to the immediate destruction of striatal cells. In contrast, no diffuse reaction seems to occur in non projection area as certified by the stability of the level of GFAP transcripts in the contralateral side. Therefore, the degeneration of terminals which propagates to projection areas seems to be able, by itself, to promote the astrocytic reaction. Our results confirm and extend the observations of Stromberg et al. (1986) and Isacson et al. (1987), who found a short delayed increase in GFAP immunoreactivity both at the injection site and in the projection areas after 6-OHDA nigral lesion by infusion of ibotenate in the striatum, respectively. Moreover, the time course of the progressive reduction of GFAP immunoreactivity in projection areas is identical to the one we have described for mRNA levels. In addition, no modification of GFAP immunostaining was detected in nonprojection territories. As shown in the ibotenate lesion model, the increase in GFAP transcript levels shortly precedes the rise of GFAP immunoreactivity indicating that the latter, results in part from a neosynthesis and not from post-translational modifications such as phosphorylation, depolymerisation or degradation of GFAP as has been questioned (Eng et al., 1987). In this respect, it should be of interest to study if the "soluble pool" of the protein, claimed to be a native intermediary in the gliofilaments synthesis (Malloch et al., 1987) is modified in reactive astrocytosis. Nonetheless, the turn-over of the polymerized GFAP is low (De Armond et al., 1986 ; Smith et al., 1984) and the corresponding immunoreactivity can still persist for a long time even after the mRNA has returned to control levels.

Clearly, not all the cytoskeletal proteins are involved in reactive astrocytosis. Using the model of nigral lesion by 6-OHDA, we were unable to detect any modifications of a α-tubulin mRNA levels in the striatum. Since modulations of the expression of this mRNA seems to be mainly associated with axonal growth or sprouting and synaptogenesis (Bhattacharya et al., 1987), such a phenomenon probably does not occur in this model. Therefore, this mRNA was used as an internal standard in our RNA quantitations. A quite different time course was observed for the changes in the level of PPE transcripts. It elevated more slowly and for a longer period of time, remaining still increased 4 months after the lesion. It is attractive to hypothesize that the signal which triggers this phenomenon does not have the same origin as that which promotes reactive gliosis. Met-enkephalinergic neurons in the striatum are well-documented targets of the nigrostriatal dopaminergic pathway (Kubota et al., 1986). Dopamine exerts on met-enkephalinergic neurons a tonic inhibitory influence (Sivam et al., 1986 ; Tang et al., 1983) ; the disinhibition activates the neuronal firing and neurotransmitter release. The resulting increased metabolism of this short half-lived peptide is thus accompanied by a corresponding elevation of mRNA amounts. In addition, these modulations of neuronal activity and metabolism participate in the adaptive processes following neuronal degeneration. A long loop compensatory mechanism is thought to occur in this experimental situation, which is able to lower met-enkephalinergic neurons'hyperactivity as has been demonstrated for other behavioural and biochemical index (Marshall et al., 1985).

We have examined more systematically the modulations of GFAP mRNA levels and the corresponding immunoreactivity at short and long term intervals in the two models of lesions. Although GFAP mRNA levels (and immunoreactivity as demonstrated in Strömberg et al., 1986) return to normal in the striatum after the 6-OHDA nigra lesion, it is probably not the case at the lesion site. A long-lasting GFAP immunoreactivity increase is also described at the lesion site after ibotenate injection (Isacson et al., 1987). This phenomenon can be deleterious to the neuronal survival creating thereby a self-sustained process similar to that thought to occur in Huntington's disease, or at the chronic stage of mechanical lesions. Whether astrocytes have to be in a reactive state to affect axonal growth and neuronal survival, needs to be clarified. In this respect, we have observed, by *in situ* hibridization, the translocation of GFAP transcripts along the astrocyte processes. This phenomenon is probably related to the cell shape modification and attendant loss of motility which occur in reactive astrocytes.

In situ hybridization should also help us to gain insights as to the heterogeneity of reactive astrocyte populations and the mechanisms which promote them. Although much more work is needed to be able to propose new therapeutic strategies for CNS repair, some clues are already available. The reaction of astrocytes to lesion, as monitored by GFAP mRNA quantitation, seems to be quasi-immediate. Therefore an early intervention in the lesional centre might avoid deleterious effects of glial scarring on neuronal regeneration and axonal remyelination which occurs not only near the lesion site but also in the projection area.

ACKNOWLEDGEMENTS

The authors gratefully acknowledge J. Mallet, N. Faucon-Biguet, F. De Vitry, P. Poulat, L. Marlier, and N. Rajaofetra for their help during the course of this investigation, J.R. Teilhac for photography and C. Bernié for typing the manuscript.

This work was supported in part by a grant from IRME, and by grants from CNRS, INSERM, and Rhône-Poulenc Santé. P. Rataboul was a fellow of Association Claude Bernard and P. Vernier is a fellow of the Fondation de l'Industrie Pharmaceutique pour la Recherche.

REFERENCES

Balcarek, J. M., and Cowan, N. J., 1985, Structure of the mouse glial fibrillary acidic protein gene : implications for the evolution of the intermediate filament multigene family, Nucleic Acid Res., 13:5527-5543.

Bhattacharya, B., Mandal, C., Basu, S., and Sarkar, P. K., 1987, Regulation of α- and β-tubulin mRNAs in rat brain during synaptogenesis, Mol. Brain Res., 2:159-162.

Bigbee, J. W., Bigner, D. D., Pegram, C., and Eng, L. F., 1983, Study of glial fibrillary acidic protein in a human glioma cell line grown in culture and as a solid tumor, J. Neurochem., 40:460-467.

Dahl, D., and Bignami, A., 1973, Glial fibrillary acidic protein from normal human brain. Purification and properties, Brain Res., 57:343-360.

De Armond, S. J, Fajardo, M., Naughton, S. A., and Eng, L.F., 1983, Degradation of glial fibrillary acidic protein by a calcium dependent proteinase : an electroblot study, Brain Res., 262:275-282.

De Armond, S. J., Lee, Y. L., Kretzschmar, H. A., and Eng, L. F., 1986, Turnover of glial filaments in mouse spinal cord, J. Neurochem., 47:1749-1753.

Dupouey, P., Benjelloun-Touini, S., and Gomes, D., 1985, Histochemical demonstration of an organized cytoarchitecture of the radial glia in the CNS of the embryonic mouse. Dev. Neurosci., 7:81-93.

Eng, L. F., and De Armond, S. J., 1982, Immunocytochemical studies of astrocytes in normal development and disease. Adv. Cell Neurobiol., 3:145-171.

Eng, L. F., Gerstl, B., and Vanderhaeghen, J. J., 1970, A study of proteins in old multiple sclerosis plaques, Trans. Amer. Soc. Neurochem., 1:42.

Eng, L. F., Reier, P. J., and Houle, J. D., 1987, Astrocyte activation and fibrous gliosis : glial fibrillary acidic protein immunostaining of astrocytes following intraspinal cord grafting of fetal CNS tissue, in: "Progress in brain research", F. J. Seil, E. Herbert, B. M. Carlson, eds, Elsevier Science Publishers B. V., Biomedical Division, Amsterdam, Vol. 71, pp 439-455.

Eng, L. F., Vanderhaegen, J. J., Bignami, A., and Gerstl, B., 1971, An acidic protein isolated from fibrous astrocytes, Brain Res., 28:351-354.

Faucon-Biguet, N., Buda, M., Lamouroux, A., Samolyk, D., and Mallet, J., 1986, Time course of the changes of TH mRNA in rat brain and adrenal medulla after a single injection of reserpine, EMBO J., 5:287-291.

Fedoroff, S., 1986, Prenatal ontogenesis of astrocytes, in: "Astrocytes", S. Fedoroff, and Vernadakis eds, Academic Press, Orlando, San Diego, New York, Austin, Boston, London, Sydney, Tokyo, Toronto, Vol. 1, pp 35-67.

Fedoroff, S., Neal, J., Opas, M., and Kalnius, V. I., 1984, Astrocyte cell lineage. III The morphology of differentiating mouse astrocytes in colony culture, J. Neurocytol., 13:1-20.

Fulcrand, J., and Privat, A., 1977, Neuroglial reactions secondary to Wallerian degeneration in the optic nerve of the postnatal rat : Ultrastructural and quantitative study. J. Comp. Neurol., 176, 189-221.

Geisler, N., and Weber, K., 1981, Comparison of the proteins of two immunologically distinct intermediate-sized filaments by amino acid sequence analysis : Desmin and vimentin, Proc. Natl. Acad. Sci. USA, 78:4120-4123.

Geisler, N., and Weber, K., 1982, The amino acid sequence of chicken muscle desmin provides a common structural model for intermediate filament proteins, EMBO J., 1:1649-1656.

Goldman, J. E., Schaumburg, H. H., and Norton, W. T., 1978, Isolation and characterization of glial filaments from human brain, J. Cell. Biol., 78:426-440.

Hain, R. F., Rieke, W. O., and Everett, N. B., 1960, Evidence of mitosis in neuroglia as revealed by radioautography employing tritiated thymidine, J. Neuropathol. Exp. Neurol., 19:147-148.
Hanahan, D., and Meselson, M., 1980, Plasmid screening at high colony density, Gene, 10:63-67.
Isacson, O., Fischer, W., Wictorin, K., Dawbarn, D., and Björklund, A., 1987, Astroglial response in the excitotoxically lesioned neostriatum and its projection areas in the rat, Neuroscience, 20:1043-1056.
Kitamura, T., Nakanishi, K., Watanabe, S., Endo, Y., and Fujita, S., 1987, GFA-protein gene expression on the astroglia in cow and rat brains, Brain Res., 423:189-195.
Kubota, Y., Inagaki, S., Kito, S., Takagi, H., and Smith, A. D., 1986, Ultrastructural evidence of dopaminergic input to enkephalinergic neurons in rat neostriatum, Brain Res., 367:374-378.
Latov, N., Nilaver, G., Zimmerman, E. A., Johnson, W. G., Silverman, A. J., Defendini, R., and Cote, L., 1979, Fibrillary astrocytes proliferate in response to brain injury : A study combining immunoperoxidase technique for glial fibrillary acidic protein and radioautography of tritiated thymidine, Develop. Biol., 72:381-384.
Lazarides, E., 1980, Intermediate filaments as mechanical integrators of cellular space, Nature (London), 283:249-256.
Lewis, S. A., Balcarek, J. M., Krek, V., Shelanski, M., and Cowan, N. J., 1984, Sequence of a cDNA clone encoding mouse glial fibrillary acidic protein : Structural conservation of intermediate filaments, Proc. Natl. Acad. Sci. USA., 81:2743-2746.
Lomedico, P. T., and Saunders, G. F., 1976, Preparation of pancreatic mRNAS cell-free translation of an insulin-immunoreactive peptide, Nucleic Acids Res., 2:381-391.
Malloch, G. D. A., Clark, J. B., and Burnet, F. R., 1987, Glial fibrillary acidic protein in the cytoskeletal and soluble protein fractions of the developing rat brain, J. Neurochem., 48:299-306.
Marshall, J. F., 1985, Neural plasticity and recovery of function after brain injury, in: "International review of neurobiology, J. R. Smythies, R. J. Bradey, eds, Academic Press, London, Vol. 26, pp 201-247.
Nathaniel, E.J.H., and Nathaniel DR, 1981, The reactive astrocyte, in: "Advances in Cellular Neurobiology", S. Federoff and L. Hertz, eds, Academic Press, New York, Vol. 2, pp 249-301.
Newcombe, J., Glynn, P., and Cuzner, M. L., 1982, The immunological identification of brain proteins on cellulose nitrate in human demyelinating disease, J. Neurochem., 38:267-274.
Patel, A. J., Weir, M. D., Hunt, A., Tahourdin, C. S. M., and Thomas, D. G. T., 1985, Distribution of glutamine synthetase and glial fibrillary acidic protein and correlation of glutamine synthetase with glutamate decarboxylase in different regions of the rat central nervous system, Brain Res., 331:1-9.

Paxinos, G., and Watson, C., 1982, "The rat brain in stereotaxic coordinates", Academic Press, Sydney.
Pelham, H. R. B., and Jackson, R. J., 1976, An efficient mRNA-dependent translation system from reticulocyte lysates, Eur. J. Biochem., 67:247-256.
Quax-Jeuken, Y. E. F. M., Quax, W. J., and Bloemendal, H., 1983, Primary and secondary structure of hamster vimentin predicted from the nucleotide sequence, Proc. Natl. Acad. Sci. USA, 80:3548-3552.
Rataboul, P., Faucon-Biguet, N., Vernier, P., De Vitry, F., Boularand, S., Privat, A., and Mallet, J., 1988, Identification of a human GFAP cDNA : A tool for the molecular analysis of reactive gliosis in the mammalian CNS, J. Neurosci. Res., 20:165-175.
Rigby, P. W. J., Dieckmann, M., Rhodes, C., and Berg, P., 1977, Labelling deoxyribonucleic acid to high specific activity *in vitro* by nick translation with DNA polymerase I, J. Mol. Biol., 113:237-251.
Schlaepfer, W. W., and Zimmerman, V. J. P., 1981, Calcium mediated breakdown of glial filaments and neurofilaments in rat optic nerve and spinal cord, Neurochem. Res., 6:243-255.
Sivam, S. P., Strunck, C., Smith, D. R., and Hong, J. S., 1986, Proenkephalin-A gene regulation in the rat striatum : influence of lithium and haloperidol. Mol. Pharmacol., 30:186-191.
Smith, M. E., Perret, V., and Eng, L. F., 1984, Metabolic studies *in vitro* the CNS cytoskeletal proteins : synthesis and degradation, Neurochem. Res., 9:1493-1507.
Steinert, P. M., Steven, A. C., and Roop, D. R., 1985, The molecular biology of intermediate filaments, Cell, 42:411-419.
Sternberger, I. A., 1979, "Immunocytochemistry", 2nd ed., J. Wiley, ed., New-York, pp 104-170.
Strömberg, I., Björklund, H., Dahl, D., Jonsson, G., Sundström, E., and Olson, L., 1986, Astrocyte responses to dopaminergic denervations by 6-hydroxydopamine and 1-methyl - 4 - phenyl - 1,2,3,6 - tetrahydropyridine as evidenced by glial fribrillary acidic protein immunohistochemistry, Brain Res. Bull., 17(2):225-236.
Tang, F., Costa, E., and Schwartz, J. P., 1983, Increase of proenkephalin mRNA and enkephalin content of rat striatum after daily injection of haloperidol for 2 to 3 weeks, Proc. Natl. Acad. Sci. USA, 80:3841-3844.
Vernier, P., Julien, J. F., Rataboul, P., Fourrier, O., Feuerstein, C., and Mallet, J., 1988, Similar time course changes in striatal levels of glutamic acid decarboxylase and proenkephaline mRNA following dopaminergic deafferentiation in the rat, J. Neurochem., 51:1375-1380.
Weir, M. D., Patel, A. J., Hunt, A., and Thomas, D. G. T., 1984, Developmental changes in the amount of glial fibrillary acidic protein in three regions of the rat brain, Dev. Brain Res., 15:147-154.
Yoshikawa, K., Williams, C., and Sabol, S.L., 1984, Rat brain preproenkephalin mRNA : cDNA cloning, primary structure, and distribution in the central nervous system, J. Biol. Chem., 259:14301-14308.

REGULATION OF THE GLIAL FIBRILLARY ACIDIC PROTEIN (GFAP) AND OF ITS ENCODING mRNA IN THE DEVELOPING BRAIN AND IN CULTURED ASTROCYTES

M. Tardy, C. Fages, G. Le Prince, B. Rolland, and J. Nunez

INSERM U-282 Hôpital Henri Mondor
94010 Créteil, France

INTRODUCTION

The Glial Fibrillary Acidic Proctein (GFAP) is the monomer of a well characterized type of intermediary filaments, the fliofilaments, structurally identified as 10nm in diameter and which are essential components of the cytoskeletal architecture of the astrocyte (see Eng 1980 for a review). The expression of GFAP has been found to be highly specific of this cell type (Eng et al. 1971; Bignami et al. 1972; Uyeda et al. 1972; Gilden et al.1976: Ludwin et al. 1976; Ludwin et al. 1976; Lach and Weinmander 1978) and may therefore be used as an exclusive marker of astroglial cells.

The monomeric form of GFAP has a molecular weight of 49 Kdaltons (see Eng and DeArmond 1983 for a review); GFAP immunologically-related entities with lower molecular weights (47-28 Kdaltons) have been also described and are probably produced by stepwise proteolytic breakdown which occur "in vivo" but is accelerated "in vitro" in the presence of Ca^{2+} ions (Schaepfer and Zimmerman 1981, Bighee et al. 1983, DeArmond et al. 1983)., Charge heterogeneity of GFAP has been also described (Bighee and Eng 1982) but is not clear whether these GFAP variants are produced only by post translational reactions such as phosphorylation or if, at least, some of them are different gene products. Southern blot analysis have shown that there are at least two genes encoding GFAP in the mouse genome (Lewis et al. 1984).

Only limited sequence data have been obtained by direct protein sequencing (Hong and Davison 1981). Recently, more than 97% of the amino acid sequence of GFAP has been established from that of a cloned GFAP cDNA (Lewis et al. 1984). The major feature of the sequence (Lewis et al. 1984) is an extensive homology of GFAP with that of vimentin (67%) and other intermediate filament proteins (geisler and Weiber 1982; Geisler et al. 1982). Such a homology was expected since GFAP and vimentin coassemble "in vitro" (Quinlan and Francke 1983a,b), as well as "in vivo" and in cultured astrocytes (Eng and Kosek 1974; Sotelo et al. 1980; Yen and Fields 1981; Chiu et al. 1981; Osborn et al. 1981; Schnitzer et al. 1981; Sharp et al. 1982; Wang et al. 1984). Yet, GFAP homopolymers have been demonstrated to be present in intact cells (Virtanen et al. 1981), in isolated gliofilament fractions (Dahl

Molecular Aspects of Development and Aging of the Nervous System
Edited by J.M. Lauder
Plenum Press, New York, 1990

and Bignami 1979; Goldman et al. 1978; Liem et al. 1978; Liem 1982; Paetau et al. 1985; Schlaepfer and Zimmerman 1981) and to be produced by "in vivo" assembly (Lucas et al. 1980; Rueger et al. 1979). Vimentin homopolymers are found at early developmental stages whereas filaments made up with GFAP alone are present in differentiated and mature brain astrocytes (Chiu et al. 1981; Dahl 1981; Dahl et al. 1981; Herpers et al. 1986). However in primary astroglial cultures the expression of vimentin remains high even in the morphologically mature cells (Schnitzer et al. 1981).

During the time course of "in vitro" astroglial cell differentiation, the organization and accumulation of gliofilaments increase in parallel to the concentration of GFAP (Sensenbrenner et al. 1980). GFAP immunofluorescence is detected in the perinuclear region of the immature cells before the rest of the cytoplasms is labeled whereas in the mature cells most of the GFAP staining is localized in the branched processes.

Factors such as neurotransmitters (McCarthy et al. 1985; Pollenz and McCarthy 1986), drugs which activate the adenylate cyclase and cyclic AMP derivatives (Sensenbrenner et al. 1980; Duffy 1983; Goldman et al. 1984a,b; Bridoux et al. 1985), phorbol esters which activate protein kinase C (Honneger 1986) hormones (Morrison et al. 1986) and a glia maturation factor (Suzuki et al. 1986) seem to increase GFAP immunoreactivity and the number of gliofilaments in the cell processes of the differentiating astrocytes. However, the growth factors AGF_1 and AGF_2 (Pettman et al. 1985), which induce a change in morpohology different from that observed in the presence of cyclic AMP derivatives (Weibel et al. 1987) increase the apparent number of gliofilaments without changing the concentration of GFAP (Weibel et al. 1985).
The bulk of information summarized above suggests that the expression of GFAP and its polymerization to gliofilaments are important hallmarks of the process of astroglial morphological differentiation and that these parameters are differently modulated depending on the regulatory signal. However, how astrocytes control at the molecular level the expression of its major filamentous protein is still poorly understood. The aim of our work has been to know whether transcriptional or/and post transcriptional mechanisms are responsible for the changes in GFAP expression and assembly seen during proliferation and differentiation of the astroglial cells.

GFAP SYNTHESIS AND EXPRESSION OF ITS ENCODING mRNA DURING BRAIN DEVELOPMENT

It has been previously shown that the GFAP content increases markedly (but unevenly depending on the brain region) during post natal development as illustrated. for instance, by immunohistochemical techniques. Recently the GFAP content has been evaluated in the developing rat (Malloch et al., 1987) and mouse (Le Prince et al., unpublished results) brain. GFAP could not be detected at day 17 post conception in the brain of the mouse, was present in low but significant amounts at birth and increased approximately 30 fold from birth until day 23 postnatal (Fig.1a). Such a developmental increase in GFAP content might result from a number of parameter : 1) increase in the number of astrocytes relative to other cell types due to astroglial cell proliferation; 2) increase in the proportion of mature astrocytes expressing higher levels of GFAP than the dividing glioblasts; 3) progressive accumulation of GFAP since this protein is very stable in normal CNS (DeArmond et al. 1986), etc.

Although the increase in GFAP content in whole brain probably results from the combination of all these, and possibly other parameters one may assume that "in vitro" translation of total RNAs extracted from the brain from a late fetal stage until adulthood would

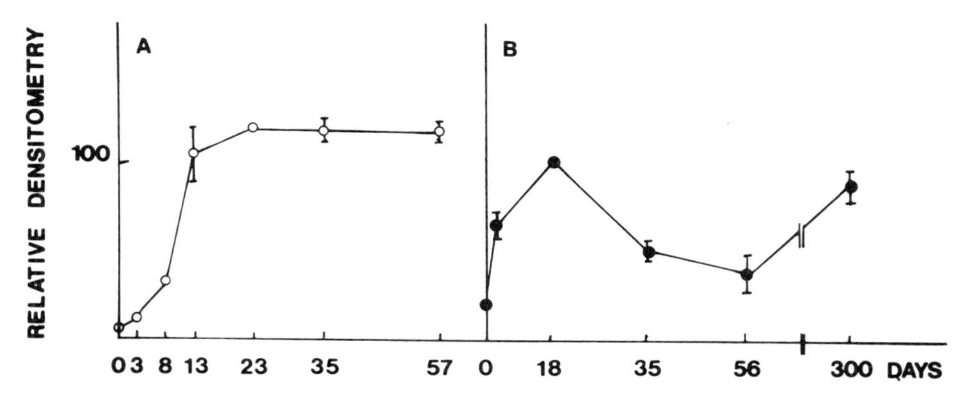

Fig.1. Densitometric evaluation of the expression of GFAP (A) and its encoding mRNA (B) in postnatal brain tissue.
GFAP concentrations were measured by immunoblots and GFAP mRNA determined by densitometric analysis of Northern blots. Results are the means ± SEM of three experiments. The value are expressed as percentage of the maximal measured values.

reflect at least partially the proportion of GFAP mRNA present at each stage. Such experiments (Tardy et al. 1989) showed that the GFAP translated from the same amount of total brain RNAs and immuno precipitated by a GFAP antibody increased several fold from 16 days of gestation until day 15 postnatal. Thus, during this developmental period there is a positive correlation between the brain total GFAP content and the concentration and translational activity of its encoding mRNA.

At later stages, however, a marked discrepancy between these two parameters was observed : the amount of GFAP immuno precipitated from the translation products of the RNAS prepared after day 15 postnatal progressively decreased with development reaching at day 36 values similar to those observed at early postnatal stages. Such a biphasic evolution suggests that: 1) either the GFAP mRNA content or its translational activity varies during development or both; 2) depending on the stage there is or not a positive correlation between the expression of GFAP mRNA and that of the encoded protein.

To answer the first question a direct evaluation of the GFAP mRNA content was performed by quantitative Northern blot analysis (Tardy et al. 1988). A GFAP-cDNA probe (Rataboul et al. 1988) was used which specifically hybridizes with the 2,7 kb GFAP as also previously shown by Lewis et al. (1984). Such a Northern blot analysis of the RNAs prepared at different stages of brain development revealed again (Fig.1b) a biphasic evolution: from birth, the earlier stage at which GFAP mRNA could be measured, the GFAP mRNA content increased steadily until day 15 postnatal (3 fold) and then declined progressively reaching two fold lower values at day 56 compared to day 15. Thus, the results of the Northern blots are consistent with the assumption that the biphasic changes in GFAP mRNA translational activity seeen during brain development result from a decrease in GFAP mRNA content relative to total brain RNA.

A second important question. i.e. the marked divergence, beginning after the second postnatal week, between the evolution of the developmental pattern of the GFAP content and that of its encoding mRNA, will be discussed below.

DEVELOPMENTAL PATTERN OF GFAP AND OF ITS ENCODING mRNA IN PRIMARY CULTURES OF ASTROGLIAL CELLS

It is not clear from the results described above whether the expression of GFAP mRNA is biphasic as the result of an intrinsic program of the astroglial cell or if it depends on the spatio temporal heterogeneity of brain development, For instance the glioblasts present in the different brain regions probably do not begin to proliferate and to differenciate at the same stages. One of the advantages of primary cultures of astrocytes is that, when prepared from new born brain, they actively proliferate during the first two weeks in our conditions of culture; after that period, when they reach confluency, the cells stop to divide, change in morphology and the population appears to be constituted by either round or short processed or spindle shaped cells. At later stages of the culture long and branched processes develop and the cells aquire the morpholoy characteristic of the mature astrocyte. Thus, although these cultures are not synchronized, it is possible to follow separately the expression of GFAP during proliferation and differentiation respectively. Another major advantage of these cultures is that they allow to know whether the biphasic developmental expression of GFAP mRNA "in vivo" depends on the intrinsic program of the astrocyte i.e. is independent of the neuronal neighbourhood.

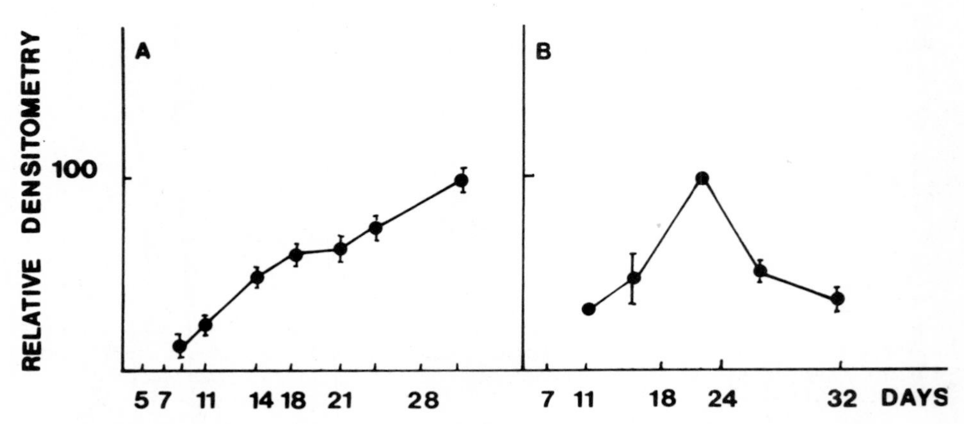

Fig.2. Time course analysis of the expression of GFAP (A) and of its encoding mRNA (B) in cultured astrocytes.
GFAP concentrations were measured by relative quantitative immunoblots. Relative amounts of GFAP mRNA were determined by densitometric analysis of Northern, blots.
Resuls are the means ± SEM of three experiments. The different values are expressed as percentages of the maximal measured values,

Immunohistochemical staining of the dividing cultured astrocytes revealed that GFAP is present at this stage but that it is perhaps organized and/or expressed differently than in the mature cells. The first assumption might be supported by the immunohistochemical data: staining by the antibody is less intense but significant in the dividing glioblasts. GFAP immunofluorescence was detected in the perinuclear region of the cells before the rest of the cytoplasm was labeled; double labeling with antisera specific for the centriole indicated that the perinuclear site corresponds to the centriolar region (Kalnins et al. 1985). At later stages, an array of fine perinuclear filaments stained by anti GFAP extends partially from this perinuclear region to the cytoplasm. It is not clear, however, whether the perinuclear localization of GFAP filaments in the immature astrocyte reflects initiation of assembly of the gliofilaments present in the cytoplasm and the processes of the mature cell or if other types of GFAP polymers with different functions are assembled during cell division.

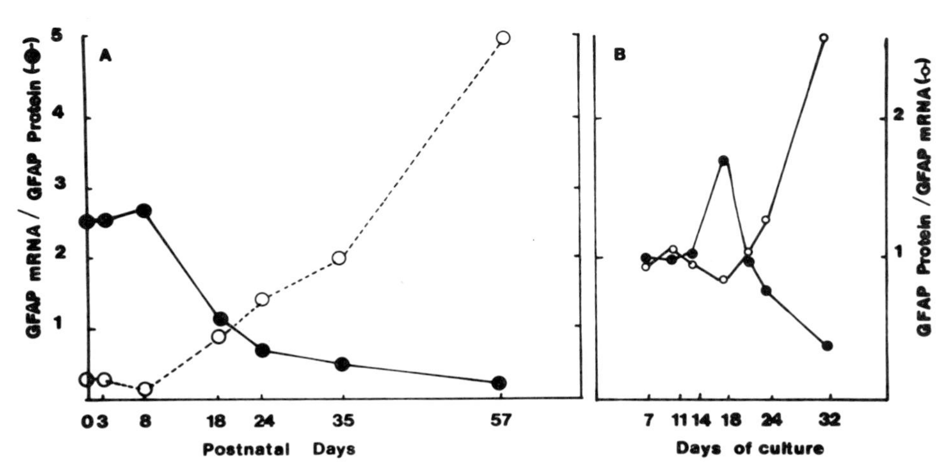

Fig.3. Changes in the ratios between GFAP and its encoding mRNA during "in vivo" (A) and "in vitro" (B) astroglial cell development.

The GFAP content of the cultured astrocytes was therefore measured by quantitative Western blot analysis (Le Prince et al., unpublished results) and found to increase 8 fold during the time course of the culture (Fig.2a). Approximately half of this increase was seen during the stages at which the cells actively proliferate suggesting either that a significant percentage of the pool of GFAP required to build up the gliofilaments of the mature astrocytes is present before the onset of morphological differentiation or that GFAP gliofilaments have a specific function in the imature dividing cell. A second marked increase in GFAP content was seen to begin during the third week of culture and to continue thereafter. Thus, during all the time course of the culture, the GFAP content increased steadily reaching several fold higher values in the mature cells compared to the immature glioblasts.

Such an evolution contrasts with that of the mRNA encoding GFAP as estimated both by "in vitro" translational experiments and by direct Northern blot analysis (Fig.2b) of the RNAs prepared at different stages of the culture. "In vitro" as "in vivo", a biphasic curve was obtained; during the two-three first weeks of the culture the GFAP mRNA content increased steadily and paralleled the increase seen for the encoded protein suggesting the existence of a positive correlation between these two parameters during the stages of astroglial cell proliferation.

After day 18 of the culture the mRNA content and its translational activity markedly decreased reaching at day 32 a value three times lower than at day 18. In other words, "in vitro" as "in vivo" the accumulation of GFAP into the cell during the stages of morphological differentiation does not seem to depend only on the transcriptional activity of the GFAP gene and/or on mechanisms leading to the accumulation of GFAP mRNA. Fig.3 shows, for instance, that the evolution of the GFAP protein/mRNA and mRNA/protein ratios is similar "in vivo" (left panel) and "in vitro" (right panel). For instance, the protein/RNA ratio markedly increases at late developmental stages both "in vivo" and "in vitro" suggesting that the divergence between the expression of the message and that of the encoded protein does not result from the spatio-temporal heterogeneity of different astroglial cell populations of the brain during development but reflects an intrinsic feature of the program of differentiation of this cell type. In contrast, the evolution of the same ratio is different "in vivo" and "in vitro" at early stages and this might account for the heterogeneity of the brain since the timing of the transition between proliferation and differentiation probably differs for the various astroglial cell populations present in the different brain regions.

EFfECTS OF CYCLIC AMP ON THE EXPRESSION OF ASTROGLIAL PROTEINS AND ON MORPHOLOGICAL DIFFERENTIATION

The conversion of immature polygonal astroglial cells to process-bearing astrocytes is accelerated following treatment of the cultures with dibutyryl cyclic AMP and agents known to increase intracellular cyclic AMP levels. Cyclic AMP has a dual effect on most cell types and these effects are triggered by two types of mechanisms i.e. phosphorylation-dephosphorylation reactions catalyzed by protein kinases and phosphatases and actions taking place probably at the transcriptional level. The existence of both fast reversible and slow permanent effects of cyclic AMP on the morphology of the astrocyte are suggested by experiments made in the presence of forskolin (Bridoux et al. 1986). Other biochemical data are also consistent with such a possibility: 1) GFAP is phosphorylated when astroglial (McCarthy et al. 1985, Pollenz and McCarthy, 1986) and C6 glioma cells (Browning and Ruina 1984) are exposed to dibutyryl cyclic AMP, forskolin or norepinephrine, 2)The same agonists stimulate or repress the synthesis of a small number of astroglial proteins (Bridoux et al. 1986). However, these agonists had little or no effect on the expression of GFAP and of its encoding mRNA (Tardy et al., unpublished results) when added for 24 to 48 hours to the cultures. Moreover, although cyclic AMP probably exerts both post translational and transcriptional effects on this cell type the functional significance of these effects remains unknown; it has not been tested, for instance, whether GFAP phosphorylation modifies either its ability to polymerize, or the stability of gliofilaments or the interactions between these structures and other components of the cytoskeleton, etc. Similarly, it is not clear whether the unidentified proteins which are induced or repressed by the cyclic AMP agonists participate or not to the changes in astroglial morphology.

DISCUSSION AND CONCLUSIONS

The different features of the expression of GFAP and of its encoding mRNA described in this work raise a number questions which might be connected to the programs of astroglial proliferation and differentiation.

The first general observation is that the concentration of GFAP and of its encoding mRNA follow a similar developmental pattern "in vivo" and "in vitro". This suggests that morphological maturation of the astrocyte does not depend, at least quantitatively and for the expression of these markers, on the presence of any neuronal neighbourhood. The signaling factors produced by the neurons are probably modulators of the programs of development of the astrocyte but their mechanism of action is still poorly understood.

The second major finding is that the developmental pattern of expression of the mRNA encoding GFAP follows a biphasic profile both "in vivo" during postnatal development and "in vitro" in primary cultures of astroglial cells. It has not been determined so far whether such an evolution is related to direct transcriptional regulatory mechanisms taking place at the level of the GFAP gene(s) or, indirectly, to changes in the stability of GFAP mRNA. The latter assumption implies that the half life of GFAP mRNA begin to decrease both in the developing brain, approximately at day 15 postnatal, and in culture when the astroglial cells begin to differentiate.

Similar situations have been described for other cytoskeletal components. For instance, the mRNAs encoding actin and tubulin (Bond and Farmer 1983), and the microtubule associated protein Tau (Couchie et al. 1988) undergo a very marked decrease in concentration during brain development whereas the level of the corresponding proteins remains essentially unchanged or decreases at much slower rates. Recently, we have shown that the decrease in brain tubulin mRNA concentration can be ascribed essentially to the neurons whereas this parameter remains constant in cultured astrocytes (Bertrand-Charriere et al., unpublished results).

Post transcriptional mechanisms have been proposed to explain the changes in stability of tubulin mRNA which are seen when the number of free tubulin subunits is increased either by colchicine treatment or by injection of tubulin monomers into the cells (Cleveland et al. 1986; Yen et al. 1988). Similarly, one may assume that the pool of free GFAP monomers is larger at late developmental stages both "in vivo" and "in vitro" and that such an increase negatively regulates the stability of GFAP mRNA. However, the data available (Malloch et al. 1987), demonstrate that, whatever the stage, most of the GFAP is assembled and that the free pool of this protein remains constant throughout development. Thus, the concentration of GFAP mRNA is probably not regulated by changes in the pool of free GFAP.

"In vitro" transcription assays, performed with isolated nuclei prepared at different stages of brain development, are in progress and will probably allow to know whether the biphasic evolution of GFAP mRNA seen in our experiments depends on transcriptional or post transcriptional regulatory mechanisms. Such answers might be of importance since a positive correlation between the expression of the GFAP mRNA and that of the protein was seen during the stages of astroglial cell proliferation whereas these parameters diverged in opposite directions during "in vitro" morphological differentiation of the astrocytes.

These data also suggest that the increase in GFAP concentration and the decrease in GFAP mRNA during astroglial cell maturation are regulated independently. Several mechanisms might explain an increase in GFAP stability: 1) GFAP might be less accessible to proteases when polymerized; 2) proteolytic activity might be more abundant in the cell body than in the cell processes which develop after confluency. It has been suggested, for instance, that turnover of gliofilaments is controlled by a Ca^{2+} protease (DeArmond et al. 1983); 3) GFAP homolymers might be more stable than the GFAP-vimentin heteropolymers which are present only at early developmental stages "in vivo" and perhaps in lower proportion in the cultured mature astrocyte compared to the proliferating cells; 4) At late stages of astroglial maturation gliofilaments might be stabilized because of their tighter interaction with other cytoskeletal elements. Such an interaction is indirectly suggested by the observation that colchicine, a drug which induces microtubule depolymerization, also produces in this cell type disassembly of the gliofilaments (Goetschy et al. 1984).

Most of these assumptions imply that the half life of GFAP changes during development and increases in the mature compared to the dividing cell. DeArmond et al. (1986) have reported that GFAP turns over relatively slowly "in vivo" in the mouse spinal cord: after injection of a radioactive amino acid forty per cent of the radioactivity incorporated into this protein was still present in cytoskeletal GFAP at 9 weeks. GFAP turnover has also been studied in 2 weeks old cultured astrocytes and found to be biphasic (Chiu and Goldman 1984): in the initial phase a fast decaying pool with a half life of 12-18 hours contributed about 40% of the total protein whereas a major portion, about 60% decayied much more slowly exhibiting a half life of about 8 days. The mechanism(s) leading to the existence of these two GFAP pools differing in turn over rates remains to be determined; Chiu and Goldman (1984) suggested that "possible post translational modifications or differences in turnover in different regions of the cell (cell body versus processes for example) may contribute to differrent turnover rates; they also proposed that "turnover in rapidly growing cells might differ from that in confluent cultures". To answer these questions work is in progress in our laboratory to determine the half life of GFAP at different stages of "in vitro" astroglial proliferation and differentiation.

Little is known on the precise function of gliofilaments during proliferation and differentiation of the astroglial cells. Their minimal role is probably to specify and to maintain, together with other cytoskeletal components such as microtubules, the shape of this cell type and the integrity of the astroglial cell processes in the mature brain. It is not excluded also that different GFAP polymers, perhaps differently organized, are formed during proliferation and maturation. The variations observed in various pathological conditions (Bignami and Dahl 1976; Duffy et al. 1980; Schechter et al. 1981; Amaducci et al. 1981; Smith et al. 1984; Goldmuntz et al. 1986; Graeber and Kreutzberg 1986) are also interesting since in all these cases increased staining of GFAP density has been noted. Large amounts of GFAP are also present in gliosed tissues (Goldman et al. 1978; Selkoe et al. 1982; Paetau et al. 1985; Newcombe et al. 1986).

Recently we have found (Tardy et al. 1989) that a marked incrase in GFAP mRNA occurs in the brain of older (300 days) mice with values as high as those observed during the initial period of postnatal brain development. This would suggest that the GFAP gene is more actively transcribed both in the developing and in the gliosed brain i.e. in conditions characterized by intensive astroglial cell proliferation. Different and apparently independent regulatory mechanisms would, in contrast, be responsible for a decrease in GFAP mRNA transcription and for an increase in GFAP levels in the normal and mature astrocytes.

REFERENCES

Amaducci, L., Forno, K.I., and Eng, L.F., 1981, Glial fibrillary acidic protein in cryogenic lesions of the rat brain, Neurosci. Lett., 21:27-32.

Bignami, A., and Dahl, D., 1976, The astroglial response to stabbing-immunofluorescence studies with antibodies to astrocyte specific protein (GFAP) in mammalian and submalian vertebrates, Neuropath. Appl. Neurobiol., 2:99-110.

Bigbee, J.W., Bigner, D.D., Pegram, C., Eng, L.F., 1983, Study of glial fibrillary acidic protein in a human glioma cell line grown in culture and as solid tumor, J. Neurochem., 40: 460-467.

Bigbee, J.W., and Eng, L.F., 1982, Analysis and comparison of in vitro synthesized glial fibrillary acidic protein with rat CNS intermediary filament proteins, J. Neurochem., 38:130-134.

Bignami, A., Eng, L.F., Dahl, D., and Uyeda, C.T., 1972, Localization of the glial fibrillary acidic protein in astrocytes by immunofluorescence, Brain Res., 43:429-435.

Bond, J.F., and Farmer, S.R., 1983, Regulation of tubulin and actin mRNA production in rat brain: expression of a new beta-tubulin mRNA with development, Mol. Cell. Biol., 3:1333-1342.

Bridoux, A.M., Fages, C., Couchie, D., Nunez, J., and Tardy, M. 1986, Protein synthesis in astrocytes: spontaneous and cyclic AMP-induced differentiation, Dev. Neurosci., 8:31-43.

Browning, E.T., and Ruina, M., 1984, Glial fibrillary acidic protein: norepinephrine stimulated phosphorylation in intact C6 glioma cells, J. Neurochem., 42:718-726.

Chiu, F.C.. and Goldman, J.E., 1984, Synthesis and turnover of cytoskeletal proteins in cultured astrocytes. J. Neurochem., 42:166-174.

Chiu, F.C., Norton, W.T., and Fields, K.L., 1981, The cytoskeleton of primary astrocytes in culture contains actin, glial fibrillary acidic protein and fibroblast-type filament protein in vimentin, J. Neurochem.. 37:147-155.

Cleveland, D.W., 1986, Molecular mechanisms controlling tubulin synthesis, in: Cell and Molecular Biology of the Cytoskeleton (Shay, J.W., ed,) 203-227, Plenum Publishing Corp. N.Y.

Couchie, D., Charriere-Bertrand, C., and Nunez, J., 1988, Expression of the mRNA for TAU proteins during brain development and in cultured neurons, J. Neurochem., 50:1894-1899.

Dahl, D., 1981, The vimentin-GFA protein transition in rat neuroglia cytoskeleton occurs at the time of myelination, J. Neurosci. Res., 6:741-748.

Dahl, D., Rueger, D.C., Bignami, A., Weber, K. and Osborn, :,, 1981, vimentin, the 57,000 molecular weight protein of fibroblast filaments in the major cyhtoskeletal component in immature glia, Eur. J. Cell. Biol., 24:191-196.

Dahl, D., and Bignami, A., 1979, Astroglial and axonal proteins in isolated brain filaments, Isolation of glial fibrillary acidic protein and of an immunologically active cyanogen bromide peptide from brain filament preparations of bovine white matter, Biochem. Biophys. Acta. 578:305-316.

De Armond, S.J., Fajardo, M., Naughton, S.A., and Eng, L.F., 1983, Degradation of glial fibrillary acidic protein by a calcium dependent-proteinase; an electroblot study, Brain Res., 262:275-282.

De Armond, S.J., Lee Yuen-Luig, Kretzchmar, H.A., and Eng, L.F., 1986, Turnover of glial filaments in mouse spinal cord,, J. Neurochem., 47:1749-1753.

Duffy, P.E., Huang, Y.Y., Rapport, M.M., and Graf, L., 1980, Glial fibrillary acidic protein in giant cell tumors of brain and other gliomas, Acta Neuropathol., 52:51-57.
Duffy, P.E., 1983, Astrocytes: Normal, reactive and neoplastic, Raven Press (New York).
Eng, L.F., 1980, The glial fibrillary acidic (GFA) protein, In: Proteins of the nervous system, second ed. Ed. R.A. Bradshaw, D.M. Schneider, p.85-117, Raven Press, N.Y.
Eng, L.F., Vauderhaegen, J.J., Bignami, A., and Gerstl, B., 1971, An acidic protein isolated from fibrous astrocytes, Brain Res., 28:351-354.
Eng, L.F., and De Armond, S.J., 1983, Immunochemistry of the glial fibrillary acidic protein, Progress in Neuropath, V;5 Ed, Zimmerman, H.M. Ed. p,19-39, Raven Press, N.Y.
Geisler, N., and Weber, K., 1982, The aminoacid sequence of chicken muscle desmin provides a common structural model for intermediate filament proteins, EMBO J., 2:1649-1656.
Geisler, N., Kaufmann, E., and Weber, K., 1982a, Protein chemical characterization of three structurally distinct domains along the protofilament unit of desmin 10nm filaments, Cell, 30:277-286.
Geisler, N., Plessmann, U., and Weber, K., 1982b, Related amino acid sequences in neurofilaments and non-neuronal intermediate filaments, Nature, 296:448-450.
Gilden, D.H., Wroblewska, Z., Eng, L.F., and Rorke, L.B., 1976, Human brain in tissue culture, J. Neurol. Sci. 29:177-184.
Goetschy, J.F., Ulrich, G., Aunis, D., and Cireselski-Treska, I., 1986, The organization and soluility properties of intermediate filaments and microtubules of cortical astrocytes in culture, J. Neurocytol., 15:375-387.
Goldman, J.E., and Chiu, F.C., 1984, Dibutyryl cyclic AMP causes intermediate filament accumulation and actin reorganization in astrocytes, Brain Res,, 306:85-95.
Goldman, J.E., Schaumburg, H.H., and Norton, W.T., 1978, Isolation and characterization of glial filaments from human brain, J. Cell. Biol., 78:426-440.
Goldmuntz, E.A., Brosman, C.F., Chiu, F.C., and Norton, W.T., 1986, Astrocyte reactivity and intermediate filament metabolism in experimental autoimmune encephalomyelities: the effect of suppression with prazosine, Brain Res., 379:16-26.
Graeber, M.B., and Kreutzberg, G.W., 1986, Astrocytes incrase in glial fibrilary acidic protein during retrograde changes of facial motor neurons, J. Neurocytol., 15:363-373.
Herpers, M.J.H.M., Ramaehers, F.C.S., Aldewareldt, F., Moesher, O., and Slooff, J., 1986, Coexpression of glial fibrillary acidic protein and vimentin-type intermediate filaments in human astrocytomas, Acta Neuropathol., 70:333-339.
Honegger. P., 1986, Protein kinase C - activating tumor promoters enhance the differentiation of astrocytes in aggregating fetal brain cell cultures, J. Neuroch., 46:1561-1566.
Hong, B.S., and Davison, P.F., 1981, Isolation and characterization of a soluble, immunoreactive peptide of glial fibrillary acidic protein, Biochim. Biophys, Acta. 670:139-145.
Kalnins, V.I., Subrahmanyan, L., and Fedoroff, S., 1985, Assembly of glial intermediate filament protein is initiated in the centriolar region, Brain Res., 345:322-326.
Lach, B., and Weimander, H., 1978, Glia specific antigen in the intracranical tumors, Immunofluorescence study, Acta Neurograth. (Berl) 4 (1) 9-15.
Lewis, S.A., Balcarek, J.M., Krek, V., Shelanski, M., and Cowan, N.J., 1984, Sequence of a cDNA clone-encoding mouse glial fibrillary acidic protein: structural conservation of intermediate filaments, Proc. Natl. Acad. Sci., USA, 81:2743-2746.

Liem, R.K.H., and Shelanski, M.L., 1978, Identity of the major protein in "native" glial fibrillary, Brain Res., 145:196-201.
Liem, R,, 1982, Simultaneous separation and purification of neurofilament and glial filament proteins from brain, J. Neurochem., 38:142-150.
Lucas, C.V., Bensch, K.G., and Eng, L.F., 1980, In vitro polymerization of glial fibrillary acidic protein (GFA) extracted from multiple sclerosis (MS) brain,, Neurochem. Res,, 5:247-255.
Ludwin, S.K., Kosek, J.C., and Eng, L.F., 1976, The topographical distribution of S-100 and GFA proteins in the adult rat brain: An immunohistochemical study using horseradish peroxidase-labeled antibodies, J. Comp, Neurol., 165:197-208.
Mc Carthy, K.D., Prime, J., Harmon, T., and Pollenz, R.S., 1985, Receptor mediated phosphorylation of astroglial intermediate filament proteins in culture, J. Neurochem., 44:723-730.
Malloch, G.D.A., Clark, J.B., and Burnet, F.R., 1987, Glial fibrillary acidic protein in the cytoskeletal and soluble protein fractions of the developing rat brain, J. Neurochem., 48:299-306.
Morrison, R.S., De Vellis, J., Lee, Y.L., Bradshaw, R.A., and Eng, L.F., 1985, Hormone and growth factors induce the synthesis of glial fibrillary acidic protein in rat brain astrocytes, J. Neurosci. Res., 14:167-176.
Newcombe, J., Woodroofe, M.N., and Cuzner, M.L., 1986, Distribution of glial fibrillary acidic protein in gliosed white matter, J. Neurochem., 47:1713-1719.
Osborn, M., Caselitz, J., and Weber, K., 1981, Heterogeneity of intermediate filament expression in vascular smooth muscle: A gradient in desmin positive cells from the rat aortic arch to the level of the arteria iliaca communes, Differentiation. 20:196-202.
Paetau, A., Virtanen, F., Steinman, S., Kurki, P., Linder, E., Vaheri, A., Westermark, B., Dahl, D., and Haltia, M., 1979, Glial fibrillary acidic protein and intermediate filaments in human glioma cells, Acta Neuropathol., 47:71-74.
Pettmann, B., Weibel, M., Sensenbrenner, M., and Labourdette, G., 1985, Purification of two astroglial growth factors from bovine brain, FEBS Lett., 189:102-108.
Pollenz, R.S., and Mc Carthy, K.D., 1986, Analysis of cyclic AMP-dependent changes in intermediate filaments protein phosphorylation and cell morphology in cultured astroglia, J. of Neurochem., 47:9-17.
Quinlan, R.A., and Franke, W.W., 1983, Molecular interactions in intermediate sized filaments revealed by chemical cross-linking heteropolymers of vimentin and glial filament protein in cultured human glioma cells, Eur. J. Biochem., 132:477-484.
Rataboul, P., Faucon-Biguet, N., Vernier, P., De Vitry, F., Boularaud, S., Privat, A., and Mallet, J., 1988) Identification of a human GFAP-cDNA, a tool for the molecular analysis of reactive glyosis in the mammalian CNS, J. Neurosc. Res., 20:165-175.
Schechter, R., Yen, S.H.C., and Terry, R.D., 1981, Fibrous astrocytes in senile dementia of the Alzheimer's type, J. Neuropath. Exp. Neuro., 40:95-101.
Schlaepfer, W.W., and Zimmerman, U.J.P., 1981, Calcium mediated breakdown of glial filaments and neurofilaments in rat optic nerve and spinal cord, Neurochem. Res., 6:243-255,
Schnitzer, J., Franke, W.N., and Schachner, M., 1981, Immunocytochemical demonstration of vimentin in astrocytes and ependymal cells of developing and adult mouse nervous system, J. Cell. Biol., 90:435-447.
Selkoe, D.J., Ihara, Y., and Salazar, F., Alzheimer's disease: insolubility of partially purified paired helical filaments in sodium dodecyl sulfate and urea, Science, 215:1243-1245.

Sensenbrenner, M,, Devilliers, G., Bock, K., aaaand Porte, A., 1980, Biochemical and ultrastructural studies of cultured rat astroglial cells, Effect of brain extract and dibutyryl AMP on glial fibrillary acidic protein and glial filament, Differentiation, 17:51-61.

Sharp, G; Osborn, M., and Weber, K., 1982, Occurence of two different intermediate filament proteins in the same filament in situ within a human glioma cell line, Exp. Cell. Res., 141:385-395.

Smith, M.E., Perret, V., and Eng, L.F., 1984, Metabolic studies in vitro in the CNS cytoskeletal proteins: synthesis and degradation, Neurochem. Res.. 9, 1493-1507.

Sotelo, J., Toh, B.H., Lolait, S.J., Yildiz, A., Sung, D. and Holobrow, E.J., 1980, Cytoplasmic intermediate filaments in cultured glial cells, Neuropath. Appl. Neurobiol., 6:291-298.

Tardy, M., Fages, C., Riol, H., Le Prince, G., Rataboul, P., Charriere-Bertrand, C. and Nunez, J., 1989, Developmental expression of the GFAP-mRNA in the central nervous system and in cultured astrocytes, J. Neurochem. 52, 1, 162-167.

Uyeda, C.T., Eng, L.F., and Bignami, A., 1972, Immunological study of the glial fibrillary acidic protein, Brain Res., 37:81-89.

Virtanen, I., Lehto, V.P., Lehronen, E., Vartio, T., Stenman, S., Kurki, P., Wayer, O., Small, J.V., Dahl, D., and Badley, R.A., 1981, Expression of intermediate filaments in cultured cells, J. Cell. Sci., 50:45-63.

Wang, E., Cairncross, J.G., and Liem, R.K.H., 1984, Identification of glial filament protein and vimentin in the same intermediate filament system in human glioma cells, Proc. Natl. Acad. Sci., 81:2102-2106.

Weibel, M., Fages, C., Belakebi, M., Tardy, M., and Nunez, J., 1987, Astroglial growth factor 2 (AGF2) increases alpha-tubulin in astroglial cells cultured in a defined medium, Neurochem. Inter,., 11:223-228.

Weibel, M., Pettmann, B., Labourdette, G., Miche, M., Bock, E., and Sensenbrenner, M., 1985, Morphological and biochemical maturation ofrat astroglial cells grown in a chemically defined medium: influence of an astroglial growth factor, Int. J. Dev. Neurosci., 3:617-630.

Yen, S.H., and Fields, K., 1981, Antibodies to neurofilaments, glial filaments and fibroblast intermediate filament-proteins bind to different cell types of the nervous systesm, J. Cell. Biol., 88:115-126.

Yen, T.J., Machlin, P.S., and Cleveland, D.W., 1988, Autoregulated instability of beta-tubulin mRNAs by recognition of the nascent amino terminus of beta-tubulin, Nature, 334:580-585.

CELL LINEAGE STUDIES IN AVIAN NEURAL CREST ONTOGENY

Anne BAROFFIO, Elisabeth DUPIN and Nicole M. LE DOUARIN

Institut d'Embryologie cellulaire et moléculaire
du CNRS et du Collège de France,
94736 NOGENT-sur-MARNE CEDEX (France)

INTRODUCTION

The neural crest of the Vertebrate embryo is a transitory structure arising during the closure of the neural tube and lying on its dorsal aspect. From it originates most of the peripheral nervous system (PNS), including neurons of all the sympathetic, parasympathetic and enteric ganglia and of the majority of sensory ganglia as well as glial cells of all these ganglia. In addition, the neural crest gives rise to Schwann cells of the peripheral nerves and to many other cell types, such as endocrine cells (e.g. adrenomedullary and calcitonin-producing cells), melanocytes and, in the head region, to the mesectoderm (see Le Douarin, 1982, for review).

The wide variety of cell types derived from the neural crest raises the question of how and when the different lineages to which they belong become segregated.

An analysis of the origin of peripheral ganglia using embryonic quail-chick chimaeras (Le Douarin and Teillet, 1974) has revealed distinct regions on the neural axis characterized by different fates (e.g. sympathetic, enteric, sensory ganglionic types). The developmental potentials of crest cells were further tested by heterotopic grafting of definite fragments of the neural primordium between quail and chick embryos. This showed that at each axial level neural crest cells are potentially able to give rise to the various cell types of the PNS and that the microenvironment in which they migrate and settle plays a major role in their phenotypic expression.

The problem is then raised as to whether neural crest cells are pluripotent or already partly or fully committed when leaving the neural tube. In the former case, the role of environmental factors will be instructive, whereas in the latter case, the selection of certain precursor cells by specific factors will result in expression of a given phenotype in a ganglion.

It has been shown that the neural crest is composed of a heterogeneous population of cells as soon as migration begins. Monoclonal antibodies raised against peripheral ganglia recognize antigenic determinants on subpopulations of neural crest cells as soon as they leave the neural tube (Barbu et al., 1986) and during migration (Ciment and Weston, 1982) or after in vitro culture (Barald,

1982). Moreover, some cells seem to be already committed to a neuronal fate before they leave the neural tube at the mesencephalic (Ziller, 1983) and vagal (Payette et al., 1984) levels or after the onset of migration in the trunk (Girdlestone and Weston, 1985).

In culture, it is possible to obtain the differentiation of neural crest cells into sensory (Ziller et al., 1983, 1987 ; Sieber-Blum et al., 1988) and autonomic neurons, the latter containing catecholamines (Cohen, 1977 ; Loring et al., 1982 ; Howard and Bronner-Fraser, 1985 ; Maxwell and Forbes, 1987) and tyrosine hydroxylase (TH) (Christie et al., 1987 ; Fauquet and Ziller, 1988). These two types of neuronal cells appear to originate from distinct precursors that have different trophic requirements (Ziller et al., 1987). Thus, the neural crest appears to be a heterogeneous population of cells more or less restricted in their developmental potentialities, although this fact does not preclude the possibility that such progenitors are derived from a common ancestor. However, a direct demonstration of the presence of pluripotent and committed cells in the neural crest remained to be made. This is what we undertook by means of the in vitro clonal analysis described below, our more distant goal being to establish lineage relationships between the various neural crest derivatives.

Cell lineages have been established in the ontogeny of the nervous system of certain Invertebrates, like Caenorhabditis elegans (Sulston et al., 1983) where this can be done by simple visual observation. In Vertebrates, this approach is not feasible and different techniques have been developed to follow the progeny of individual cells. These include retrovirus-mediated gene transfer (Price et al., 1987 ; Turner and Cepko, 1987 ; Cepko, 1988) or vital dye injection (e.g. Wetts and Fraser, 1988). Another approach consists in analyzing in vitro the progeny of single cells, allowing, as for example for the optic nerve, the identification of bipotential progenitors (Raff et al., 1983) and of factors involved in each differentiation step (Hughes et al., 1988 ; Raff et al., 1988). The in vitro culture approach is particularly suitable for disclosing the potentialities of a single cell retrieved from its natural environmental context. In 1975, Cohen and Konigsberg (1975) initiated a study of clonal potencies of quail truncal neural crest cells. By cultivating the cells using the limit-dilution method, they observed three types of colonies, one formed by melanocytes, another by unpigmented cells and mixed colonies containing both pigmented and non-pigmented cells. This study was extended to a search for catecholamine-containing cells, which were found in unpigmented and mixed colonies (Sieber-Blum and Cohen, 1980). This was the first clear evidence for bipotent progenitors for melanocytes and adrenergic cells in the neural crest. However, this study involved a very limited phenotypic analysis of neural crest cells compared to the extraordinary range of phenotypes they are able to express both in vivo and in vitro. For this reason we decided to undertake a clonal analysis of neural crest cell differentiation by using different culture conditions (Baroffio et al., 1988).

RESULTS

1) In Vitro Cultures of Dissociated Quail Mesencephalic Neural Crest Cells

First, we removed the neural crest at the mesencephalic level since, at this site, the crest is easy to dissect free of contamination by the neural tube (Ziller et al., 1983) while having a large array of derivatives including mesectoderm, besides different types of neurons, glial cells and melanocytes (see Le Douarin, 1982). Second, single cells were seeded individually under microscopic control. Third, culture conditions were devised that allow differentiation of many of the cell types mentioned above. Culture was carried out on a feeder-layer of growth-inhibited mouse 3T3 fibroblasts (Todaro and Green, 1963 ; Rheinwald and Green,

1975) in a complex medium containing foetal calf serum (10%) and chick embryo extract (CEE, 2%) along with hormones and growth factors (Baroffio et al., 1988).

In these culture conditions, dissociated neural crest cells proliferate strikingly and differentiate in a time-dependent manner into various cell types as defined by specific markers (see Table 1).

Table 1. Markers used to define the cell types differentiating in cultures of neural crest cells on 3T3 feeder-layers.

CELL TYPE		MARKERS
Neuronal cells	adrenergic	TH
	non-adrenergic	NF, SP, VIP
	neuron-like	HNK1
Non-neuronal cells		HNK1
Melanocytes		pigment
Mesectoderm	cartilage	toluidine blue

After about 1 day bipolar neurons appear (approximately 100 per neural crest), all of them possessing immunoreactivity for neurofilament proteins (NF) and about half of them substance P (SP)-like immunoreactivity. Morphologically and immunocytochemically, these neurons are similar to those appearing in serum-free medium and are likely of sensory type (Ziller et al., 1983, 1987 ; Bader et al., 1983). They disappear gradually from the 3rd day of culture and very few of them are still present after 6 days.

From the 4th day TH-containing adrenergic cells (about 1000 per neural crest) differentiate, 10-15% of which possess NF immunoreactivity. It must be stressed that, to our knowledge, this is the first report of the differentiation of adrenergic cells from early migrating neural crest cells cultivated in the presence of only low amounts of CEE (2% compared to at least 10% in other studies, see e.g. Howard and Bronner-Fraser, 1985 ; Maxwell and Forbes, 1987 ; Fauquet and Ziller, 1988). This could be due either to a matrix effect of the 3T3 cell feeder-layer or to substances produced by these cells. Moreover, these are the first culture conditions reported for early migrating neural crest cells permitting the development of TH-containing cells that possess a mature neuronal marker, NF, (see, e.g., Christie et al., 1987 ; Fauquet and Ziller, 1988). Some of these TH-positive cells contain either SP or vasoactive intestinal peptide (VIP). Another type of NF-positive cells differentiating in our cultures lacks TH but contains SP or VIP. It seems difficult to categorize these cells into the sensory or autonomic lineages. In the avian PNS, TH and catecholamines are found in sympathetic neurons and chromaffin cells, SP in sensory spinal ganglia and VIP in sympathetic and enteric ganglia (Black, 1982 ; Fontaine-Pérus, 1984 ; New and Mudge, 1986 ; Garcia-Arraras et al., 1987) but SP has also been reported to occur in sympathetic neurons of rat (Kessler et al., 1981) and cells with both SP and TH differentiate in cultures of quail truncal neural crest (Fauquet and Ziller, 1988) and rat superior cervical ganglia (Bohn et al., 1984).

The HNK1 epitope (Abo and Balch, 1981), which is carried by all glial cells and by some neurons of the PNS (Tucker et al., 1984) is expressed by non-neuronal glial cells from the first day in culture on 3T3 cells. In some instances, this marker also enables the visualization of neuron-like cells of otherwise unknown phenotype.

Finally, neural crest cells cultivated on 3T3 cells give rise to melanocytes and to cartilage.

These experiments show that neural crest cells cultured on 3T3 feeder-layer differentiate into a large array of cell types including neurons and non-neuronal cells, melanocytes and mesectodermal derivatives.

2) Clonal cultures of quail mesencephalic neural crest cells

In addition to the wide range of phenotypes expressed by the neural crest cells, this system presents the advantage of allowing clonal culture without altering differentiation ability. We obtained about 400 clones after 7 to 16 days of culture, with a mean cloning efficiency of 35% (maximum 58% in one series).

Clones of different sizes, ranging from one cell to several thousand, were regularly found, irrespective of the duration of the culture (Figure). To analyze whether these differences were due to an earlier arrest of cell proliferation in the smaller clones, cultures were given ^{3}H-thymidine 24 hours before fixation at 7, 10, 12, 15 and 16 days of culture. In the largest clones (more than 10^{3} cells), the percentage of dividing cells gradually decreased from 80-100% of ^{3}H-thymidine-labelled cells after 7 days to 0-10% after 16 days, indicating that most cells divide during the first 7 days, at the rate of about one division every 15 hours. The smaller clones (less than 100 cells) contained dividing cells at all times, but in very variable proportions, ranging from 10-70% at 7 days and from 1-20% at 16 days. This suggests that in small clones either cells divide slowly or, more likely, that only a proportion of the cells divide from the beginning. Migrating mesencephalic neural crest cells thus appear heterogeneous as far as their proliferation potential is concerned.

Diversity was also revealed by an analysis of the cell types that had differentiated. In 277 clones obtained after 7 to 12 days in culture, eleven different combinations of phenotypes were found (Table 2). Therefore, the neural crest cells have variable differentiating abilities, thus confirming and extending the heterogeneity revealed by the monoclonal antibody studies mentioned above. Neural crest cells were able to give rise to up to 4 different cell types in a single clone as identified by the chosen markers. Fifty-two percent of the founder cells were demonstrated to be pluripotent since they gave rise to clones containing more than one identifiable cell type. Eight percent of the cells yielded both neural and melanocytic lineages. Neurons and non-neuronal glial cells can arise from common precursors, since they were colocalized in 48% of the clones. Non-neuronal cells were also found associated with other derivatives, such as melanocytes and cartilage (3%), or alone (47%). The potential of neural crest cells to give rise to non-neuronal cells concerns 98% of the migratory crest cells at the mesencephalic level.

DISCUSSION

Pluripotency of neural crest cells, established in vitro, has recently been confirmed by an in vivo study (Bronner-Fraser and Fraser, 1988) in which a

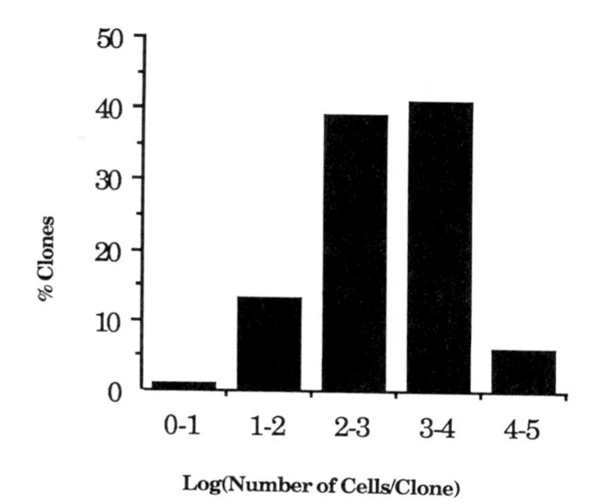

Figure 1 Distribution of 10-day clonal cultures (n = 226) according to their cell number

lysinated rhodamine dextran was injected into presumptive neural crest cells in the dorsal neural tube at the trunk level, and the location of fluorescent daughter cells was analyzed in different neural crest derivatives. Sixty percent of the clones obtained were constituted of cells distributed in multiple sites ; pluripotency is thus not confined to crest cells at the mesencephalic level, but extends to premigratory truncal neural crest cells as well. Nevertheless, this finding does not exclude the possibility that some neural crest cells are

Table 2. Types of clones as defined by the different combinations of cell types found after 7 to 12 days in culture.

CELL TYPE	TYPE OF CLONE										
N adrenergic	+	+		+	+						
N non-adrenergic	+		+	+		+					
N neuron-like										+	
nN non-neuronal	+	+	+	+	+	+	+	+	+		
M melanocytes	+	+	+				+				
C cartilage								+			+
% (n=277)	1.5	1.5	2.5	32	7	1.5	2.5	3	47	1	1

restricted in their developmental potentials quite early, i.e. at the beginning of their migration or even before leaving the neural tube. In fact, cloning experiments both with mesencephalic (Baroffio et al., 1988) and truncal (Bronner-Fraser and Fraser, 1988) crest cells suggest the existence of some precursors with a limited developmental potential, since they give rise to a unique cell type. In particular, we observed clones that were constituted of neuron-like cells or of cartilage. The former could correspond to the early committed neuronal progenitors already revealed in mass cultures of neural crest (Ziller et al., 1983, 1987). The latter confirm previous observations suggesting that most cells of the mesectodermal lineage are segregated early during neural crest migration (Le Douarin and Teillet, 1974 ; Hall and Tremaine, 1979).

The results described above lead us to speculate about possible filiations between these progenitors. In particular, the precursors of several types of clones we have obtained seem to originate from the members of a same lineage family. This is in good agreement with the hypothesis proposing that neural crest cells undergo a progressive restriction of their potentialities during the course of their migration (Le Douarin, 1986) finally to yield precursors for the diverse cell types at the sites of differentiation, some of them only being able to survive in each type of ganglion. Neural crest derivatives would therefore differentiate from a mixture of different kinds of precursors, whose survival and further development would depend on specific trophic factors.

There is now good evidence for the existence at the sites of gangliogenesis of two different types of precursors for the sensory and autonomic lineages respectively (see Le Douarin, 1986). The existence of "sensory" precursors was deduced from the back-transplantation of peripheral ganglia into neural crest migration pathways (Ayer-Le Lièvre and Le Douarin, 1982 ; Schweizer et al., 1983). These precursors are present in spinal (see also Rohrer et al., 1985 ; Ernsberger and Rohrer, 1988), but not in autonomic sympathetic ganglia and seem to survive only in response to factors produced by the neural tube (Teillet and Le Douarin 1983 ; Kalcheim and Le Douarin, 1986) and particularly to brain-derived neurotrophic factor (Kalcheim et al., 1987). There are strong indications that these precursors exist already in the early migrating neural crest (Ziller et al., 1983, 1987) or even before migration (Bronner-Fraser and Fraser, 1988).

Other precursors of an "autonomic" type are largely distributed among different types of ganglia (Ayer-Le Lièvre and Le Douarin, 1982 ; Schweizer et al., 1983). They are present in sensory ganglia, where they stay in a resting state, but can be triggered to differentiate into a variety of autonomic derivatives in culture (Xue et al., 1985, 1987, 1988a ; Rohrer et al., 1986) in the presence of CEE or insulin (Xue et al., 1988b). Moreover, there is now good evidence for the presence of bipotential progenitors, whose choice of differentiation pathway is determined by environmental factors. In particular, it has been shown that sympathetic and adrenal lineages arise from a common precursor (Doupe et al., 1985a, b). In vitro, this sympathoadrenal precursor differentiates either into sympathetic neurons in the presence of NGF, or into chromaffin cells under the influence of glucocorticoids (Anderson and Axel, 1986). Even differentiated, these cell types are still able to interconvert (Unsicker et al., 1978 ; Doupe et al., 1985a). Recently, another type of bipotential progenitor for both enteric neurons and parafollicular cells has been found (Barasch et al., 1987).

In conclusion, in vitro and in vivo cell lineage studies clearly demonstrate the presence in the avian neural crest of pluripotent cells and suggest the existence of fully committed precursors. Our results as well as these from other studies, fit with the idea already developed by one of us (Le Douarin, 1986) of a progressive restriction in the developmental potentials of these pluripotent neural crest cells during migration, that would yield different types of precursors at peripheral sites, where only those able to respond to specific growth factors would develop.

Abbreviations

CEE : chick embryo extract
NF : neurofilament proteins
PNS : peripheral nervous system
SP : substance P
TH : tyrosine hydroxylase
VIP : vasoactive intestinal peptide

REFERENCES

Abo, T., and Balch, C.M., 1981, A differentiation antigen of human NK and K cells identified by a monoclonal antibody (HNK-1), J. Immunol., 127:1024.
Anderson, D.J., and Axel, R., 1986, A bipotential neuroendocrine precursor whose choice of cell fate is determined by NGF and glucocorticoids, Cell, 47:1079.
Ayer-Le Lièvre, C., and Le Douarin, N.M., 1982, The early development of cranial sensory ganglia and the potentialities of their component cells studied in quail-chick chimeras, Dev. Biol., 94:291.
Bader, C.R., Bertrand, D., Dupin, E., and Kato, A.C., 1983, Development of electrical membrane properties in cultured avian neural crest. Nature, 305:808.
Barald, K.F., 1982, Monoclonal antibodies to embryonic neurons. Cell-specific markers for chick ciliary ganglion. in: "Neuronal Development", N.C. Spitzer, ed, Plenum Press, New York.
Barasch, J.M., Mackey, H., Tamir, H., Nunez, E.A., and Gershon, M.D., 1987, Induction of a neural phenotype in a serotonergic endocrine cell derived from the neural crest, J. Neurosci., 7:2874.
Barbu, M., Ziller, C., Rong, P.M., and Le Douarin, N.M., 1986, Heterogeneity in migrating neural crest cells revealed by a monoclonal antibody, J. Neurosci., 6:2215.
Baroffio, A., Dupin, E., and Le Douarin, N.M., 1988, Clone-forming ability and differentiation potential of migratory neural crest cells, Proc. Natl. Acad. Sci. USA, 85:5325.
Black, I.B., 1982, Stages of neurotransmitter development in autonomic neurons, Science, 215:1198.
Bohn, M.C., Kessler, J.A., Adler, J.E., Markey, K., Goldstein, M., and Black, I.B., 1984, Simultaneous expression of the SP-peptidergic and noradrenergic phenotypes in rat sympathetic neurons, Brain Res., 298:378.
Bronner-Fraser, M., and Fraser, S.E., 1988, Cell lineage analysis reveals multipotency of some avian neural crest cells, Nature, 335:161.
Cepko, C., 1988, Immortalization of neural cells via oncogene transduction, Trends in Neurosci, 11:6.
Christie, D.S., Forbes, M.E., and Maxwell, G.D., 1987, Phenotypic properties of catecholamine-positive cells that differentiate in avian neural crest cultures, J. Neurosci., 7:3749.

Ciment, G., and Weston, J.A., 1982, Early appearance in neural crest and crest-derived cells of a antigenic determinant present in avian neurons, Dev. Biol., 93:355.
Cohen, A.M., and Konigsberg, I.R., 1975, A clonal approach to the problem of neural crest determination, Dev. Biol., 46:262.
Cohen, A.M., 1977, Independent expression of the adrenergic phenotype by neural crest cells in vitro, Proc. Natl. Acad. Sci. USA, 74:2899.
Doupe, A.J., Landis, S.C., and Patterson, P.H., 1985a, Environmental influences in the development of neural crest derivatives : glucocorticoids, growth factors and chromaffin cell plasticity, J. Neurosci., 5:2119.
Doupe, A.J., Patterson, P.H., and Landis, S.C., 1985b, Small intensely fluorescent cells in culture: role of glucocorticoids and growth factors in their development and interconversions with other neural crest derivatives, J. Neurosci., 5:2143.
Ernsberger, U., and Rohrer, H., 1988, Neuronal precursor cells in chick dorsal root ganglia: differentiation and survival in vitro, Dev.Biol., 126:420.
Fauquet, M., and Ziller, C., 1988, Tyrosine hydroxylase-positive cells, differentiated in vitro from neural crest, exhibit neuronal features and peptide-like immunoreactivities, J. Histochem. Cytochem., (submitted).
Fontaine-Pérus, J., 1984, Development of VIP in the peripheral nervous system of avian embryo, Peptides, 5:195.
Garcia-Arraras, J.E., Chanconie, M., Ziller, C., and Fauquet, M., 1987, In vivo and in vitro expression of vasoactive intestinal polypeptide-like immunoreactivity by neural crest derivatives, Dev. Brain Res., 33:255.
Girdelstone, J., and Weston, J.A., 1985, Identification of early neuronal subpopulations in avian neural crest cultures, Dev. Biol., 109:274.
Hall, B.K., and Tremaine, R., 1979, Ability of neural crest cells from the embryonic chick to differentiate into cartilage before their migration away from the neural tube, Anat. Rec., 194:469.
Howard, M.J., and Bronner-Fraser, M., 1985, The influence of neural tube-derived factors on differentiation of neural crest cells in vitro. I. Histochemical study on the appearance of adrenergic cells, J. Neurosci., 5:3302.
Hughes, S.M., Lillien, L.E., Raff, M.C., Rohrer, H., and Sendtner, M., 1988, Ciliary neurotrophic factor induces type-2 astrocyte differentiation in culture, Nature, 335:70.
Kalcheim, C., and Le Douarin, N.M., 1986, Requirement of a neural tube signal for the differentiation of neural crest cells into dorsal root ganglia, Dev. Biol., 116:451.
Kalcheim, C., Barde, Y.-A., Thoenen, H., and Le Douarin, N.M., 1987, In vivo effect of brain-derived neurotrophic factor on the survival of neural crest precursor cells of the dorsal root ganglia, EMBO J., 6:2871.
Kessler, J.A., Adler, J.E., Bohn, M.C., and Black, I.B., 1981, Substance P in principal sympathetic neurons : regulation by impulse activity. Science, 214:335.
Le Douarin, N.M. 1982, "The Neural Crest", Cambridge University Press, Cambridge.
Le Douarin, N.M., and Teillet, M.-A., 1974, Experimental analysis of the migration and differentiation of neuroblasts of the autonomic nervous system and of neurectodermal mesenchymal derivatives, using a biological cell marking technique, Dev. Biol., 41:162.

Le Douarin, N.M., 1986, Cell line segregation during peripheral nervous system ontogeny, Science, 231:1515.
Loring, J., Glimelius, B., and Weston, J.A.,1982, Extracellular matrix materials influence quail neural crest cell differentiation in vitro. Dev. Biol., 90:165.
Maxwell, G.D., and Forbes, E., 1987, Exogenous basement membrane-like matrix stimulates adrenergic development in avian neural crest cultures, Development, 101:767.
New, H.V., and Mudge, A.W., 1986, Distribution and ontogeny of SP, CGRP, SOM and VIP in chick sensory and sympathetic ganglia, Dev. Biol., 116:337.
Payette, R.F., Bennett, G.S., and Gershon, M.D., 1984, Neurofilament expression in vagal neural crest-derived precursors of enteric neurons, Dev. Biol., 105:273.
Price, J., Turner, D., and Cepko, C., 1987, Lineage analysis in the vertebrate nervous system by retrovirus-mediated gene transfer, Proc. Natl. Acad. Sci. USA, 84:156.
Raff, M.C., Miller, R.H., and Noble, M., 1983, A glial progenitor cell that develops in vitro into an astrocyte or an oligodendrocyte depending on culture medium, Nature, 303:390.
Raff, M.C., Lillien, L.E., Richardson, W.D., Burne, J.F., and Noble, M.D., 1988, Platelet-derived growth factor from astrocytes drives the clock that times oligodendrocyte development in culture, Nature, 333:562.
Rheinwald, J.G., and Green, H., 1975, Serial cultivation of strains of human epidermal keratinocytes : the formation of keratinizing colonies from single cells, Cell, 6:331.
Rohrer, H., Henke-Fahle, S., El-Sharkawy, T., Lux, H.D., and Thoenen, H., 1985, Progenitor cells from embryonic chick dorsal root ganglia differentiate in vitro to neurons : biochemical and electrophysiological evidence, EMBO J., 4:1709.
Rohrer, H., Acheson, A.L., Thibault, J., and Thoenen, H., 1986, Developmental potential of quail dorsal root ganglion cells analyzed in vitro and in vivo, J. Neurosci., 6:2616.
Schweizer, G., Ayer-Le Lièvre, C., and Le Douarin, N.M., 1983, Restrictions of developmental capacities in the dorsal root ganglia during the course of development, Cell Diff., 13:200.
Sieber-Blum, M., and Cohen, A.M., 1980, Clonal analysis of quail neural crest cells : they are pluripotent and differentiate in vitro in the absence of non crest cells, Dev. Biol., 80:96.
Sieber-Blum, M., Kumar, S.R., and Riley, D.A., 1988, In vitro differentiation of quail neural crest cells into sensory-like neuroblasts, Dev. Brain Res., 39:69.
Sulston, J.E., Schierenberg, E., White, J.G., and Thomson, J.N., 1983, The embryonic cell lineage of the nematode Caenorhabditis elegans, Dev. Biol., 100:64.
Teillet, M.A., and Le Douarin, N.M., 1983, Consequences of neural tube and notochord excision on the development of the peripheral nervous system in the chick embryo, Dev. Biol., 98:192.
Todaro, G.J., and Green, H., 1963, Quantitative studies of mouse embryo cells in culture and their development into cell line, J. Cell Biol., 17:299.
Tucker, G.C., Aoyama, H., Lipinski, M., Tursz, T., and Thiery, J.P., 1984, Identical reactivity of monoclonal antibodies HNK-1 and NC-1: conservation in vertebrates on cells derived from the neural primordium and on some leukocytes, Cell Diff., 14:223.
Turner, D.L., and Cepko, C.L., 1987, A common progenitor for neurons and

glia persists in rat retina late in development, Nature, 328:131.
Unsicker, K., Krisch, B., Otten, U., and Thoenen, H., 1978, Nerve growth factor-induced fiber outgrowth from isolated rat adrenal chromaffin cells: impairment by glucocorticoids, Proc. Natl. Acad. Sci. USA, 75:3498.
Wetts, R., and Fraser, S.E., 1988, Multipotent precursors can give rise to all major cell types of the frog retina, Science, 239:1142.
Xue, Z.G., Smith, J., and Le Douarin, N.M., 1985, Differentiation of catecholaminergic cells in cultures of embryonic avian sensory ganglia, Proc. Natl. Acad. Sci. USA, 82:8800.
Xue, Z.G., Smith, J., and Le Douarin, N.M., 1987, Developmental capacities of avian embryonic dorsal root ganglion cells : neuropeptides and tyrosine hydroxylase in dissociated cell cultures, Dev. Brain Res., 34:99.
Xue, Z.G., and Smith, J., 1988a, High affinity uptake of noradrenaline in quail dorsal root ganglion cells that express tyrosine hydroxylase immunoreactivity in vitro, J. Neurosci., 8:806.
Xue, Z.G., Le Douarin, N.M., and Smith, J., 1988b, Insulin and insulin-like growth factor-I trigger the differentiation of catecholaminergic precursors in cultures of dorsal root ganglia, Cell Diff., 25:1.
Ziller, C., Dupin, E., Brazeau, P., Paulin, D., and Le Douarin, N.M., 1983, Early segregation of a neuronal precursor cell line in the neural crest as revealed by culture in a chemically defined medium, Cell, 32:627.
Ziller, C., Fauquet, M., Kalcheim, C., Smith, J., and Le Douarin, N.M., 1987, Cell lineages in peripheral nervous sytem ontogeny : medium-induced modulation of neuronal phenotypic expression in neural crest cell cultures, Dev. Biol., 120:101.

CHARACTERIZATION OF TROPHIC FACTORS STORED AND SECRETED BY NEURONS

K. Unsicker, D. Blottner, D. Gehrke, C. Grothe, D. Heymann, F. Stögbauer and R. Westermann

Department of Anatomy and Cell Biology
University of Marburg
D-3550 Marburg

INTRODUCTION

Naturally occurring neuronal cell death during development is generally accepted to be regulated in part by neurotrophic factors (NTFs) (Berg 1982; Cunningham 1982; Oppenheim 1985). Beyond neural development NTFs are also involved in the maintenance and regeneration of the peripheral and central nervous system (Levi-Montalcini 1972; Johnson et al. 1980; Gage et a. 1986).

Mainly from studies with nerve growth factor (NGF), the best established NTF (Thoenen and Barde 1980; Levi-Montalcini 1987), a concept has evolved implying that target cells provide trophic molecules for their innervating neurons (Fig. 1). In addition to NGF several other neurotrophic proteins have been described during the last few years whose physiological relevance is still obscure. These factors, supporting different sets of peripheral (PNS) and central nervous system (CNS) neurons include brain-derived neurotrophic factor (BDNF, Barde et al. 1982), ciliary neuronotrophic factor (CNTF, Barbin et al. 1984), acidic and basic fibroblast growth factor (aFGF and bFGF, Morrison et al. 1986; Walicke et al. 1986; Unsicker et al. 1987), purpurin (Schubert et al. 1986), and seminal vesicle-derived neurotrophic factor (SVNF, Hofmann and Unsicker, 1987). Sources for characterization and isolation of these NTFs have been non-neuronal tissues or the complex nervous tissue, where their distribution and cellular localization is still widely unknown. **Glial cells** are an established source for NTFs (Rudge et al. 1985; Westermann et al. 1988). The only evidence for NTFs being present in **neurons** comprises the demonstration of NGF mRNA (Ayer Le Lievre et al. 1988) and bFGF-like immunoreactivity in hippocampal neurons (Pettmann et al. 1986) and neurons in cell culture (Janet et al. 1988).

In order to investigate whether NTFs are present in neurons we chose the sympathetic neuron-like adrenal chromaffin and neuroblastoma cells.

Molecular Aspects of Development and Aging of the Nervous System
Edited by J.M. Lauder
Plenum Press, New York, 1990

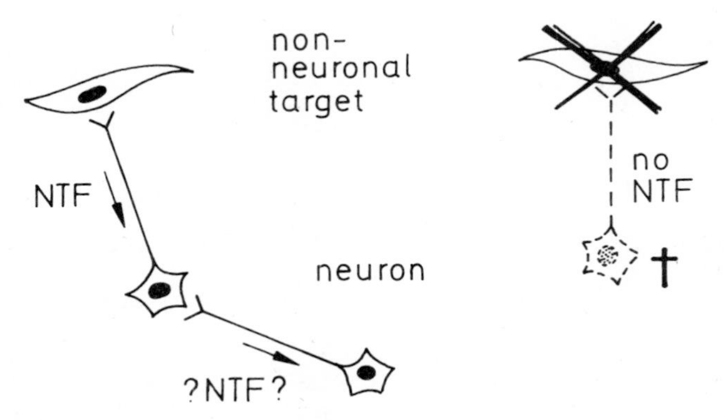

Fig. 1. The neurotrophic hypothesis. Neurons innervating non-neuronal target tissues are maintained by NTFs present in and released from these tissues. Absence or removal of the target causes death of neurons. Since most neurons are connected to other neurons rather than non-neuronal tissues, NTFs should also be present in neurons.

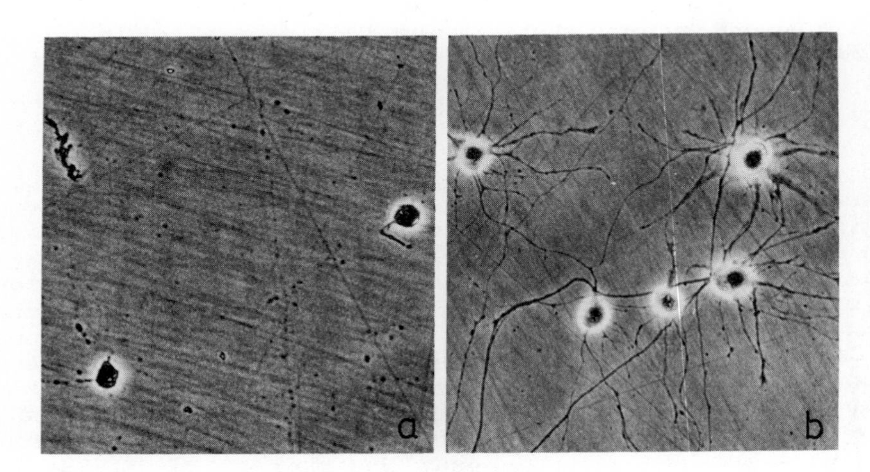

Fig. 2. Bioassay for the detection and quantitation of NTF activities. For the detection of NTFs, several neuronal tissues (see Table 1) are dissociated, neurons enriched by preplating and cultured for 24 or 48 h. In the absence of a trophic support, most neurons die during 24 h (a). In the presence of saturating amounts of an adequat NTF most neurons survive (b). Different dilutions of substances to be assayed for survival promoting activity are added to the cultures. Numbers of surviving neurons are a measure for quantifying NTF activities.

NEUROTROPHIC FACTORS ARE PRESENT IN ADRENAL CHROMAFFIN CELLS

Adrenal chromaffin cells have distinct advantages in studies on neuronal NTFs: (i) they are available in large quantities (Livett et al. 1987), (ii) they represent a relatively homogeneous and well characterized population of modified neurons, (iii) their secretory organelles, the chromaffin granules, are well characterized and can be easily isolated in large quantities (Winkler and Smith 1975), and, (iv) modes of release of putative NTFs can be studied with purified cultured cells (Unsicker et al. 1989).

Adrenal medullary cells store neurotrophic factors

NTF activities were assayed in vitro by their capacity to promote survival of neurons selected from various areas of the CNS and PNS (for methods see Barbin et al. 1984 and Hofmann and Unsicker 1987) (Fig. 2).

In order to demonstrate NTF activities of adrenal medullary components we first assayed extracts obtained from high speed supernatants of rat adrenal medullae. These preparations promoted survival of several distinct PNS and CNS neuron populations, indicating that adrenal medullary cells contain NTFs (Blottner and Unsicker 1988a) (Tab. 1). High speed supernatants from purified bovine chromaffin cells also contained NTF activities suggesting the chromaffin cell as one source of adrenal medullary NTFs.

Chromaffin cells release NTFs

If chromaffin cells contain NTFs responsible for the survival of their presynaptic neurons, these factors should be released. We therefore analyzed the presence of neurotrophic activities in media conditioned (CM) by cultured bovine chromaffin cells. Table 1 shows that CM contains neurotrophic activities addressing a similar set of PNS and CNS neurons as medullary extracts. In order to further clarify the mode of release we investigated whether these activities were liberated by Ca^{2+}-dependent cholinergic stimulation in vitro.

Catecholamines (CA) and chromogranin A (CHR A), the main soluble constituents of chromaffin vesicles (Winkler and Smith 1975), are coreleased under these conditions (Winkler et al. 1988). As shown in Table 2, similar ratios of neurotrophic activities, CA, and CHR A were released during stimulation with the cholinergic agonist carbachol, inhibition of carbachol-induced exocytosis with the Ca^{2+}-channel blocker verapamil, and in unstimulated control cultures. These results suggest that NTFs are likely to be released from chromaffin granules.

Chromaffin granules contain NTFs

Analyzes of the soluble chromaffin granule contents in terms of neuronal survival promoting effects indicated the precence of NTF activities addressing various PNS and CNS neuron populations (Table 1). The molecular properties of the NTFs present in chromaffin granules were further investigated by different chromatographic procedures. Results are compiled in Table 3.

Table 1. Adrenal NTF activities towards various CNS- and PNS-neurons.

	RAT ADRENAL MEDULLARY EXTRACT	BOVINE CHROMAFFIN CELL CONDITIONED MEDIUM	BOVINE CHROMAFFIN GRANULE CONTENT
cCG8	+	+	+
cDRG8	+	+	+
cDRG10	+	+	+
cSG11	+	+	+
cSC6	+	n.d.	+
rHIPP18	n.d.	n.d.	+
rSCGp1	n.d.	+	+
rNGp1	n.d.	-	-
rCCp8	n.d.	+	+

Rat or bovine medullary extracts (a), conditioned medium from cultured bovine chromaffin cells (b), or soluble proteins from bovine chromaffin granules (c) were tested for neurotrophic activities in bioassays. Abbreviations for neurons supported: Embryonic chick ciliary, dorsal root and lumbar sympathetic ganglia at day 8, 10 and 11, respectively (cCG8, cDRG8, cDRG10, cSG11), and spinal cord at day 6 (cSC6); embryonic rat day 18 hippocampus (rHIPP18); newborn rat superior cervical ganglia (rSCGp1), nodose ganglia (rNGp1) and chromaffin cells (rCCp8).

Table 2. NTFs released by cultured bovine chromaffin cells.

	NTF ACTIVITY	CATECHOLAMINES	CHR A
CARBACHOL	100%	100%	100%
CARBACHOL + VERAPAMIL	50%	42%	44%
CONTROL	4%	2%	3%

Cultures were stimulated with carbachol. Catecholamines and CHR A were quantified by HPLC/electrochemical detection and ELISA, respectively. NTF activities released into the culture medium were assayed on cCG8 neurons. Values obtained in response to carbachol were set to 100%. Comparable reductions for all three substances were observed following inhibition of exocytosis with the calcium channel blocker verapamil or in unstimulated control cultures.

Separation of bovine chromaffin vesicle proteins by gel filtration (HPLC, TSK-W 3000) revealed a single fraction with NTF activity, addressing embryonic chick ciliary ganglion neurons of day 8 (cCG8), but not embryonic chick dorsal root ganglion neurons of day 8 (cDRG8), with an apparent molecular weight of about 20- 24 kDa. An established NTF of comparable molecular weight and target neurons being addressed is CNTF. CNTF also retains its activity in the presence of sodium dodecyl sulphate (SDS) as shown by cell blots (Carnow et al. 1985). This method allows the estimation of molecular weights of NTFs in crude extracts after separation by SDS-polyacrylamidegelelectrophoresis and subsequent transfer to nitrocellulose, which is then used as a substrate for CG8 bioassays (Carnow et al. 1985; Heymanns and Unsicker, 1987). Chromaffin granule proteins in such experiments yielded an activity of about 24 kD, corresponding to the molecular weight of CNTF from rat sciatic nerve (Manthorpe et al. 1986).

Since CHR-A is the major soluble protein in bovine chromaffin granules (Winkler and Smith 1975) we expected a considerable enrichment of NTFs by the removal of this protein. Surprisingly, immunoaffinity-purified CHR A revealed an activity addressing cDRG8 neurons. In contrast, CHR A purified according to the established biochemical procedure (generous gift from Dr. R. Hogue-Angeletti) revealed no NTF activity. In addition, no survival activity could be observed for several peptides derived from the CHR A sequence including pancreastatin (Tatemoto et al. 1986; Eiden 1987). Thus, it is yet unclear how this cDRG8 activity relates to Chr A.

Several growth factors have been shown to specifically bind to heparin (Lobb et al. 1986). We therefore subjected bovine granule proteins to heparin affinity-chromatography. Elution by 0.6 M NaCl yielded an activity addressing cCG8 and cDRG8 neurons. A second activity supporting cCG8 neurons could be eluted from heparin at 1 to 3M NaCl. This chromatographic behaviour might suggest a relationship with bFGF which has a molecular weight of 16 to 18 kDa (Gospodarowicz et al. 1984) and promotes the in vitro survival of cCG8 neurons (Unsicker et al. 1987). The 1 to 3M NaCl heparin-binding fraction revealed also an 18kD protein band. To further investigate the possible presence of bFGF in chromaffin vesicles we used polyclonal antisera against this growth factor in immunological studies. Application of antibodies against bFGF revealed immunoreactivity in chromaffin cells in situ (Fig. 3) and in vitro. In addition, Western blot analysis of bovine adrenal medullary extracts with anti-bFGF antisera yielded an 18kDa band. Moreover the NTF activity of both medullary extracts and vesicle content was inhibited by anti-bFGF antibodies. Finally mitogenic activity similar to bFGF was detected in chromaffin granule extracts using an endothelial cell proliferation assay.

Taken together these results suggest the presence of several immunologically, biochemically and biologically distinct NTFs in chromaffin cells and granules.

Table 3. Characteristics of chromaffin granule NTFs.

	RESPONSIVE NEURONS	MW (kD)
FGF-like	cCG8	18
CNTF-like	cCG8	24
CHR A-related	cDRG8	n.d.
0.6 M Heparin-binding	cDRG8 + cCG8	n.d.

(Abbreviations for neurons see Table 1).

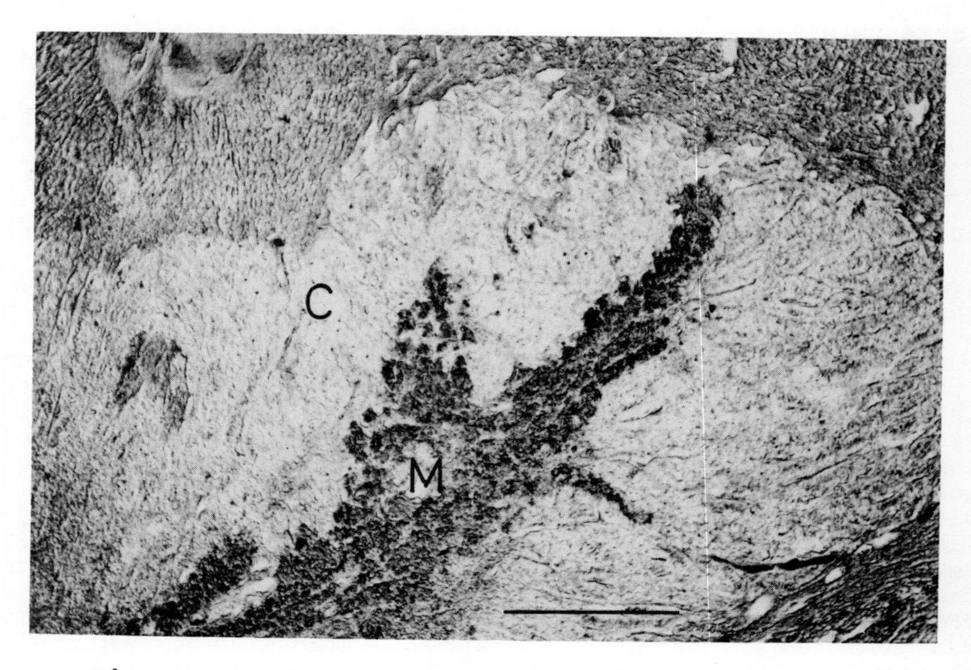

Fig. 3. Basic FGF-like immunoreactivity in bovine adrenal gland. Cryostat sections of bovine adrenals were subjected to peroxidase-anti-peroxidase immunohistochemistry. Chromaffin cells in the medulla (M) show strong immunoreactivity for bFGF, while the inner cortical layer (C) is not stained. (bar=0,25mm)

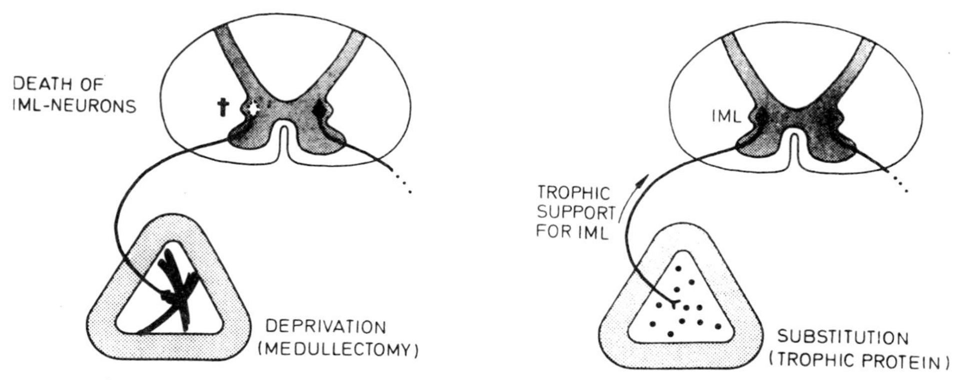

Fig. 4. Neurotrophic requirements of IML neurons: Target deprivation and substitution of the target by NTFs. Adrenal chromaffin cells are mostly innervated by neurons located in the intermedio-lateral column (IML) of the spinal cord. According to the neurotrophic hypothesis (Fig. 1) medullectomy should lead to IML-neuron death whereas substitution of the target should prevent this neuronal cell death.

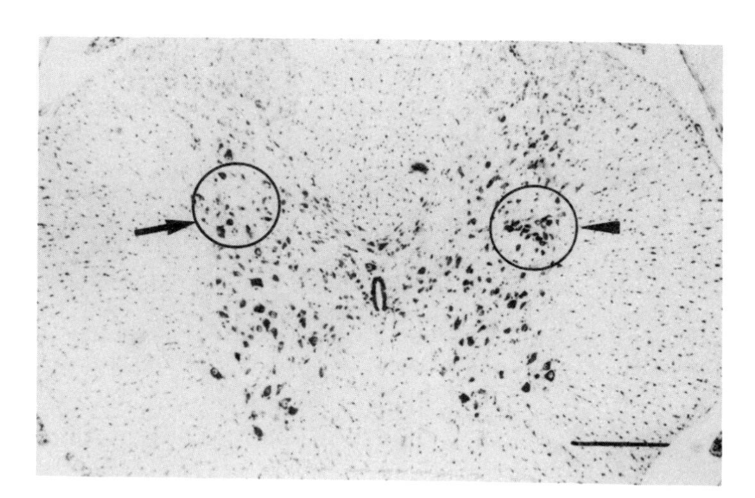

Fig. 5. IML-neuron death following adrenomedullectomy. Cresyl-violet stained transverse-sections of rat spinal cord showing cell losses in the ipsilateral IML (arrow) after medullectomy as compared to the contalateral unlesioned side (arrowhead). IML neurons are circled. (bar=0,25mm)

NTF ACTIVITIES IN NEUROBLASTOMA CELLS

Another homogeneous neuron population also derived from the sympatho-adrenal cell lineage are neuroblastoma cells. Extracts of the mouse C 1300 and the human IMR 32 cell lines have been shown to contain neurotrophic activities directed towards several embryonic neuron populations. In accordance with chromaffin cells one of these activities seems to be related to CNTF in terms of molecular weight as determined by cell blots and responding embryonic neurons (Heymanns and Unsicker 1987).

In analogy to chromaffin cells, IMR 32 neuroblastoma cells also contain several heparin-binding NTFs. Immunological studies suggest a possible identity of one of these heparin-binding proteins with bFGF (Heymann and Unsicker 1988).

POSSIBLE IN VIVO RELEVANCE OF NEURON DERIVED NTFs

In order to clarify the in vivo relevance of neuron-derived NTFs we used one of the putative chromaffin cell and neuroblastoma NTFs - bFGF - for in vivo experiments.

Chromaffin cells in the adrenal medulla are innervated by preganglionic neurons located in the intermediolateral (IML) of the lateral horn in the spinal cord (Schramm et al. 1975). According to the neuron-target neurotrophic hypothesis (Fig. 1) adrenomedullectomy should result in IML neuron death (Fig. 4). Unilateral adrenomedullectomy in the rat in fact caused the disappearance of about 25% IML neurons on the ipsilateral side (Fig. 5). Substitution of chromaffin tissue by implanted gel foams soaked with bFGF almost totally prevented IML neuron death (Blottner and Unsicker 1988b). These results therefore suggest that bFGF, possibly present in chromaffin granules, is able to rescue preganglionic IML neurons in vivo. Neurotrophic effects of bFGF have previously been shown in two other CNS lesion-paradigms (Anderson et al. 1988; Otto et al. 1988; Sievers et al. 1987).

CONCLUSIONS

Data from our studies summarized above provide evidence for the presence of neurotrophic activities in neurons using adrenal medullary chromaffin cells and neuroblastoma cells as model systems. In chromaffin cells neuronal survival promoting activities are stored in and released from transmitter storing granules. Basic FGF or an immunologically related protein is present in chromaffin as well as in neuroblastoma cells. A putative physiological role of bFGF in chromaffin cells has been shown in lesion experiments.

Isolation and characterization of these NTFs and availability of antibodies will lead to a better understanding of trophic neuron-neuron interactions during development, maintenance and diseases of the nervous system. NTFs from adrenal medullary chromaffin cells may also be responsible for the functional recovery foolowing transplantation of adrenal tissue in Parkinsonian rats, mice and humans (Olson et al. 1987; Bohn et al. 1987; Lewin 1987).

ACKNOWLEDGEMENTS

We thank D. Gospodarowicz for a generous gift of bovine bFGF, PROGEN (Heidelberg) for providing recombinant bFGF and P. Böhlen for performing proliferation-assays.

The excellent technical assistance of H. Hlawaty, M. Johannsen, P. Lattermann, W. Lorenz, H. Reichert-Preibsch, K. Thomassen and K. Zachmann is greatfully acknowledged.

This work was supported by a grant from the German Research Foundation (Un 34/11-3).

REFERENCES

Anderson K.J., Dam D., Lee S., Cotman C.W. (1988): Basic fibroblast growth factor prevents death of lesioned cholinergic neurones in vivo; Nature 332, 360-1

Ayer-LeLievre C., Olson L., Ebendal T., Seiger A., Persson H (1988): Expression of the nerve growth factor gene in hippocampal neurons; Science 240, 1339-41

Barbin G., Manthorpe M., Varon S. (1984): Purification of the chick eye ciliary neuronotrophic factor; J. Neurochem. 43, 1468-78

Barde Y.A., Edgar D., Thoenen H. (1982): Purification of a new neurotrophic factor from mammalian brain; EMBO J. 4, 549-53

Berg D.K. (1982): Cell death in neuronal development: regulation by trophic factors; In: N.C. Spitzner (ed.): Neuronal Development: Regulation by Trophic Factors. New York, Plenum, pp. 297-331

Blottner D. and Unsicker K. (1988a): Spatial and temporal patterns of neurotrophic activities in rat adrenal medulla and cortex; submitted

Blottner D., Westermann R., Grothe C., Böhlen P., Unsicker K. (1988b): Basic fibroblast growth factor in the adrenal medulla: possible trophic role for preganglionic neurons in vivo; submitted

Bohn M.C., Cupit L., Marciano F., Gash D.M. (1987): Adrenal medulla grafts enhance recovery of striatal dopaminergic fibers; Science 237, 913-6

Carnow T.B., Manthorpe M., Davis G.E., Varon S. (1985): Localized survival of ciliary ganglionic neurons identifies neuronotrophic factor bands on nitrocellulose blots; J. Neurosci. 5, 1965-71

Cunningham T.J. (1982): Naturally occuring neuron death and its regulation by developing neural pathways; Int. Rev. Cytol., Vol. 74, 163-86

Eiden L.E. (1987): Is chromogranin a prohormone?; Nature 325, 301

Gage F.H., Wictorin K., Fisher W., Williams L.R., Varon S., Bjoerklund A. (1986): Retrograde cell changes in medial septum and diagonal band following fimbria-fornix transection: quantitative temporal analysis; Neurosci. 17, 241-55

Heymann D. and Unsicker K. (1988): Cell blotting and isoelectric focusing of neuroblastoma-derived heparin-binding neurotrophic activities; submitted

Heymanns J. and Unsicker K. (1987): Neuroblastoma cells contain a trophic factor sharing biological and molecular properties with ciliary neurotrophic factor; Proc. Natl. Acad. Sci. 84, 7758-62

Hofmann H.D. and Unsicker K. (1987): Characterization and partial purification of a novel neuronotrophic factor from bovine seminal vesicle; J. Neurochem. 48, 1425-33

Janet T., Grothe C., Pettmann B., Unsicker K., Sensenbrenner M. (1988): Immunocytochemical demonstration of fibroblast growth factor in cultured chick and rat neurons; J. Neurosci. Res. 19, 195-201

Johnson E.M., Gorin P.D., Brandeis L.D., Pearson J. (1980): Dorsal root ganglion neurons are destroyed by exposure in utero to maternal antibody to nerve growth factor; Science 210, 916-8

Levi-Montalcini R. (1972): The morphological effects of immunosympathectomy; In: G. Steiner and E. Schoenbaum (ed.): Immunosympathectomy. Amsterdam: Elsevier, pp. 55-78

Levi-Montalcini R. (1987): The nerve growth factor: thirty-five years later; EMBO J. 6, 1145-54

Lewin R. (1987): Dramatic results with brain grafts; Science 237, 245-7

Livett B.G., Mitchelhill K.I., Dean D.M. (1987): Adrenal chromaffin cells - their isolation and culture; In: A.M. Poisner and J.M. Trifaro (eds.): In Vitro Methods for Studying Secretion; Secretory Process, Vol.3, pp. 171-7

Lobb R.R., Harper J.W., Fett J.W. (1986): Purification of heparin-binding growth factors; Anal. Biochem. 154, 1-14

Manthorpe M., Skaper S.D., Williams L.R., Varon S. (1986): Purification of adult rat sciatic nerve ciliary neuronotrophic factor; Brain Res. 367, 282-6

Morrison R.S., Sharma A., De Vellis J., Bradshaw R.A. (1986): Basic fibroblast growth factor supports the survival of cerebral cortical neurons in primary culture; Proc. Natl. Acad. Sci. 83, 7537-41

Olson L., Backlund E.O., Gerhardt G., Hoffer B., Lindvall O., Rose G., Seiger A., Stromberg I. (1987): Nigral and adrenal grafts in parkinsonism: recent basic and clinical studies; Adv. Neurol. 45, 85-94

Oppenheim R.W. (1985): Naturally occuring cell death during neural development; TINS 8, 487-93

Otto D., Frotscher M., Unsicker K. (1988): Basic fibroblast growth factor and nerve growth factor administered in gel foam rescue medial septal neurons after fimbria fornix transection; J. Neurosci. Res., in press

Pettmann B., Labourdette G., Weibel M., Sensenbrenner M. (1986): The brain fibroblast growth factor (FGF) is localized in neurons; FEBS Lett. 118, 195-9

Rudge J.S., Manthorpe M., Varon S. (1985): The output of neuronotrophic and neurite-promoting agents from rat brain astroglial cells: a microculture method for screening potential regulatory molecules; Dev. Brain Res. 19, 161-72

Schramm L.P., Adair J.R., Stribling J.M., Gray L.P. (1975): Preganglionic innervation of the adrenal gland of the rat: A study using horseradish peroxidase; Exp. Neurol. 49, 540-53

Schubert D., LaCorbiere M., Esch F. (1986): A chick neural retina adhesion and survival molecule is a retinol binding protein; J. Cell Biol. 102, 2295-301

Sievers J., Hausmann B., Unsicker K., Berry M. (1987): Fibroblast growth factors promote the survival of adult rat retinal ganglion cells after transection of the optic nerve; Neurosci. Lett. 76, 157-62

Tatemoto K., Efendic S., Mutt V., Makk G., Feistner G.J., Barchas J.D. (1986): Pancreastatin, a novel pancreatic peptide that inhibits insulin secretion; Nature 324, 476-8

Thoenen H. and Barde Y.A. (1980): Physiology of nerve growth factor; Physiol. Rev. 60, 1284-336

Unsicker K., Reichert-Preibsch, Schmidt R., Pettmann B., Labourdette G., Sensenbrenner M. (1987): Astroglial and fibroblast growth factors have neurotrophic functions for cultured peripheral and central nervous system neurons; Proc. Natl. Acad. Sci. 84, 5459-63

Unsicker K., Reichert-Preibsch H., Hofmann H.-D. (1989): Chromaffin cell cultures and their use for assaying effects of nerve growth factor and other growth factors; In: R.A. Rush (ed.): Nerve Growth Factors. London: Wiley, pp.149-67

Walicke P., Cowan W.M., Ueno N., Baird A., Guillemin R. (1986): Fibroblast growth factor promotes survival of dissociated hippocampal neurons and enhances neurite extension; Proc. Natl. Acad. Sci. 83, 3012-6

Westermann R., Hardung M., Meyer D.K., Ehrhard P., Otten U., Unsicker K. (1988): Neuronotrophic factors released by C6 glioma cells; J. Neurochem. 50, 1747-58

Winkler H. and Smith D. (1975): The chromaffin granule and the storage of catecholamines; In: Handbook of Physiology 7/VI. Washington: Am. Physiol. Soc., pp. 321-39

Winkler H., Fischer-Colbrie R., Schober M., Rieker S., Weiler R. (1988): Catecholamine-storing cells secrete a blended mixture of catecholamines, neuropeptides and chromogranins; In: N.A. Thorn et al. (eds.): Molecular Mechanisms in Secretion, Alfred Benzon Symposion 25, pp. 347-59

NEUROPEPTIDES AS POSITIVE OR NEGATIVE NEURONAL GROWTH REGULATORY FACTORS: EFFECTS OF ACTH AND LEU-ENKEPHALIN ON CULTURED SEROTONERGIC NEURONS

Martha I. Davila-Garcia and Efrain C. Azmitia

New York University
Washington Square Center for Neuroscience
New York, NY 10003

INTRODUCTION

In general, neuronal growth regulatory factors can be divided into three categories with respect to their effects *in vitro*. The first category consists of the survival factors, which increase the number of neurons surviving after a certain time *in vitro* (as typified by NGF). The second category is composed of the neurite promoting factors which increase the proportion of neurons growing neurites (such as laminin, fibronectin, collagen and polycations). The third category includes developmental growth factors (also called instructing or specifying factors) which influence the type of growth (cell shape, transmitter content, neurotransmitter release, etc.) for specific neuronal cell types. Evidence would suggest that neurotransmitters, including neuropeptides, can be placed within any of these categories, being either positive or negative factors on neuronal growth.

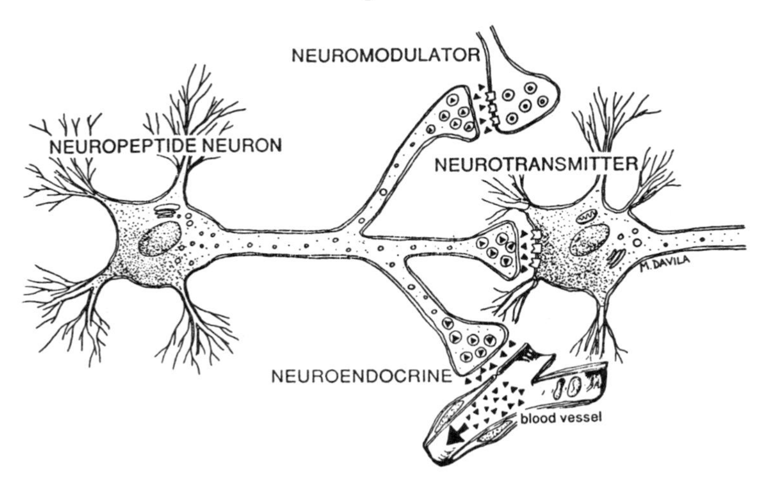

Fig.1 Actions of neuropeptides on the CNS. Neuropeptides may act as neurotransmitters, neuromodulators or as neuroendocrine hormones.

Molecular Aspects of Development and Aging of the Nervous System
Edited by J.M. Lauder
Plenum Press, New York, 1990

Since their discovery in the 1920's, peptide hormones were shown to act as neuroendocrine factors. Synthesized in the pituitary or hypothalamus, they are released into the portal blood system where they affect distant organs. A few decades later they were found to also act as neuromodulators in synaptic activity. Most recently, these peptide hormones have been shown to be active in the brain as neurotransmitters and hence named neuropeptides (Fig.1). Finally, neuropeptides are present early in fetal development (Table I), when active neurogenesis and differentiation are taking place, thus a direct influence of these peptides as neuronal growth regulatory factors could be expected.

Table I. Neuropeptide Ontogeny. Neuropeptides appear very early during development and before they play a role as neurotransmitters or neuromodulators.

NAME OF NEUROPEPTIDE	FIRST DAY OF IMMUNOCYTOCHEMICAL DETECTION	REFERENCE
ACTH	16DG	Chatelain et al., 1979
Bombesin	0DPN	McGregor et al., 1982
CRF	18DG	Daikoku, 1986
		Bugnon et al., 1982, 1984
CRF receptors	17DPN	Insel et al., 1988
LHRH	20DG	Daikoku et al., 1978
Neuropeptide Y	13DG	Foster and Shultzberg, 1984
Neurophysin (vaso-oxy prec)	16DG	Wolff and Sterba, 1978
Neurotensin	<10DPN	Aperia et al., 1982
Opiates in T.C.	13-15DG	Neale et al., 1978
Opiate receptors	15DG	Coyle and Pert, 1976
(in spinal cord)	16DG	Kirby, 1984
Oxytocin	17DG	Sinding et al., 1980
POMC peptides	15DG	Brayon et al., 1979
		Swaab and Martin, 1981
Somatostatin	14DG	McGregor et al., 1982
Substance P	14DG	McGregor et al., 1982
TRHRH	18DG	Conklin et al., 1973
Vasopressin	14DG	Sinding et al., 1980
Vasopressin fibers	16DG	Boer et al., 1980, 1985
		Buijs, et al., 1980
VIP	0DPN	McGregor et al., 1982

Neuropeptides have been found to be neuronal growth regulatory factors during development[16,19,47,78,86,87], in primary neuronal cell cultures[11,36,37,39,40,70] and in regeneration[15,40,68,69]. Their effects include regulation of morphogenesis, mitosis, cell death, neurite extension, and synaptogenesis. The evidence indicates that the mechanism involves the stimulation or inhibition of specific receptors located both pre- and post-synaptically, that the effects are dose and time dependent, and most importantly, that they have their strongest effect during "critical periods" in development. Furthermore, the action of these peptides on their corresponding receptors and the subsequent activation of second messenger systems, which trigger a cascade of events involving phosphorylation of proteins as well as enhancement on metabolic activity, is an open field for investigation. The realization that neuropeptides function as neuronal regulatory growth factors not only in development but in regeneration as well, has revolutionized the field of neural plasticity and opened a new chapter on the study of factors that regulate neuronal maturation and differentiation.

The purpose of this section is not to review the effects of all neuropeptides on development, nor to cover their behavioral effects. The aim will be to briefly review studies with ACTH and enkephalins and their best known neurotrophic/neurotropic effects as regulatory growth factors on neuronal development *in vivo* or *in vitro*, with particular emphasis on their actions on serotonergic neurons.

In our laboratory, the study of neuropeptides in neuronal development is carried out by three principal methods which we shall describe here in some detail. We believe that the same methodology may be used for any neuronal system. (1) Primary fetal cells are grown in 96 well plates for biochemical analysis (uptake, binding, protein assays). These microcultures permit fast and sensitive screening of the actions of neuropeptide drugs in serial dilutions. Furthermore, this microculture system permits many variables such as plating density, influence of target neurons and non-neuronal cells, time in culture, and drug interactions to be examined. (2) Primary fetal cells are grown on 8-chambered glass slides for morphometric analysis. The slide cultures are processed by immunocytochemistry to identify neurotransmitter specific neurons with antibodies and by radioautography to localize the site of action of radioactive ligands. The slide cultures can then be cover-slipped and studied under optimal microscopic conditions for photographic work or computer-assisted morphometric analysis. (3) Pregnant rats are administered drugs during key periods in brain development of the fetuses. The pups are studied post-natally for alterations in the neuroanatomy of selected neurotransmitter systems, the expression of related receptor(s) and on the behavioral repertoire of the developing rat. Usually these three approaches will provide complimentary data on the role of a particular neuropeptide on a selected neurotransmitter system.

METHODS

Tissue Culture

The cultures are prepared as previously described by Azmitia and Whitaker-Azmitia[10]. Embryos are removed by Caesarean section from timed-mated pregnant rats certified pathogen-free (Taconic Farms, Germantown NY) and the age of each litter verified by measurement of the crown-to-rump length. The brains are removed and placed in a solution containing 0.8% NaCl, 0.04% KCl, 0.006% $Na_2HPO_4 \cdot 12H2O$, 0.003% KH_2PO_4, 1% glucose, 0.00012% phenol red, 0.0125% penicillin G, and 0.02% streptomycin. Rostral brain stem cells are obtained from 14 day gestation (14DG) fetuses by dissecting between the mesencephalic and pontine flexures. The tectum is cut through the midline to expose the cerebral aqueduct and two lateral cuts are made at the ventral part of the tectum. The tegmentum is cut at both sides (0.25 mm from the midline), leaving a block of tissue which contains most of the midbrain (ascending) serotonergic cell bodies[7]. The meninges and the adjacent mesenchymal tissue are carefully removed. Target cells from spinal cord (SC) are obtained from 14DG fetuses by extracting the entire spinal cord. Hippocampal cells (HIPP) are obtained from 18DG fetuses using the landmarks provided by Banker and Cowan[12]. The hemispheres are carefully separated from the thalamus. The hippocampal formation is exposed and separated from the rest of the cortex at its septal and temporal poles by sectioning the anterior and posterior end of the lateral ventricle. The hippocampal cortex is separated as a quarter-moon shaped tissue containing the dentate gyrus, the hippocampus, and part of the subiculum. The meninges and choroid plexus are then carefully removed. The tissues are transferred to 1 ml of complete neuronal medium (CNM) which contains: Eagle's minimal essential medium (MEM), 0.5% glucose, non-essential amino acids, and 7.5% fetal calf serum (all from Gibco); pH 7.4. Each tissue is gently repipetted using a pasteur pipette with a fire polished tip (>1 mm). A second repipetting is done with a smaller diameter pipette (<0.5 mm) until the tissue is completely dissociated. To these cell suspensions, 5 ml of CNM are added and the cell density is determined using a hemocytometer. The cells are plated in 96 well plates (NUNCLON, NUNC (InterLab) Denmark) or 8 well chambered slides previously coated with poly-D-lysine (25 ug/ml) and washed with CNM just before plating. Initial plating densities are specified for each experiment between 800,000 to 1 million cells/cm^2 (70 ug of protein/well) with a final volume of 200 ul of CNM per well. Co-cultures of raphe and target cells are plated in a 1:1 ratio. The plates are incubated for 3 or 5 days (3DIC or 5DIC) in a CO_2 incubator at 37°C, with 95% humidity, and 5% CO_2.

Measure of High Affinity (^{3}H)5-HT Uptake

Monoamine reuptake and storage has been used by several workers as a model system to study and monitor neuronal maturation and innervation density[10,11,55,61]. In culture, the high-affinity uptake of (^{3}H)5-HT is linearly related to the number of plated 5-HT-immunoreactive (5-HT-IR) neurons and to the days in culture[10]. The method for measuring uptake is briefly described here. The cultures are washed with MEM containing 1% glucose at 37°C. The cultures are incubated with 50 nM (^{3}H)5-HT (26 Ci/mmol, New England Nuclear), 1% glucose and 10^{-4} M pargyline in MEM for 20 minutes at 37°C[10]. This concentration of (^{3}H)5-HT has been shown to be selectively retained by serotonergic neurons as demonstrated by radio-autographic pharmacology[6,72] and it is five times lower than the concentration needed to be taken up by mature glia[52]. Non-specific accumulation is measured by incubating the cultures as described with 10^{-5} M fluoxetine (a gift from Ely Lilly Co.). Fluoxetine is a recognized 5-HT uptake blocker[6,46]. After the 20 min incubation period, the medium is removed and the cultures are washed twice with 0.1 M PBS. The cells are allowed to air dry before 200 ul of absolute ethanol are added to each well for 1 hr. From each well 150 ul are removed and placed in 5 ml of liquiscint (National Diagnostics) and counted for 1 minute in a Beckman liquid scintillation counter (counting efficiency 58%). The specific accumulation of (^{3}H)5-HT is calculated as the difference between the total and the non-specific accumulation.

Drug Preparation

The ACTH 1-39, ACTH peptide fragments (a gift from Organon), and Leu-enkephalin (Sigma Chemical Co., St. Louis) are dissolved in MEM at a high concentration (1 mM) and passed through a Uniflo filter (0.2 um pore). This stock concentration is aliquoted and frozen. Dilutions are prepared in CNM from the stock in 1:10 serial dilutions. The drugs are added to the cultures either at the time of plating or once daily. Because peptides are rapidly degraded by aminopeptidases in culture, the drugs are prepared with or without bacitracin 0.05 mg/ml, an aminopeptidase inhibitor.

Immunocytochemistry of Serotonin Neurons

Positive identification of serotonin-containing neurons in tissue culture is achieved using an antibody raised against serotonin conjugated to hemocyanin by formaldehyde[57]. The specificity of this antiserum was characterized, and the antiserum has been shown to exhibit little or no cross reactivity to norepinephrine, dopamine, or histamine, as determined by a variety of pre-absorption tests, immunocytochemical staining pattern, and by prior treatment with selective neurotoxins[9,44,57]. The cells grown on 8 chambered slides, are fixed after 24 hrs, 3DIC or 5DIC in culture, by removing the media and adding 200 ul of freshly prepared ice cold (4°C) 4% paraformaldehyde and 0.05% $MgCl_2$ in 0.1 M phosphate buffer (PB) pH 7.4. The cells are fixed for at least 2 hrs at 4°C. The cultures are washed carefully three times with 0.1 M PB with 0.9% saline (0.1 M PBS) pH 7.4 and three times with 0.1 M Tris buffer with 0.9% saline (0.1 M TBS) pH 7.6. The primary antibody against serotonin is prepared in 0.3% Triton X-100 and 1% normal goat serum in 0.1 M TBS (TNT3) at a 1:8,000 concentration and a small volume is added (100 ul) to each well. The plates are incubated overnight at 4°C. The cultures are then washed gently three times with 0.1 M TBS and incubated at room temperature in biotynilated antibody against rabbit (ABC Vector Stains) for 1 hr and gently washed twice with 0.1 M TBS. The cultures are then incubated in an Avidin-biotin solution (Vector Stains) for 1 hr and washed twice with 0.1 M TBS. They are then incubated in a solution containing 0.05% 3,3'-diaminbenzidine, 0.003% H_2O_2 in 0.1 M TBS for 4 to 8 min, washed three times with 0.1 M TBS, dehydrated through a series of alcohols (70%, 95%, 100%), cleared with xylene and mounted with DPX mountant (BDH limited). Morphological analysis of the 5-HT-IR cells is achieved using a Bioquant for measuring neurite cell length, cell body area and cell density.

Pregnant Rats

Pregnant Sprague Dawley rats (Taconic Farms) are kept in separate cages at 22°C, in a 12 hr light/dark cycle and feed rat chow ad libitum. The rats are treated with the peptides from day 6 to day 21 of gestation (DG). An untreated and a saline group are always used as controls. At birth the pups are cross fostered with untreated mothers in groups of eight. Uptake of (^{3}H)5-HT is determined using the method by Azmitia et al[8] with slight modifications. The pups are sacrificed at 26 days postnatally (26DPN). Their brains are removed, and the midbrain, hippocampus, and spinal cord are dissected out and placed in cold (4°C) 0.32 M sucrose; the volume of sucrose being ten times the tissue weight. The samples are homogenized into subcellular organelles including mitochondria, membrane fractions and myelin, using a Cole Palmer motor drive set at 1000 rev/min. The homogenates are centrifuged for 10 min at 1000 X g in a Sorvall centrifuge and the supernatants (S1) removed and saved on ice. The pellets (P1) are resuspended in 2 ml of cold 0.32 M sucrose and centrifuged again for 10 min at 1000 X g. The resultant supernatants (S2) are combined with their respective S1 and spun at 14,000 X g for 10 min. The resulting pellets (P2) are the crude synaptosomal preparation and are resuspended in cold MEM with 1% glucose at a volume equal to ten times the tissue weight. From this suspension, 20 ul of the tissue are placed in a 96 well plate (NUNC) and incubated with 50 nM (^{3}H)5-HT in MEM with 10^{-4} M pargyline and 1 % glucose for 200 ul total volume per well. Each sample is run in quadruple. For non specific uptake 10^{-5} M cold 5-HT is added to 4 wells/condition. The plates are incubated at 37°C for 5 minutes and harvested into G/B Whatman filter paper, and washed for 15 sec with cold 0.1 M PBS using an automatic Titrek cell harvester (Flow Laboratories). The filters are then placed in scintillation vials with 5 ml of liquiscint (National Laboratories) and counted on a Beckman Scintillation Counter for 1 min. The remaining tissue is frozen and used for protein determination using the Lowry[58] method.

Anatomical Studies

Pups at postnatal day 26 (DPN) are perfused intracardially with 100 ml cold (4°C) 0.9% saline for 1 min followed by 200 ml cold (4°C) 4% paraformaldehyde in 0.1 M PBS with 0.05% MgCl, pH 7.4 for 15 min. The brains are removed and postfixed in cold 4% paraformaldehyde for 4 hrs and sectioned on a vibratome at 40 um intervals. The sections are washed with 0.1 M PBS three times and three times with 0.1 M TBS for 10 min each and processed for 5-HT immunohistochemistry as free floating sections following the same protocol as for tissue culture slides (see above), being careful to wash thoroughly in between antibodies. The sections are mounted on gelatinized slides, dehydrated through a series of alcohols (70%, 90%, 100%), cleared with xylene and coverslipped with DPX mountant (BHD limited). Cell number, density of fibers and cell body distribution are assessed using a computer assisted image analizer (Nikon, Joyce-Loebl Magiscan).

I. ACTH

A. ACTH IN VITRO

ACTH on 5-HT Neurons

The action of ACTH and related neuropeptides has been studied in cultured dissociated mesencephalic cells. Azmitia and de Kloet[11] showed that at the time when fetal serotonergic cells complete their neurogenesis (12E-14E) and are ready to begin differentiation and neurite extension, they are particularly sensitive to the action of these peptides. ACTH, ACTH 4-10, ACTH 1-24, ACTH 1-39, and the ACTH 4-9 analogue Org

2766 at various concentrations, stimulate the expression of (^3H)5-HT uptake by serotonergic fetal neurons cultured alone but have no effect when these neurons are co-cultured with target cells from the hippocampus. A physiological role for circulating neuropeptides might be to provide a low level of neurotrophic activation until the afferent cells grow close enough to their target, at which time growth is enhanced by neurotrophic substances secreted by the target cells. This effect is not seen with α-MSH, which had no neurotrophic effect on raphe cells. However, this peptide was able to stimulate serotonergic maturation in raphe-hippocampal co-cultures. Thus, α-MSH may affect the ability of the target cell to produce a specific neurotrophic factor, although it has no direct effect on the afferent neuron itself (Fig. 2).

ACTH and development of other systems

Enhancement of neurite outgrowth has been described by Davila and collegues[36] where ACTH 1-13 (α-MSH) (50 ng/ml) increases neurite cell length on dissociated serotonergic neurons grown in culture by 30% over control with no apparent effect on cell body area. This enhancement of neurite outgrowth has also been described by Richter-Landsberg and co-workers[66]. They investigated the role of ACTH 1-24 and ACTH 4-10 in the neuronal growth and development of cultured rat cerebral cells. At low concentrations (10^{-8} M) these peptides increase acetylcholinesterase activity but not protein synthesis, increase the density of the neuritic network, promote neurite fasciculation and increase the size and number of neuronal aggregates. Others have observed similar trophic effects of α-MSH (0.1 nM) and ACTH 4-10 (0.001-0.01 nM) on spinal cord explant cultures[82] and of ACTH 1-24 on chick cerebral neurons[35]. These studies taken together argue strongly in favor of the trophic capabilities of ACTH fragments on various central neuronal cell groups. The mechanism by which these ACTH fragments promote and enhance neurite outgrowth, however, is not known. Receptors for ACTH have not been fully characterized. Nevertheless, ACTH is known to influence brain protein kinase activity and to affect phosphorylation of brain B-50 protein in synaptic plasma membranes[62,90], although, it has not been shown to play a direct role in phosphorylation of cytoskeletal proteins[42] as one might predict given its neurite enhancing effects.

B. ACTH IN VIVO

ACTH and 5-HT Neurons

We have examined the postnatal effects of prenatally administered ACTH 4-10 (10 ug/Kg), on high affinity (^3H)5-HT uptake by brainstem, hippocampus and spinal cord[38]. Pregnant rats were injected daily from 6DG - 21DG with the peptide. The pups were sacrificed at 26 days postnatal (DPN) and the brainstem, hippocampus and spinal cord dissected. A synaptosomal preparation from these tissues was obtained as described above to measure the specific high affitity uptake during a 20 min incubation. The ACTH 4-10 treated group showed a significant increase in (^3H)5-HT uptake in brainstem and hippocampus compared to untreated controls and an increase in hippocampus compared with the saline treated group, suggesting that the ascending serotonergic system is affected by prenatal administration of this peptide most effectively at the terminal region in the hippocampus. Even mild stressors such as saline injections have been shown to produce early neurogenesis of 5-HT-IR neurons[57]. This effect is probably due to the stress-induced release of endogenous ACTH by the daily IP injection to the pregnant rats. In our study, however, only ACTH prenatal treatment but not saline injections showed an increase in (^{3}H)5-HT uptake expression in brainstem.

This ACTH family of neuropeptides has been shown to also influence the biosynthetic pathway of serotonin. Ramaekers et al[65] have shown that the ACTH 4-10 and ACTH 4-10 (7D-Phe) fragments, cause an increase in the levels of serotonin present in the adult hippocampus. Moreover, corticosterone, which is released form adrenal glands after stress-induced-ACTH release, is required for the normal development of tryptophan hydroxylase[80] and is essential for homotypic collateral sprouting of serotonergic fibers in the adult hippocampus[89]. Finally, circulating adrenal steroids have been found to increase the activity of tryptophan hydroxylase[5,74] and change the 5-HT levels in hypothalamus and midbrain of adult rat[56]. Conversely, adrenalectomy, which increases ACTH levels and decreases corticosterone leads to decreased levels of tryptophan hydroxylase activity[3] and serotonin turnover[4,41].

ACTH and Development

Neurogenic effects of ACTH peptides during development are in general, poorly characterized. The administration of ACTH peptides to pregnant rats has produced permanent alterations in the motor behavior[13,53] and sexual behavior[71] of offspring. Postnatal chronic administration of ACTH to rat pups one or two weeks of age causes heightened motor behavior[1] and accelerates motor development[1].

Studying the development of the neuromuscular junction, Rose and Strand[67] have shown that chronic administration of the ACTH peptide fragment ACTH 4-10 to pregnant rats between 3DG and 12DG, before neuromuscular innervation occurs, accelerates the maturation of the hindlimb _extensor digitorum longus_ muscle in the resulting pups, indicating that the ACTH peptide has, at least _in utero_, myotrophic effects as well. Moreover, the neuromuscular junctions of neonatal rats treated with the peptide fragments ACTH 4-10 and ORG 2766 (the tri-substituted analogue of ACTH 4-9) increase in size, complexity and extent of internal nerve terminal branching[45], again an indication of accelerated maturation.

II. OPIOIDS AND OPIATES

A. OPIOIDS IN VITRO

Opioid Peptides, Opiates and 5-HT Neurons

In contrast to the positive growth effects of ACTH peptides, opioids act as negative growth factors in neuronal development. Leu-enkephalin, for example, has been shown to have potent inhibitory effects on serotonergic maturation in tissue culture, as measured by high affinity (^{3}H)5-HT uptake. This inhibition does not involve neuronal death since cell density remains unchanged[37]. Chronic (daily dose) administration of the peptide for 3DIC or 5DIC inhibited uptake by 40% with maximal effects observed at a concentration of 18 nM[39]. Acute administration (one dose at the time of plating), however, stimulated uptake at 1.8 fM at 3DIC. This paradoxical stimulation of the opioid may be due to a rebound mechanism of receptors. Not only is it reversed by addition of bacitracin (0.05 mg/ml), an aminopeptidase inhibitor, but the stimulatory effect is transient since the increase in uptake returns to control levels after 5DIC[39]. Opiate receptors have been shown to exist on serotonergic neurons[64] and to be present early in development[27]. Studies have shown that chronic application of opioid peptides down-regulates striatal opiate receptors in the adult rat[77], while their antagonists up-regulate opioid receptors in cultured fetal mouse spinal cord-ganglion explants in a dose dependent manner[81]. Thus, dose and time of receptor occupancy seems to be a major factor affecting opioid activity. Moreover, continuous blockade of opioid receptors with the opiate antagonist naltrexone has been shown to enhance growth, whereas temporary receptor blockade inhibits growth[87]. These observations are also consistent with the finding that administration of naloxone to serotonergic neurons grown in culture stimulates their growth[37]. Furthermore, high levels of leu-enkephalin inhibit and low levels stimulate neurite length of 5-HT-IR neurons.[37].

Opiates and Development on Other Systems

This duality of opiate effects has also been reported in electrophysiological studies by Crain and collegues. They have shown that opioid peptides selectively depress sensory-evoked dorsal horn responses in dorsal root ganglion (DRG)-spinal cord (DRG-cord) cultures from 13 day old fetal mice[28] and that this effect is stereospecific, naloxone reversible, and dose dependent[29]. After chronic exposure to morphine (1 uM) or opioid peptides for 2-3 days, for example, these sensory evoked dorsal horn responses recover and can be elicited by DRG stimuli even in the presence of opioids at concentrations 10 - 100 fold higher than required to depress a naive explant. Their electro-physiological analyses have been correlated with binding assays showing high levels of opiate receptors in dorsal spinal cord regions of DRG-cord cultures[49] and in the DRG neuritic outgrowths[48]. Furthermore, they have shown that increasing concentrations of DADLE elicit dual modulatory effects on the duration of the action potential (APD) of DRG neurons on DRG-cord explants[32]: low concentrations (nM) prolongue APD, while high concentrations (uM) shorten APD (Shen and Crain, in press). Chronic exposure to DADLE on DRG-cord explants markedly decreases the proportion of DRG neurons showing APD shortening effects while increasing the proportion of DRG cells showing APD prolongation[32]. Similar dual opioid modulation of APD occurs in dissociated DRG neurons in monolayer cultures[25]. These authors suggest that the inhibitory effects may be mediated by down-regulation of inhibitory opiate receptors and an up-regulation of excitatory opiate receptor subtypes. An adenylate cyclase/cAMP/protein kinase A second messenger system appears to mediate the excitatory effects via positive coupling to a specific (high affinity) opiate receptor subtype[25,29,30].

Besides having negative growth effects on cultured serotonergic neurons and modulating sensory-evoked dorsal horn responses on DRG-cord explants, opiates have been shown to influence enzyme activity. Sakellaridis et al[70] demonstrated that morphine (10^{-5} M), applied from 4DIC to 6DIC, decreases choline acetyltransferase (ChAT) activity in neuron enriched cultures derived from 6 day old chick embryos. In addition, application of 10^{-5} M and 10^{-6} M morphine from 4DIC to 6DIC or 6DIC to 8DIC, decreases neurite length and increases the number of flat cells (glia) and the number and degree of neuronal aggregations. On the other hand, administration of methadone (10^{-6} M) to these cultures from 4DIC to 6DIC or 6DIC to 8DIC also decreases ChAT activity, but has no associated morphological changes. These studies suggest that methadone may be exerting a more specific effect compared to a more general effect by morphine[70].

B. OPIATES IN VIVO

Opiates and Serotonin Neurons

At 26 days postnatally, rats treated prenatally with leu-enkephalin (1 ug/ml) show marked enhancement on (^{3}H)5-HT high affinity uptake by spinal cord synaptosomal preparations[38]. At the same time, this prenatal treatment has no effect on uptake expression by brainstem or hippocampal synaptosomes. Morphological studies using immunohistochemistry for 5-HT demonstrate that prenatal exposure to this opioid, results in a sharp modification of 5-HT cell migration, with a decrease in the number of dorsal raphe cells and an inconsistent segregated group of large 5-HT-IR cells just above B9 area[38]. These results, in combination with the studies on prenatal exposure to ACTH 1-39 and ACTH 4-10[38] suggest a specific, time dependent effect of these peptides not only on the expression of serotonin uptake but on the subsequent development of 5-HT cell groups.

Opiates and Opioid Peptides in Development

Zagon and collegues have shown that daily injections of the opioid antagonist naltrexone to preweaning rats at doses (1 mg/Kg) that block opiate receptors only transiently (acute administration), inhibit brain[87] and cerebellar[86] growth and development. In contrast, high doses of naltrexone (50 mg/Kg) increase brain, organ, and body weight[86,87] and accelerate the appearance of physical characteristics as well as the maturation of spontaneous motor and sensory motor behaviors[88]. Prenatal exposure to exogenous opiates such as morphine, heroin and methadone have also been shown to alter neuronal growth of humans and laboratory animals[47,63,88]. The mechanism of this growth inhibition, however, is unknown but suppression of cell division and alteration in polyamines, nucleic acid and protein synthesic systems are thought to be involved[50,75,83,85]. Furthermore, these inhibitory effects seen <u>in vivo</u> are blocked by concomitant administration of opiate antagonists[76,86,87] such as naloxone or naltrexone.

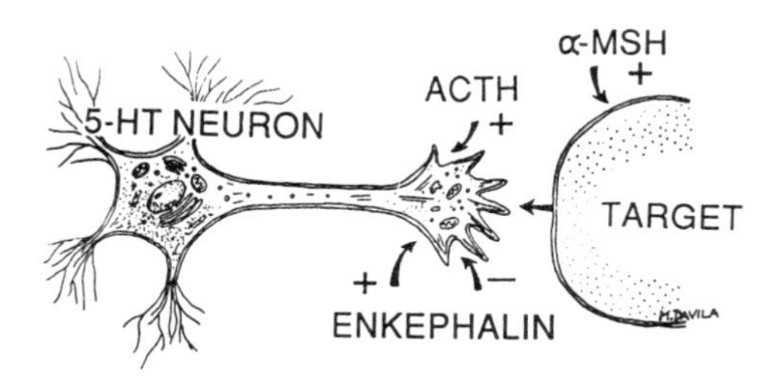

Fig.2 Positive and negative effects of neuropeptides on developing 5-HT neurons.

SUMMARY

In summary, we have presented evidence that neuropeptides can function as either positive or negative growth regulatory factors during development. The ACTH family of peptides appear to act predominantly as a positive growth regulatory factor - enhancing neurite outgrowth, cell survival, biochemical maturation and behavioral expression. These effects of ACTH are most pronounced prior to the time the afferent cell has reached its target. Thus, ACTH may act as a low level general neurotrophic growth regulatory factor. The opioids have the opposite effect. These neuropeptides inhibit neurite extension, cell survival, and biochemical maturation. The effects of these negative growth regulatroy factors are observed even when the afferents have reached their targets. The action of the opioids is thought to occur through specific receptors and known second messenger systems. Thus, CNS neuropeptide levels can have important actions in regulating the development of a variety of CNS systems, and permanently influencing the structure and function of the brain.

References

1. Acker, G.R., Berran, J. and Strand, F.L. (1985) ACTH neuromodulation of the developing motor system and neonatal learning in the rat. Peptides 6, 41-49.

2. Aperia, A., Herin, P., Eklov, A.-C. and Johnsson, V. (1982) Importance of AVP for blood pressure control during development: a study in the Brattleboro rat. Ann. NY. Acad. Sci. 394, 350-360.

3. Azmitia, E.C. and McEwen, B.S. (1969) Corticosterone regulation of tryptophan hydroxylase in midbrain of the rat. Science 166, 1274-1276.

4. Azmitia, E.C., Algeri, S. and Costa, E. (1970) In vivo conversion of ^{3}H-L-tryptophan into ^{3}H-L-serotonin in brain areas of adrenalectomized rats. Science 9, 633-637.

5. Azmitia, E.C. and McEwen, B.S. (1974) Adrenalcortical influence on rat brain tryptophan hydroxylase activity. Brain Res. 78, 291-302.

6. Azmitia, E.C. and Marovitz, W.F. (1980) In vitro hippocampal uptake of tritiated serotonin (5HT) a morphological, biochemical and pharmacological approach to specificity. J. Histochem. Cytochem. 28, 636-644.

7. Azmitia, E.C., Perlow, M.J., Brennan, M.J. and Lauder, J.M. (1981) Fetal raphe and hippocampal transplants in adult and aged C57BL/6N mice; an immunocytochemical study. Brain Res. Bull. 7, 703-710.

8. Azmitia, E.C., Brennan, M.J. and Quartermain, D. (1983) Adult development of the hippocampal-serotonin system of C57BL/6N mice; analysis of high affinity uptake of ^{3}H-5HT. Intern. J. Neurochem. 5, 39-44.

9. Azmitia, E.C. and Gannon, P. (1983) The ultrastructural localization of serotonin immunoreactivity in myelinated axons within the medial forebrain bundle of rat and monkey. J. Neurosci. 3, 2083-2090.

10. Azmitia, E.C. and Whitaker-Azmitia, P.M. (1987) Target cell stimulation of dissociated serotonergic neurons in culture. J. Neurosci. 20, 47-63.

11. Azmitia, E.C. and de Kloet, R. (1987) ACTH neuropeptide stimulation of serotonergic neuronal maturation in tissue culture: modulation by hippocampal cells. Prog. Brain Res. 72, 311-318.

12. Banker, G.A. and Cowan, W.M. (1977) Rat hippocampal neurons in dispersed cell culture. Brain Res. 126, 397-425.

13. Beckwith, B.E., O'Quinn, R.K., Petro, M.S., Kastin, A.J. and Sandman, C.A. (1977a) The effects of neonatal injections of a-MSH on the open field behavior of juvenile and adult rats. Physiol. Psychol. 5, 295-299.

14. Beckwith, B.E., Sandman, C.A., Hothersall, D., and Kastin, A.J. (1977b) The influence of neonatal injections of a-MSH on learning, memory and attention. Physiol. Behav. 18, 63-71.

15. Bijlsma, W.A., Jennekins, F.G.I., Schotman, P. and Gispen, W.H. (1984) Neurotropic factors and regeneration in the peripheral nervous system. Psychoneuroendocrinology 9, 199-215.

16. Boer, G.J., Swaab, D.F., Uylings, H.B.M., Boer, K., Buijs, R.M. and Velis, D.N. (1980) Neuropeptides in rat brain development, Prog. Brain Res. 53, 201-227.

17. Boer, G.J., Buijs, R.M., Swaab, D.F. and De Vries, G.J. (1980) Vasopressin and the developing rat brain. Peptides 1 (Suppl 1), 203-209.

18. Boer, G.J. (1985) Vasopressin and brain development: studies using the Brattleboro rat. Peptides 6, 49-62.

19. Boer, G.J. and Swaab, D.F. (1985) Neuropeptide effects on brain development to be expected from behavioral teratology. Peptides 6 (Suppl 2), 21-28.

20. Brayon, A., Shoemster, W.J., Bloom, F.F., Manns, A. and Guillemin, R. (1979) Perinatal development of the endorphin- and enkephalin- containing systems in the rat brain. Brain Res. 179, 93-101.

21. Buijs, R.M., Velis, D.N., Swaab, D.F. (1980) Ontogeny of vasopressin and oxytocin in the fetal rat: early vasopressinergic innervation of the fetal brain. Peptides 1, 315-324.

22. Bugnon, C., Fellmann, D., Gouget, A. and Cardot, J. (1982) Ontogeny of the corticoliberin neuroglandular system in rat brain. Nature 298, 159-161.

23. Bugnon, C., Fellmann, D., Gouget, A., Bresson, J.L., Clavequin, M.C., Hadjiyassemis, M. and Cardot, J. (1984) Corticoliberin neurons: cytophysiology, phylogeny and ontogeny. J. Steroid Biochem. 20, 183-195.

24. Chatelain, A., Dupouy, J.P. and Dubois, M.P. (1979) Ontogenesis of cells producing polypeptide hormones (ACTH, MSH, LPH, GH, Prolactin) in the fetal hypophysis of the rat: influence of the hypothalamus. Cell Tissue Res. 196, 409-427.

25. Chen, G.-G., Chalazonitis, A., Shen, K.-F. and Crain, S.M. (1988) Inhibitor of cyclic AMP-dependent protein kinase blocks opioid-induced prolongation of the action potential of mouse sensory ganglion neurons in dissociated cell cultures. Brain Res. 462, 372-377.

26. Conklin, P.M., Schindler, W.J. and Hull, S.F. (1973) Hypothalamic thyrotropin releasing factor. Activity and pituitary responsiveness during development of the rat. Neuroendocrinology 11, 197-211.

27. Coyle, J.T. and Pert, C.B. (1976) Ontogenetic development of [^{3}H]naloxone binding in rat brain. Neuropharmacology 15, 555-560.

28. Crain, S.M., Crain, B., Peterson, E.R. and Simon, E.J. (1978) Selective depression by opioid peptides of sensory-evoked dorsal-horn network responses in organized spinal cord cultures. Brain Res. 157, 191-201.

29. Crain, S.M., Crain, B. and Peterson, E. (1986) Cyclic AMP or forskolin rapidly attenuates the depressant effects of opioids on sensory-evoked dorsal-horn responses in mouse spinal cord-ganglion explants. Brain Res. 370, 61-72.

30. Crain, S.M., Crain, B. and Makman, M.H. (1987) Pertussis toxin blocks depressant effects of opioids, monoaminergic and muscarinic agonists on dorsal horn network responses in spinal cord-ganglion cultures. Brain Res. 400, 185-190.

31. Crain, S.M. (1988) Regulation of excitatory opioid responsivity of dorsal root ganglion neurons. In:

Regulatory roles of opioid peptides, (eds. P. Iles and C. Farsang), pp. 186-201. VCH Publishers, Germany.

32. Crain, S.M., Shen, K.-F. and Chalazonitis, A. (1988) Opioids excite rather than inhibit sensory neurons after chronic opioid exposure of spinal cord-ganglion cultures. Brain Res. 455, 99-109.

33. Daikoku, S., Kanano, H., Matsumura, H. and Saito, S. (1978) In vivo and in vitro studies on the appearance of LHRH neurons in the hypothalamus of perinatal rats. Cell Tissue Res. 194, 433-445.

34. Daikoku, S. (1986) Development of the hypothalamic-hypophyseal axis in rats. In: Parsdistalis of the pituitary gland -structure, function and regulation, (eds. F. Yoshimura and A. Gorbman), pp. 21-26. Elsevier, Amsterdam.

35. Daval, J.L., Louis, J.C., Gerard, M.J., Vincendon, G. (1983) Influence of adrenocorticotropic hormone on the growth of isolated neurons in culture. Neurosci. Lett. 36, 299-304.

36. Davila, M.I., de Kloet, R. and Azmitia, E.C. (1986) Effects of neuropeptides on the maturation of serotonergic neurons in tissue culture. Soc. Neurosci. Abstr. 12, 152.

37. Davila-Garcia, M.I., Hou, X.P., Murphy, R. and Azmitia, E.C. (1987) Inhibitory effects of leu-enkephalin on ^{3}H-5HT uptake in dissociated mesencephalic raphe neurons. Soc. Neurosci. Abstr. 13, 1136.

38. Davila-Garcia, M.I., Hlibczuk, V., Akbari, H., Alves, S. and Azmitia, E.C. (1988) Postnatal effects of prenatal administration of neuropeptides on ^{3}H-5HT uptake by serotonergic neurons. Soc. Neurosci. Abstr. 14, 416.

39. Davila-Garcia, M.I. and Azmitia, E.C. (1989) Effects of acute and chronic administration of leu-enkephalin on cultured serotonergic neurons. Devl. Brain Res. (in press).

40. Dekker, A., Gispen, W.H. and De Wied, D. (1987) Axonal growth and neuropeptides. Life Sci. 41, 1667-1678.

41. De Kloet, E.R., Kovacs, G.L., Telegdy, G., Bohus, B., Versteeg, D.H.G. (1982) Decreased serotonin turnover in the dorsal hippocampus of the rat brain shortly after adrenalectomy: selective normalization after corticosterone substitution. Brain Res. 239, 659-663.

42. Edwards, P.M., Verhaagen, J., Spierings, T., Schotman, P., Jennekens, F.G.I. and Gispen, W.H. (1985) The Effects of ACTH 4-10 on protein synthesis, actin and tubulin during regeneration. Brain Res. Bull. 15, 267-272.

43. Foster, G.A. and Schultzberg, M. (1984) Immunohistochemical analysis of the ontogeny of neuropeptide Y immunoreactive neurons in foetal rat brain. Int. J. devl. Neurosci. 2, 387-407.

44. Frankfurt, M. and Azmitia, E.C. (1983) The effect of intra-

cerebral injections of 5,7-dihydroxytryptamine and 6-hydroxydopamine on the serotonin immunoreactive cell bodies and fibers in the adult rat hypothalamis. Brain Res. 261, 91-99.

45. Frischer, R.E., El-Kawa, N.M. and Strand, F.L. (1985) ACTH peptides as organizers of neuronal patterns in development: maturation of the rat neuromuscular junction as seen by scanning electron microscopy. Peptides 6, 13-19.

46. Fuller, R.W., Perry, K.W. and Molloy, B.B. (1974) Effects of an uptake inhibitor on serotonin metabolism in rat brain: studies with 3-(p-trifluoromethylphenoxy)-n-methyl-3-phenyl propylamine (Lilly 110140). Life Sci. 15, 1161-1171.

47. Handelmann, G.E. (1985) Neuropeptide effects on brain development. J. Physiol. (Paris). 80, 268-274.

48. Hiller, J.M., Simon, E.J., Crain, S.M. and Peterson, E.R. (1978) Opiate receptors in cultures of fetal mouse dorsal root ganglia (DRG) and spinal cord: predominance in DRG neurites. Brain Res. 145, 369-400.

49. Hiller, J.M., Simon, E.J., Crain, S.M. and Peterson, E.R. (1978) Opiate receptor distribution in organized cultures of fetal mouse spinal cord and dorsal root ganglia. In: Characteristics and functions of opioids, (eds. J.M. Van Ree and L. Terenius), pp.477-478. Elsevier, Amsterdam.

50. Hui, F.W., Krikun, E., Hirsh, E.M., Blaiklock, R.G. and Smith, A.A. (1976) Inhibition of nucleic acid synthesis in the regenerating limb of salamanders treated with dl-methadone or narcotic antagonists. Exp. Neurol. 53, 267-273.

51. Insel, T.R., Battaglia, G., Fairbanks, D.W. and De Souza, E.B. (1988) The ontogeny of brain receptors for corticotropin-releasing factor and the development of their functional association with adenylate cyclase. J. Neurosci. 8, 4151-4158.

52. Kimelberg, H.K. and Katz, D.M. (1985) High affinity uptake of serotonin into immunocytochemically identified astrocytes. Science 228, 889-895.

53. King, J.A. (1988) Maturational effects of ACTH and nicotine on developing rat motor and neuroendocrine systems. [PhD thesis. New York University. N.Y.].

54. Kirby, M.L. (1984) Alterations in fetal and adult responsiveness to opiates following various schedules of prenatal morphine exposure. In: Neurobehavioral Teratology, (ed. J. Yanai), pp. 235-248. Elsevier, Amsterdam.

55. Kirskey, D.F. and Slotkin, T.A. (1979) Concomitant development of (^{3}H)dopamine and (^{3}H)5-hydroxytryptamine uptake systems in rat brain regions. Br. J. Pharmac. 67, 387-391.

56. Kovac, G.L., Kishonti, J., Lissak, K. and Telegdy, G.

(1977) Dose-dependent dual effect of corticosterone on cerebral 5HT metabolism. Neurochem. Res. 2, 311-322.

57. Lauder, J.M., Wallace, J.A., Krebs, H., Petruz, P. and McCarthy, K. (1982) In vivo and in vitro development of serotonergic neurons. Brain Res. Bull. 9, 605-625.

58. Lowry, O.H., Rosebrough, N.H., Farr, A.L., Randall, R.J. (1951) Protein measurement with folin phenol reagent. J. Biol. Chem. 193, 265-275.

59. McGregor, G.P., Woodmans, P.L., O'Suaghnessy, D.J., Ghatei, M.A., Polak, J.M. and Bloom, S.R. (1982) Developmental changes in bombesin, substance P, somatostatin, and vasoactive intestinal polypeptide in the rat brain. Neurosci. Lett. 28, 21-27.

60. Neale, J.H., Barker, J.L., Uhl, G.R. and Snyder, S.H. Enkephalin-containing neurons visualized in spinal cord cell cultures. Science 201, 467-469.

61. Nomura, Y., Maitoh, F. and Segawa, T. (1976) Regional changes in monoamine content and uptake of rat brain during postnatal development. Brain Res. 101, 305-315.

62. Oestreicher, A.B., Zwiers, H. and Gispen, W.H. (1977) Synaptic membrane phosphorylation: target for neurotransmitters and neuropeptides. Prog. Brain Res. 55, 349-367.

63. Peters, M.A. (1977) The effects of maternally administered methadone on brain development in the offspring. J. Pharm. Exp. Ther. 203, 340-346.

64. Parenti, M., Tirone, F., Oligati, V.R. and Groppetti, A. (1983) Presence of opiate receptors on striatal serotonergic nerve terminals. Brain Res. 280, 317-322.

65. Ramaekers, F., Rigter, H. and Leonard, B.E. (1978) Parallel changes in behaviour and hippocampal monoamine metabolism in rats after administration of ACTH analogues. Pharmacol. Biochem. Behav. 8, 547-551.

66. Richter-Landsberg, C., Bruns, I. and Flohr, H. (1987) ACTH neuropeptides influence development and differentiation of embryonic rat cerebral cells in culture. Neuro. Res. Comm. 1, 153-162.

67. Rose, K.J. and Strand, F.L. (1988) Mammalian neuromuscular development accelerated with early but slowed with late gestational administration of ACTH peptide. Synapse 2, 200-204.

68. Saint-Come, C., Acker, G.R. and Strand, F.L. (1982) Peptide influences on the development and regeneration of motor performance. Peptides 3, 439-449.

69. Saint-Come, C. and Strand, F.L. (1985) ACTH/MSH 4-10 improves motor unit reorganization during peripheral nerve regeneration in the rat. Peptides 6 (Suppl 1), 77-83.

70. Sakellaridis, N., Mangoura, D. and Vernadakis, A. (1986) Effect of opiates on the growth of neuron-enriched cultures from chick embryonic brain. Int. J. devl. Neurosci. 4, 293-302.

71. Segarra, A. (1988) Sexual differentiation in rats is altered by perinatal nicotine or ACTH administration, [PhD Thesis. New York University. N.Y.]

72. Shaskan, E.G. and Snyder, S.H. (1970) Kinetics of serotonin accumulation into slices from rat brain: relationship to catecholamine uptake. J. Pharmacol. Exp. Ther. 175, 404-418.

73. Sinding, C., Robinson, A.G. and Seif, S.M. (1980) Neurohypophyseal peptides in the developing rat fetus. Brain Res. 195, 177-186.

74. Singh, V.H., Corley, K.C., Phan, T-H., Kalimi, M. and Boadle-Biber, M.C. (1988) Sound stress induced increases in tryptophan hydroxylase (TrpH) activity blocked by adrenalectomy (ADX). Soc. Neurosci. Abstr. 14, 447.

75. Slotkin, T.A., Seidler, F.J. and Whitmore, W.L. (1980) Effects of maternal methadone administration on ornithine decarboxylase in brain and heart of offspring: relationships of enzyme activity to dose and to growth impairment in the rat. Life Sci. 26, 861-867.

76. Smith, A.A., Hui, F.W. and Crofford, M.J. (1977) Inhibition of growth in young mice treated with D-1-methadone. Eur. J. Pharmacol. 43, 307-313.

77. Steece, K.A., DeLeon-Jones, F.A., Meyerson, L.R., Lee, J.M., Fields, J.Z. and Ritzmann, R.F. (1986) In vivo down regulation of rat striatal opioid receptors by chronic enkephalin. Brain Res. Bull. 17, 255-257.

78. Swaab, D.F. (1981) Neuropeptides and brain development - a working hypothesis. In: A multidisciplinary approach to brain development, (eds. C.Ci. Benedetta, R. Balazs and G. Gombos), pp. 181-196. Elsevier, Amsterdam.

79. Swaab, D.F. and Martin, J.T. (1981) Functions of an a-melatonin and other opiomelanocortin peptides in labour, intrauterine growth and brain development. Ciba Found. Symp. 81, 196-217.

80. Sze, P.Y. (1980) Glucocorticoids as a regulatory factor for brain tryptophan hydroxylase during development. Dev. Neurosci. 3, 217-223.

81. Tempel, A., Crain, S.M., Peterson, E.R., Simon, E.J. and Zukin, R.S. (1986) Antagonist-induced opiate receptor upregulation in cultures of fetal mouse spinal cord-ganglion explants. Dev. Brain Res. 25, 287-291.

82. Van der Neut, R., Bar, P.R., Sodaar, P. and Gispen, W.H. (1988) Trophic influences of Alpha-MSH on ACTH 4-10 on neuronal outgrowth in vitro. Peptides 9, 1015-1020.

83. Vertes, Z., Melegh, G., Vertes, M., and Kovacs, J. (1982) Effects of naloxone and D-met^2-pro^5-enkephalinamide treatment on the DNA synthesis in the developing rat brain. Life Sci. 31, 119-126.

84. Wolf, G. and Sterba, G. (1978) Development of the hypothalamic-neurohypophysial system in rats. In: Hormones and brain development, (eds. G. Dorner and M. Kawakami), pp. 217-222. Elsevier, Amsterdam.

85. Zagon, I.S. and McLaughlin, P.J. (1978) Perinatal methadone exposure and brain development; a biochemical study. J. Neurochem. 56, 49-54.

86. Zagon, I.S. and McLaughlin, P.J. (1986) Opioid antagonist (naltrexone) modulation of cerebellar development: histological and morphometric studies. J. Neurosci. 6, 1424-1432.

87. Zagon, I.S. and McLaughlin, P.J. (1986) Opioid antagonist-induced modulation of cerebral and hippocampal development: histological and morphometric studies. Dev. Brain Res. 28, 233-246.

88. Zagon, I.S. and McLaughlin, P.J. (1988) Endogenous opioid systems and neurobehavioral development. In: Endorphins, opiates and behavioral processes, (eds. R.J. Rodgers and S.J. Cooper), pp. 287-309. John Wiley and Sons, New York.

89. Zhou, F.C. and Azmitia, E.C. (1985) The effects of adrenalectomy and corticosterone on homotypic collateral sprouting of serotonergic fibers in the hippocampus. Neurosci. Lett. 34, 111-116.

90. Zwiers, H., Veldhus, D., Schotman, P. and Gispen, W.H. (1976) ACTH, cyclic neuclotides and brain protein phosphorylation in vitro. Neurochem. Res. 1, 669-677.

BASIC FIBROBLAST GROWTH FACTOR AFFECTS THE SURVIVAL AND DEVELOPMENT OF MESENCEPHALIC NEURONS IN CULTURE

Giovanna Ferrari, Maria-Cristina Minozzi, Gino Toffano
Alberta Leon and Stephen D. Skaper

Fidia Research Laboratories
Via Ponte della Fabbrica, 3/A
35031 Abano Terme, Italy

INTRODUCTION

The survival and growth of neurons within the CNS is thought to be regulated by the availability of appropriate neuronotrophic factors. Compelling evidence for this idea derives from the recent findings that Nerve Growth Factor (NGF), the classical trophic factor for neural crest-derived peripheral neurons (Levi-Montalcini, 1966; Greene and Shooter, 1980; Thoenen and Barde, 1980) is also present within the CNS, where it has been proposed to act as a neuronotrophic factor for developing and lesioned adult cholinergic neurons in the basal forebrain (Gnahn et al., 1983; Hefti, 1986; Korsching, 1986; Kromer, 1987; Williams et al., 1986). These observations have encouraged the view that neuronotrophic factors in general may play an important role, not only during development but also in the adult, in particular following brain injury. At present NGF appears to address specific and limited populations of neurons. It is thus reasonable to believe that additional trophic factors exist for yet other types of neurons.

BASIC FIBROBLAST GROWTH FACTOR: NEURONOTROPHIC ACTIVITY

The most significant recent advance in identifying such factors has come from the findings that basic fibroblast growth factor (bFGF), a well-known polypeptide growth factor recognized for its mitogenic properties on endothelial-derived cells (Esch et al., 1985; Gospodarowicz et al., 1986; Lobb et al., 1986; Thomas and Gimez-Gallego, 1986), can also act as a neuronotrophic factor for hippocampal (Walicke et al., 1986), cortical (Morrison et al., 1986), spinal cord (Unsicker et al., 1987) and cerebellar granule (Hatten et al., 1988) neurons in vitro. The additional findings that brain is a rich source of bFGF (Lobb and Fett, 1984) and expresses receptors for this protein (Imamura et al. 1988) raises the possibility that bFGF plays a key role in normal nervous system development and/or function. Here, we report that also rat mesencephalic neurons are responsive to bFGF. The present findings, in addition, provide the first evidence that bFGF can effect at least two defined CNS populations, i.e. dopaminergic and GABAergic neurons. Most of the techniques in this study have been described elsewhere (Ferrari et al., 1988).

Molecular Aspects of Development and Aging of the Nervous System
Edited by J.M. Lauder
Plenum Press, New York, 1990

MORPHOLOGICAL RESPONSE OF MESENCEPHALIC NEURONS TO bFGF

At the in vitro level, the morphological evaluation of neuronal survival and neurite outgrowth induced by bFGF has been described for cells from different brain areas. These effects, however, have never been tested on mesencephalic neurons nor associated with specific neuronal cell types (e.g. in terms of neurotransmitter-related properties). To address these questions, 14 day fetal rat dissociated mesencephalic cells were utilized and the trophic effects of bFGF first analyzed at the morphological level. Since bFGF can stimulate astroglial cell growth in vitro, it was important to verify whether bFGF acted directly on mesencephalic neurons or indirectly via a mitotic action on astroglial cells. Mesencephalic cultures grown for 20 days in the absence or presence of bFGF (3 ng/ml) were stained with antibodies to neurofilament protein or Glial Fibrillary Acidic Protein (GFAP), two well-established markers for neurons and astroglia, respectively. Fig. 1 shows that in both cases the vast majority of cells (> 99%) and their processes stained positively with neurofilament antibody, indicating that most of the surviving cells were neurons. Note that while the control culture (panel A) contained few cells with moderate neurite outgrowth, the addition of bFGF clearly increased not only the number of surviving cells but also the degree of their process branching (panel B). This effect was already evident after 10 days. Less than 1% of the cells were GFAP-positive. This small number of glial cells present in our system makes it likely that bFGF is acting directly on mesencephalic neurons, as already reported for hippocampal neurons (Walicke and Baird, 1988).

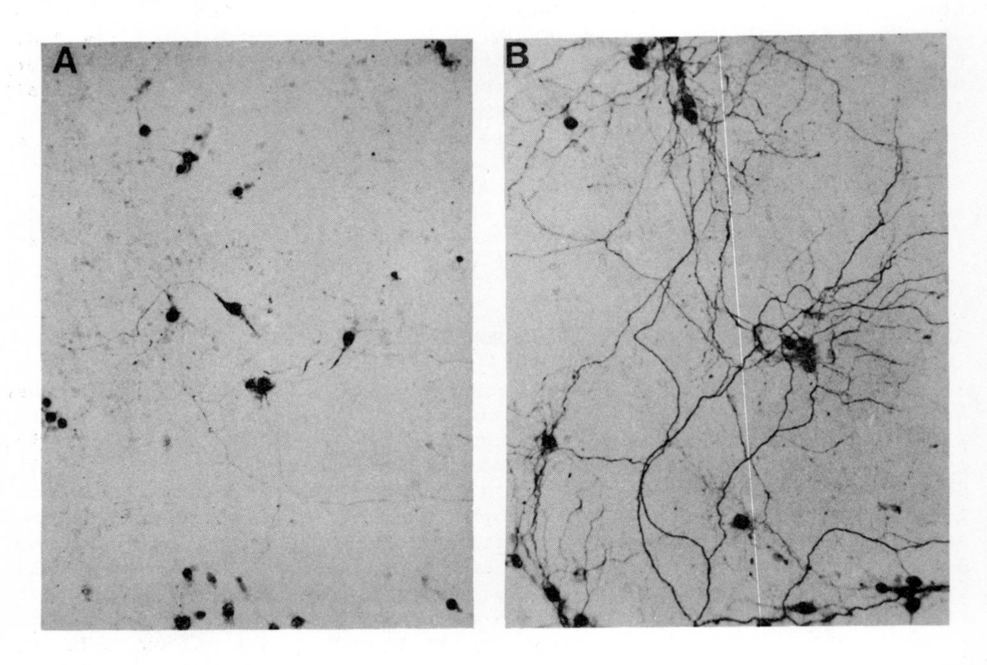

Immunohistochemical staining of rat mesencephalic cells for neurofilament. Cells were maintained for 20 days in the absence (A) or presence (B) of bFGF (3 ng/ml). Dissociated mesencephalic cells (1×10^6) were seeded on polyornithine-coated 35 mm dishes and maintained in serum-free hormone-supplemented medium. bFGF was added 3 hr after plating.

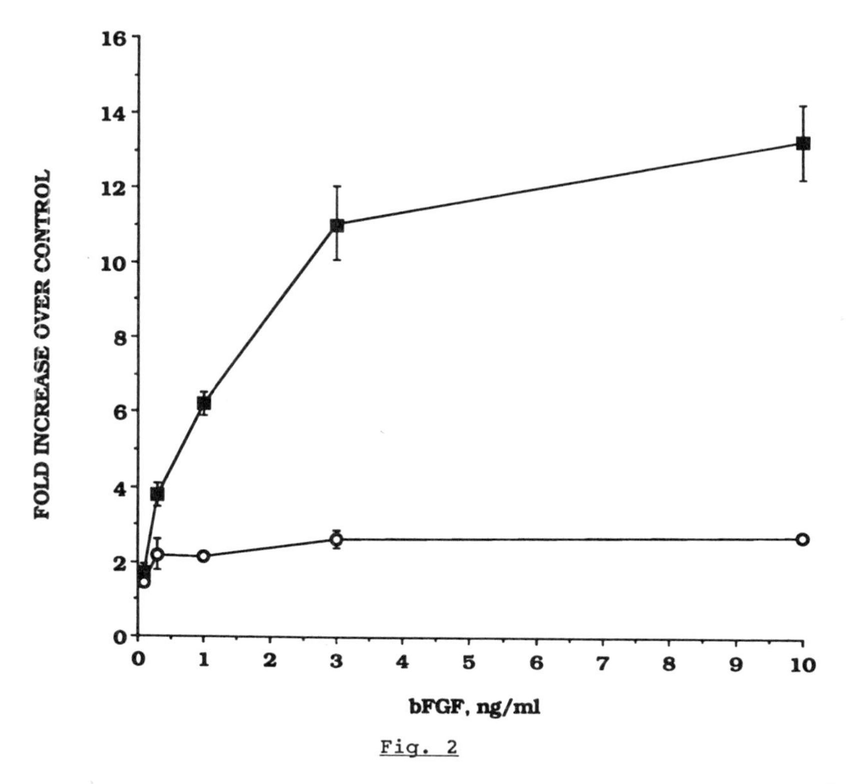

Fig. 2

Dose-response curve of bFGF on benztropine mesylate-sensitive ^{3}[H]DA uptake (■) and L-2,4 diaminobutyric acid-sensitive [^{14}C]GABA uptake (○). Measurements were performed using mesencephalic neurons at 10 days in vitro. Data are expressed as fold increase over control.

BASIC FGF ADDRESSES DOPAMINERGIC AND GABAERGIC NEURONS

Another question addressed by the present study was whether the morphological effect induced by bFGF was directed to particular neuronal cell types, in terms of neurotransmitter-related properties. For this purpose, we followed the effect of bFGF on the survival of dopaminergic and GABAergic neurons by assessing their specific ^{3}[H]dopamine (DA) and [^{14}C] γ-aminobutyric acid (GABA) uptakes, respectively. Long-term cultures were used, since under these conditions such uptake parameters have been shown to be a good index of cell survival (Dal Toso et al., 1988). Fig 2 shows that bFGF at 10 days was capable of increasing, in a dose-dependent manner, both DA and GABA uptakes, indicating an enhanced survival of the dopaminergic as well as the GABAergic neurons. Basic FGF was already active in the picogram range, with saturation levels at 3 ng/ml. Dopamine uptake increased about 10-fold, while GABA uptake was increased only 2.5-fold. This result suggests that dopaminergic neurons, at least at this developmental stage, are more sensitive to bFGF than GABAergic neurons. The effect of bFGF is highly specific, since at 0.3 ng/ml it was completely blocked by affinity-purified polyclonal antibodies against bFGF. The trophic effect of bFGF still persisted at 20 days in vitro (Table 1). At this latter time also, the stimulatory increase induced by bFGF was much greater on dopamine uptake than on GABA uptake, confirming that dopaminergic and GABAergic neurons respond differentially to bFGF.

TABLE 1

Trophic effect of bFGF on mesencephalic neurons: 20 days in vitro

	RELATIVE UPTAKE	
	DOPAMINE	GABA
Control	100	100
bFGF (3 ng/ml)	1350	400

DISCUSSION

The long-recognized mitogen for endothelial-derived cells, basic FGF, has more recently come to light as a molecule displaying neuronotrophic activity for a variety of both centrally and peripherally originating cells. These targets include hippocampal (Walicke et al., 1986), cortical (Morrison et al., 1986), spinal cord (Unsicker et al., 1987), cerebellar (Hatten et al., 1988), and parasympathetic (Unsicker et al., 1987) neurons, as well as PC12 pheochromocytoma cells (Rydel and Greene, 1987). The present results clearly show that mesencephalic cells are also to be added to the list of bFGF-responsive neurons. At 20 days in vitro, bFGF produced a striking increase in the number of neurite-bearing cells and in the degree of their neurite network.

Our findings also document, for the first time, that bFGF affects biochemically identifiable populations of CNS neurons. Basic FGF was able to reduce the death of at least the dopaminergic and GABAergic neurons derived from mesencephalic cultures, as quantified by their specific $[^3H]$DA and $[^{14}C]$GABA uptakes with time in vitro. The stimulatory effect of bFGF was always much greater on $[^3H]$DA uptake, indicative of an apparently greater sensitivity of dopaminergic neurons to bFGF. The trophic action of bFGF at 20 days in culture for dopaminergic neurons was confirmed by directly counting the number of neurite-bearing cells with glyoxylic acid-induced catecholamine fluorescence (GIF). GIF-positive neurons elaborating long and branched neurites were observed only in bFGF-treated cultures.

The concentration dependence of the effects of bFGF on mesencephalic neurons (ED50 = 1.5 ng/ml) is in good agreement with other reports. The low levels of bFGF effective in the present cultures, together with the extremely small number of GFAP-positive cells observed suggests that bFGF acts directly on the mesencephalic-derived neurons, a conclusion reached by others for the neuronotrophic action of bFGF on CNS neurons (Walicke and Baird, 1988). Nevertheless, one cannot rule out with absolute certainly that bFGF acts on glial cells, which in turn produce another molecule acting in concert with bFGF.

In the case of NGF, the innervation territories of basal cholinergic neurons, namely hippocampus and neocortex, contain the highest levels of NGF protein and its mRNA (Shelton and Reichardt, 1986; Whittemore et al., 1986). Retrograde transport of NGF to these cholinergic cell bodies occurs (Schwab et al., 1979), as well as NGF induction of choline acetyltransferase (Gnahn et al., 1983). Thus, NGF appears to function as a target-derived trophic factor for defined populations of CNS neurons. The broad spectrum of bFGF trophic action for CNS-derived neurons in vitro raises the question as to whether bFGF as well operates in vivo as a target-derived trophic protein. The answer to this question will probably require mapping of bFGF and its mRNA in those brain regions serving as presumptive innervation territories for dopaminergic and GABAergic neurons, among others.

Axotomy, which disconnects the neuronal body from its end organs will lead even in the adult to neuronal death. From this, a neuronotrophic hypothesis has been proposed for the CNS (Appel, 1981; Varon, 1985). According to this concept central neurons, as neurons in the periphery, require a continuous supply of endogenous trophic molecules for their survival through adulthood. Supporting evidence for this hypothesis has been provided by recent experiments in which the apparent loss of a large proportion of cholinergic neurons in the septal region following transection of the fimbria-fornix has been prevented by the intraventricular infusion of NGF (Hefti, 1986; Kromer, 1987; Williams et al., 1986) or bFGF (Anderson, 1988). Additional studies have shown that NGF (Carmignoto et al., 1988) and bFGF (Sievers et al., 1987) promote the survival of adult rat retinal ganglion cells after optic nerve section. Furthermore, increased levels of NGF (Gasser et al., 1986; Whittemore et al., 1987) and bFGF (Finklestein et al., 1987) occur following brain injury. If confirmed, this hypothesis can have important clinical implications, in that administration of endogenous trophic factors may reduce or interrupt the effects of a neurodegenerative process.

Clinically, various pathological states of the CNS appear to selectively attack particular populations of neurons. For example, amyotrophic lateral sclerosis strikes motor neurons, Parkinson's disease the dopaminergic neurons of the substantia nigra, and Alzheimer's disease is characterized by a deficit of the central cholinergic system. The selectivity of the neurons involved in these degenerative situations suggests the action of specific substances for that group of neurons affected, and in particular the possibility of an insufficiency of the corresponding neuronotrophic factors. In this vein, an NGF deficit has already been proposed to be involved in Alzheimer's disease (Hefti, 1983). Theoretically, the present data raise the possibility of an analogous role for bFGF in Parkinson's disease. In vivo experiments will be needed, however, to validate this hypothesis. The neuronotrophic hypothesis of the CNS and the experimental data with NGF and bFGF now offer realistic encouragement for the idea that neuronotrophic deficits are at the base of the pathological processes in these diseases.

ACKNOWLEDGEMENTS

We wish to thank Dr. Denis Gospodarowicz for the generous gifts of bFGF and bFGF antibodies and Antonia Bedeschi for typing of the manuscript.

REFERENCES

Anderson, K.J., Dam, D., Lee, S., and Cotman, C.W., 1988, Basic fibroblast growth factor prevents death of lesioned cholinergic neurons in vivo, Nature, 332:360.

Appel, S.H., 1981, A unifying hypothesis for the cause of amyotrophic lateral sclerosis, parkinsonism and Alzheimer's disease, Ann. Neurol., 10:499.

Carmignoto, G., Maffei, L., Candeo, P., Canella, R., and Comelli, C.,

1988, Effect of NGF on the survival of rat retinal ganglion cells following optic nerve section, J. Neurosci., in press.

Dal Toso, R., Giorgi, O., Soranzo, C., Kirschner, G., Ferrari, G., Favaron, M., Benvegnù, D., Presti, D., Vicini, S., Toffano, G., Azzone, G.F., and Leon, A., 1988, Development and survival of neurons in dissociated fetal mesencephalic serum-free cell cultures: I. Effects of cell density and of an adult mammalian striatal derived neuronotrophic factor (SDNF), J. Neurosci., 8:733.

Esch, F., Baird, A., Ling, N., Veno, N., Hill, F., Denorey, L., Klepper, R., Gospodarowicz, D., Bohlen, P., and Guillemin, R., 1985, Primary structure of bovine basic fibroblast growth factor (FGF) and comparison with the amino-terminal sequence of bovine acidic FGF, Proc. Natl. Acad. Sci. USA, 82:6507.

Ferrari, G., Minozzi, M.-C., Toffano, G., Leon, A., and Skaper, S.D., 1988, Basic fibroblast growth factor promotes the survival and development of mesencephalic neurons in culture, Dev. Biol., in press.

Finklestein, S.P., Apostolides, P.J.,Caday, C.G., Prosser, J., Phillips, M.F., and Klagsbrun, M., 1988, Increased basic fibroblast growth factor (bFGF) immunoreactivity at the site of focal brain wounds, Brain Res., 460:253.

Gasser, U.E.,Weskamp, G., Otten, U., and Dravid, A.R., 1986, Time course of the elevation of nerve growth factor (NGF) content in the hippocampus and septum following lesions of the septohippocampal pathway in rats. Brain Res., 376:351.

Gnahn, H., Hefti, F., Heumann, R., Schwab, M.E., and Thoenen, H., 1983, NGF-mediated increase of choline acetyltransferase (ChAT) in the neonatal rat forebrain: evidence for a physiological role of NGF in the brain? Dev. Brain Res., 9:45.

Gospodarowicz, D., Neufeld, G., and Schweigerer, L., 1986, Molecular and biological characterization of fibroblast growth factor, an angiogenic factor which also controls the proliferation and differentiation of mesoderm and neuroectoderm derived cells, Cell Diff., 19:1.

Greene, L.A., and Shooter, E.M., 1980, The nerve growth factor: biochemistry, synthesis and mechanism of action, Ann. Rev. Neurosci., 3:353.

Hatten, M.E., Lynch, M., Rydel, R.E., Sauchez, J., Joseph-Silverstein, J., Moscatelli, D., and Rifkin, D.B., 1988, In vitro neurite extension by granule neurons is dependent upon astroglial-derived fibroblast growth factor, Dev. Biol., 125:280.

Hefti, F., 1983, Is Alzheimer's disease caused by lack of nerve growth factors? Ann. Neurol., 13:109.

Hefti, F., 1986, Nerve growth factor promotes survival of septal cholinergic neurons after fimbrial transection, J. Neurosci., 6:2155.

Imamura, T., Tokita, Y., and Mitsui, Y., 1988, Purification of basic FGF receptors from rat brain, Biochem. Biophys. Res. Comm., 155:583

Korsching, S., 1986, The role of nerve growth factor in the CNS, Trends Neurosci., 9:570.

Kromer, L.F., 1987, Nerve growth factor treatment after brain injury prevents neuronal death, Science, 235:214.

Levi-Montalcini, R., 1966, The nerve growth factor: its mode of action on sensory and sympathetic nerve cells, Harvey Lect., 60:217.

Lobb, R.R. and Fett, J.W., 1984, Purification of 2 distinct growth factors from bovine neural tissue by heparin affinity chromatography, Biochemistry, 23:6925

Lobb, R.R., Harper, J.W., and Fett, J.W., 1986, Purification of heparin-binding growth factors, Anal. Biochem., 154:1.

Morrison, R.S., Sharma, A., De Vellis, J., and Bradshaw, R.A., 1986, Basic fibroblast growth factor supports the survival of cerebral cortical neurons in primary culture, Proc. Natl. Acad. Sci. USA,

83:7537.
Prochiantz, A., Di Porzio, U., Kato, A., Berger, B., and Glowinski, J., 1979, In vitro maturation of mesencephalic dopaminergic neurons from mouse embryos is enhanced in presence of their striatal target cells, Proc. Natl. Acad. Sci. USA, 76:5387.
Rydel, R.E., and Greene, L.A., 1987, Acidic and basic fibroblast growth factors promote stable neurite outgrowth and neuronal differentiation in cultures of PC12 cells, J. Neurosci., 7:3639.
Schwab, M.E., Otten, U., Agid, Y. and Thoenen, H., 1979, Nerve growth factor (NGF) in the rat CNS: absence of specific retrograde axonal transport and tyrosine hydroxylase induction in locus coeruleus and substantia nigra, Brain Res., 168:473.
Shelton, D.L., and Reichardt, L.F., 1986, Studies on the expression of the ß nerve growth factor (NGF) gene in the central nervous system: level and regional distribution of NGF mRNA suggest that NGF functions as a trophic factor for several distinct populations of neurons, Proc. Natl. Acad. USA, 83:2714.
Sievers, J., Hausmann, B., Unsicker, K., and Berry, M., 1987, Fibroblast growth factors promote the survival of adult rat retinal ganglion cells after section of the optic nerve, Neurosci. Lett., 76:156.
Thoenen, H., and Barde, Y.-A., 1980, Physiology of nerve growth factor, Physiol. Rev., 60:1284.
Thoenen, H., and Edgar, D., 1985, Neurotrophic factors, Science, 225:238.
Thomas, K.A., and Gimenez-Gallego, G., 1984, Fibroblast growth factors: broad spectrum mitogens with potent angiogenic activity, Trends Biol. Sci., 11:81.
Unsicker, K., Reichert-Preibsch, H., Schmidt, R., Pettmann, B., Labourdette, G., and Sensenbrenner, M., 1987, Astroglial and fibrobast growth factors have neurotrophic functions for cultured peripheral and central nervous system neurons, Proc. Natl. Acad. Sci. USA, 84:5459.
Varon, S., 1985, Factors promoting the growth of the nervous system, in: "Discussions in Neurosciences", Vol. 2 n° 3.
Walicke, P.A., and Baird, A., 1988, Neuronotrophic effects of basic and acidic fibroblast growth factors are not mediated through glial cells, Dev. Brain. Res., 40:71.
Walicke, P., Cowan, W.M., Ueno, N., Baird, A., and Guillemin, R., 1986, Fibroblast growth factor promotes survival of dissociated hippocampal neurons and enhances neurite extension, Proc. Natl. Acad. Sci. USA, 83:3012.
Whittemore, S.R., Ebendal, T., Lärkfors, L., Olson, L., Seiger, A., Stromberg, I., and Persson, H., 1986, Developmental and regional expression of ß-nerve growth factor messenger RNA and protein in the rat central nervous system, Proc. Natl. Acad. Sci. USA, 83:817.
Whittemore, S.R., Lärkfors, L., Ebendal, T., Holets, V.R., Ericsson, A., and Persson, H., 1987, Increased ß-nerve growth factor messenger RNA and protein levels in neonatal rat hippocampus following specific cholinergic lesions, J. Neurosci., 7:244.
Williams, L.R., Varon, S., Peterson, G.M., Victorin, K., Fischer, W., Björklund, A., and Gage, F.H., 1986, Continuous infusion of nerve growth factor prevents basal forebrain neuronal death after fimbria-fornix transection, Proc. Natl. Acad. Sci. USA, 83:9231.

ROLE OF TROPHIC FACTORS IN NEURONAL AGING

J. Regino Perez-Polo

Dept. of HBC & G
Univ. of Texas Medical Br.
Galveston, Tx. 77550-2777 USA

1. Neuronal Cell Death

Neuronal cell death is an event associated with the pathophysiology of nerve injury, stroke and aging that is also a positive regulatory element in neuronal development. Whereas in all the aforementioned, the neurite outgrowth, as a part of the spectrum of responses by the surviving neurons, has been extensively studied at the molecular level; less is known about the molecular mechanisms that determine cell survival, except for the many classical biological experiments that have established that in vertebrate development, neuronal cell death is, in large measure but not uniquely, under the control of neuronotrophic factors such as the nerve growth factor protein, NGF (Thoenen et al., 1981). Two hypotheses have been proposed to explain the phenomena resulting in neuronal death. One hypothesis is that neurons are particularly susceptible to free radical damage because of their very low basal endogenous levels of antioxidants and antioxidant enzymes and that neuronotrophic factors acting through their respective cell surface receptors can shift the oxidant-antioxidant balance through induction of antioxidant enzymes (Perez-Polo et al., 1986). The second hypothesis proposes that there are "suicide genes" whose repression by NGF results in cell survival (Martin et al., 1988). Following axotomy to peripheral sympathetic and sensory neurons or the more central projections of the basal forebrain into the hippocampal areas, NGF has also been shown to be necessary to peripheral regeneration and to sparing effects on cholinergic basal forebrain neurons following deafferentation. NGF has dramatic trophic effects on cholinergic neurons of the CNS in vitro, a regeneration paradigm (Bostwick et al, 1987; Hatanaka et al, 1988). In aged rodent and human CNS and PNS, there are reported decreases in the levels of NGF and its receptor, NGFR (Goedert et al, 1986; Angelucci et al, 1988; Uchida and Tonionaga, 1987). In the periphery, NGF has well documented effects on certain neuronal and non-neuronal cells alike whose physiological relevance is not understood in a definitive fashion (Levi-Montalcini, 1987; Lillien and Claude, 1985; Thorpe, et al., 1988).

Molecular Aspects of Development and Aging of the Nervous System
Edited by J.M. Lauder
Plenum Press, New York, 1990

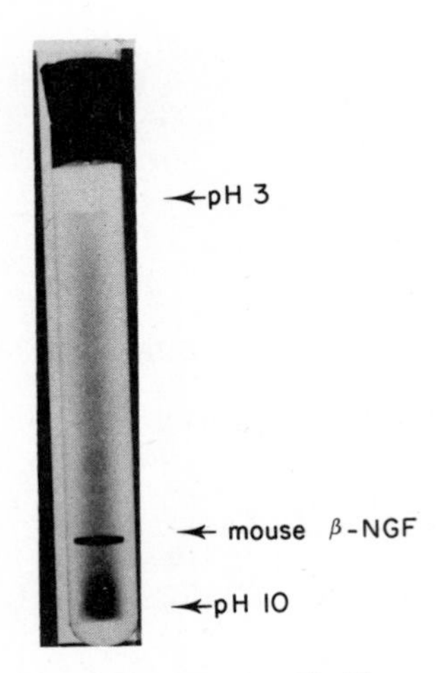

Isoelectric focusing on 7½% disc polyacrylamide gels with 3-10 pH gradient of 150 μg purified mouse β-NGF.

Fig. 1. Structure of murine NGF.

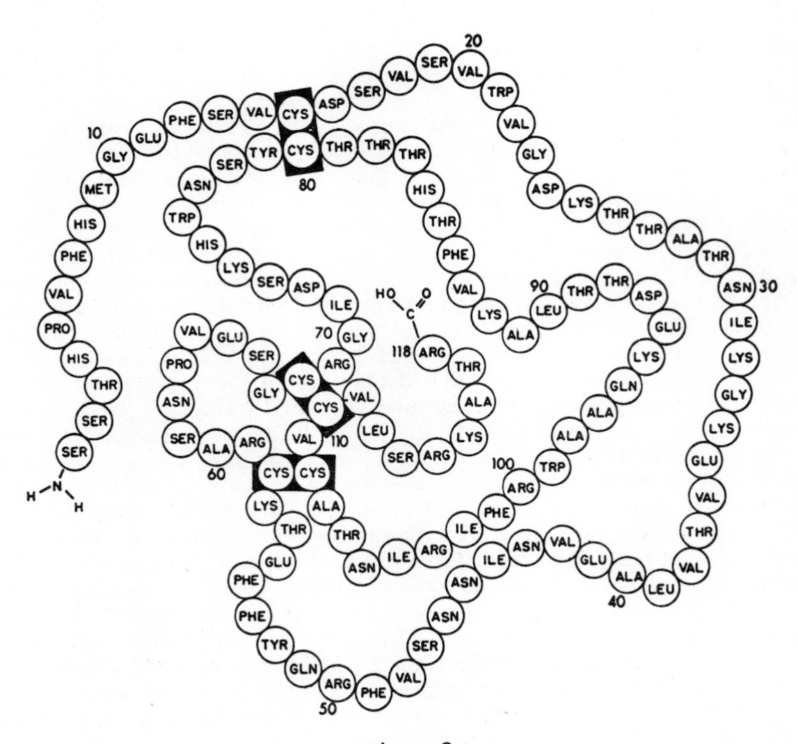

Fig. 2

The molecular events relevant to sparing effects of NGF on neuronal survival in development, following injury or associated with aging are not understood in a definitive manner. It may be that different mechanisms are at work at different stages or in different CNS compartments. It is known that a potent and widespread consequence of aging and injury to CNS is increased or cumulative oxidative stress partially manifested by an increase in free radicals followed by peroxidative events (Demopoulos et al, 1980; Agranoff, 1984; Cutler, 1985; Davies, 1987). Thus,the perturbation of oxidant-antioxidant balance disturbs energy metabolism and damages proteins, nucleic acids and membranes.

2. The Nerve Growth Factor

NGF is a neuronotrophic protein whose structural features have been well chronicled (Levi-Montalcini, 1987) Fig. 1. NGF has been purified from the submaxillary gland of mice and rats, murine saliva, several snake venoms, the guinea pig prostate, bovine seminal plasma, seminal vesicle and human term placenta (Perez-Polo, 1987). In all instances, only beta-NGF, henceforth called NGF, has nerve growth promoting activity. The amino acid sequence (Fig. 2) and the DNA sequence of the NGF gene is known for mouse, rat, bovine, human and chick NGF and all are highly conserved, (Whittemore et al., 1988). NGF mRNA levels have been determined for brain, superior cervical ganglia and spinal cord and correlated with NGF protein levels as a function of development and in response to injury (Whittemore and Seiger, 1987). Alternate splicing of NGF mRNA may have developmental significance in the periphery and perhaps also in the CNS (Whittemore and Seiger, 1987). Levels of NGF mRNA and protein in periphery and CNS correlate with the density of innervation (Whittemore and Seiger, 1987). The highest levels of NGF in the CNS are in cortex and hippocampus, the terminal axonal regions for basal forebrain cholinergic neurons (Whittemore and Seiger, 1987). This is also where NGF effects on ChAT induction and cell sparing following lesions have been best documented (Whittemore and Seiger, 1987). However, it should be emphasized that there is evidence for NGF and its receptor in non-cholinergic regions of brain and spinal cord although it is not established if these are present in neuronal cells.

3. The Effects of NGF

Here, NGF effects on non-neuronal tissues will not be discussed (Levi-Montalcini, 1987; Thorpe et al, 1988). Treatment of sensory and sympathetic ganglia at appropriate stages in development with NGF results in a decrease in neuronal cell death, neuronal hypertrophy, increased anabolic activity, neurotransmitter synthesis and exaggerated neurite outgrowth (Perez-Polo, 1985; Levi-Montalcini, 1987). In particular, the neurite outgrowth of explanted embryonic chick has long been used to assay NGF activity in vitro (Fig. 3). Depression of NGF levels in early development due to injections of anti-NGF, results in irreversible sympathectomy and neuronal cell loss in sensory ganglia (Perez-Polo, 1985; Levi-Montalcini, 1987). In CNS, NGF stimulates ChAT in the hippocampus, septum and cortex of rat neonates and adult rats

Fig. 3. Explanted chick embryonic sensory ganglia in the absence (Top) and presence (Bottom) of 10 ng/ml NGF after 24 hours in culture.

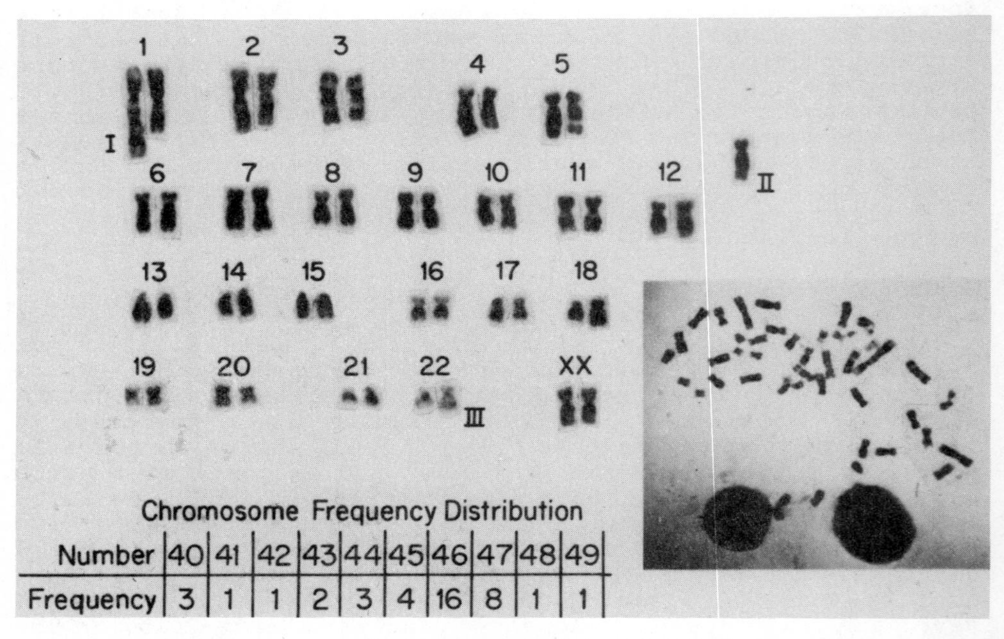

Chromosome Frequency Distribution

Number	40	41	42	43	44	45	46	47	48	49
Frequency	3	1	1	2	3	4	16	8	1	1

Fig. 4. Kayo type of SK-N-SH-SY5Y. Adopted from Perez-Polo et al, 1979.

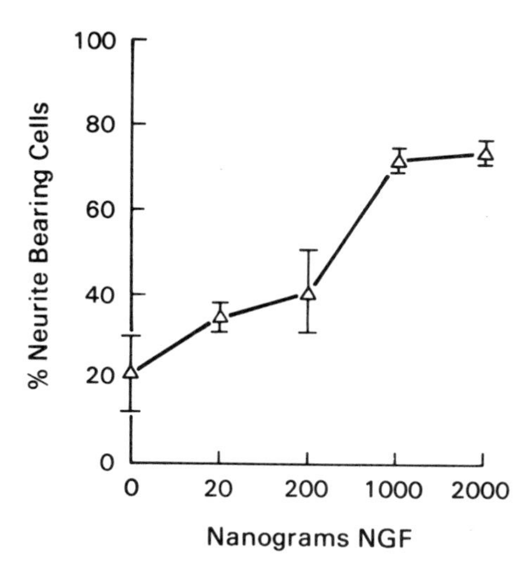

Fig. 5. Dose response curve of an SK-SH line to NGF.

following fimbrial lesions, most likely due to its sparing effect on cholinergic septal neurons (Whittemore and Seiger, 1987). NGF increases ChAT activity and neurite outgrowth in cultures of dissociated fetal basal forebrain and septal cells (Bostwick et al., 1987; Whittemore and Seiger, 1987). The NGF induction of neurotransmitters is specific in septal cells to ChAT (Bodwick et al, 1987). NGF treatment rescues basal forebrain cholinergic and non-cholinergic neurons following fimbria lesions (Whittemore and Seiger, 1987). The retrograde transport of NGF from innervated tissues to cell bodies has been demonstrated for PNS and CNS neurons (Thoenen et al., 1981). NGF in spinal cord is also retrogradely transported by central processes to sensory neurons (Yip and Johnson, 1984; Khan et al., 1987). Like other protein hormones, NGF binds to a cell surface receptor, NGFR, and is internalized (Yankner and Shooter, 1982).

4. Clonal Model Systems

There are difficulties in studying the mechanism of action of NGF in vivo. It is difficult to have homogenous cellular populations, to observe isolated cellular events and to quantitatively manipulate the cellular environment (Perez-Polo, 1987). Three neuronal cell lines that have proved useful are the PC12 rat pheochromocytoma line, the SK-N-SH-SY5Y, SY5Y, and the LAN-1 human neuroblastoma lines. PC12 display type I and type II NGFR and respond to NGF with extended neurites and increased neurotransmitter metabolism (Greene and Shooter, 1980). SY5Y and LAN-1 are nearly diploid lines (Fig. 4). The SY5Y display only NGFR I, whereas LAN-1 display both type I and II NGFR (Marchetti and Perez-Polo, 1987). Exposure of SY5Y and LAN-1 to NGF induces neurites, electrically excitable membranes, inhibits cell division and stimulates the anabolic machinery of the cell (Perez-Polo et al., 1982) (Fig. 5 & 6). For PC12 and SY5Y, NGF provides protection form 6-hydroxydopamine (Fig. 7 & 8) and hydrogen peroxide by inducing catalase and glutathione transferase activity (Perez-Polo and Werrbach-Perez, 1988).

5. The Receptor to NGF

Studies on NGF receptors, NGFR, are of three different types: binding studies, detection of NGFR protein and detection of NGFR mRNA. Binding studies use ^{125}I-NGF as a ligand to NGFR (Stach and Perez-Polo, 1987); detection of NGFR protein usually relies on immunoprecipitation of iodinated NGFR or NGF covalently linked to NGFR followed by immunoprecipitation with a monoclonal antibody to NGFR (Yau and Johnson, 1987). Immunoprecipitates have been analyzed exclusively by SDS-PAGE. NGFR mRNA has been detected by Northern analysis using two independently derived cDNA probes (Chao et al., 1986; Buck et al., 1987). PNS neurons display two NGF binding activities consisting of a high Bmax, low affinity (Kds: 10^{-9} M) binding site called type II or fast dissociating NGFR and a low Bmax, high affinity (Kds: 10^{-11} M) binding site called type I or slowly dissociating NGFR (Stach and Perez-Polo, 1987). There is some interconvertability of the two types of binding activities (Marchetti and Perez-Polo, 1987). The two binding activities can be separated using preparative electrofocusing on granulated gel (Marchetti and Perez-Polo, 1987) (Fig. 9). It appears that the type I receptor is internalized and may be required for neurite elongation by PC12 cells (Stach and Perez-Polo, 1987). The NGF binding activities described for non-neuronal cells are all of the type II variety (Thorpe et al, 1988).

If a soluble NGFR binding assay is applied to LAN-1 and PC12 cells and the properties of NGFR binding and molecular species present are compared to NGFR on dissociated chick embryo sensory ganglia, the NGFR are all very similar (Marchetti and Perez-Polo, 1987; Stach and Perez-Polo, 1987). In order to enhance the signal to noise ratio, one can carry out a partial purification by lectin chromatography followed by reparative electrofocusing on granulated gel, PEGG, that separates the high affinity from the low affinity binding activities on LAN-1. To do binding assays on ^{125}I labelled NGFR ^{131}I-NGF is used as ligand (Marchetti and Perez-Polo, 1987). It has been shown that the NGF binding activity that displays high affinity binding to NGF is associated with a protein displaying a molecular weight by SDS-PAGE of 200 KDa under conditions where the NGF binding activity that displays the low affinity binding to NGF displays a molecular weight by SDS-PAGE of 92.5 KDa (Marchetti and Perez-Polo, 1987). Although there is some variations to the actual molecular weights reported, the results are consistent with the hypothesis that the receptor to NGF is the product of one gene where expression as a high or low affinity receptor to NGF is determined by post-transnational events and the milieu of the receptor on the cell surface. That is, the affinity of the NGF receptor is modulated in vivo.

One outcome of the comparative SDS-PAGE analysis of dissociated sensory ganglia is the continued presence of a third molecular species (Lyons et al, 1983) perhaps due to the presence of NGFR on Schwann cells and astrocytes

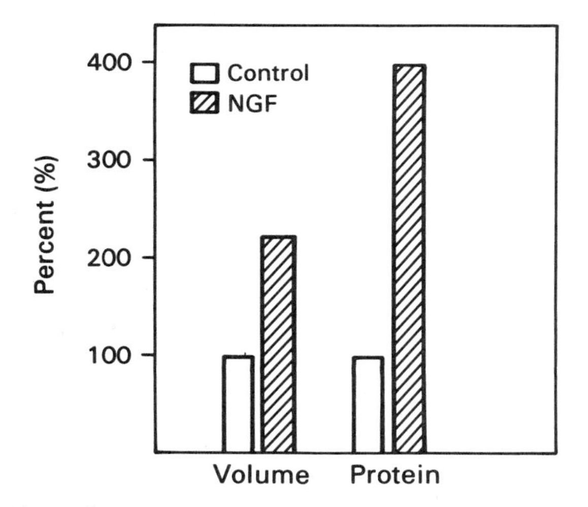

Fig. 6. Response of the SY5Y to NGF.

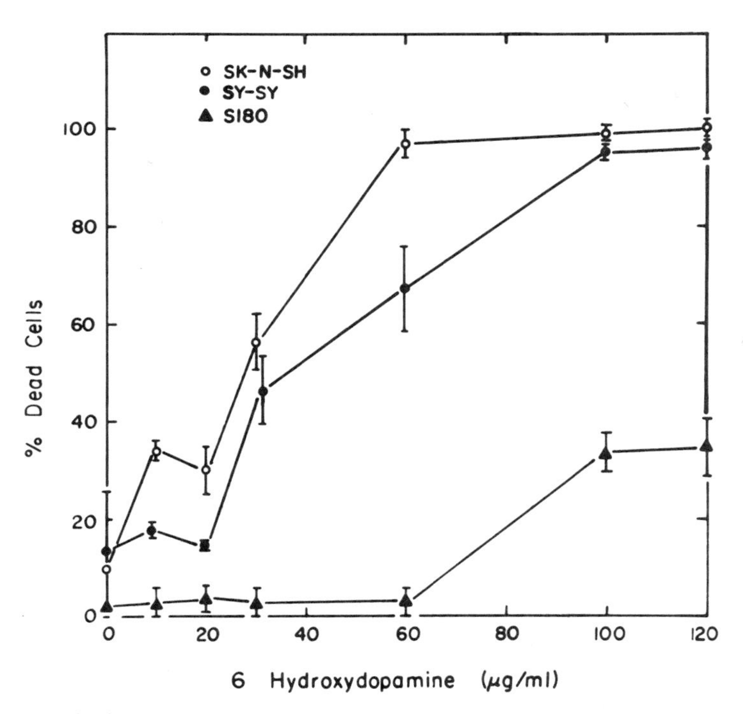

Fig. 7. Toxicity of 6-OHDA to SY5Y compared to sarcoma cells (S180).

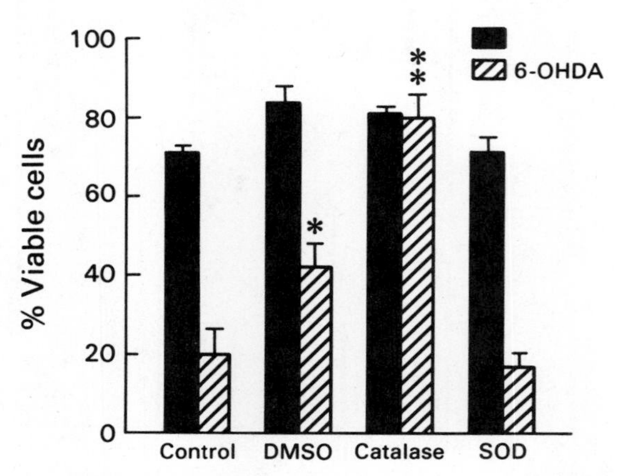

Fig. 8. Protection from 6-OHDA by anti-oxidants.

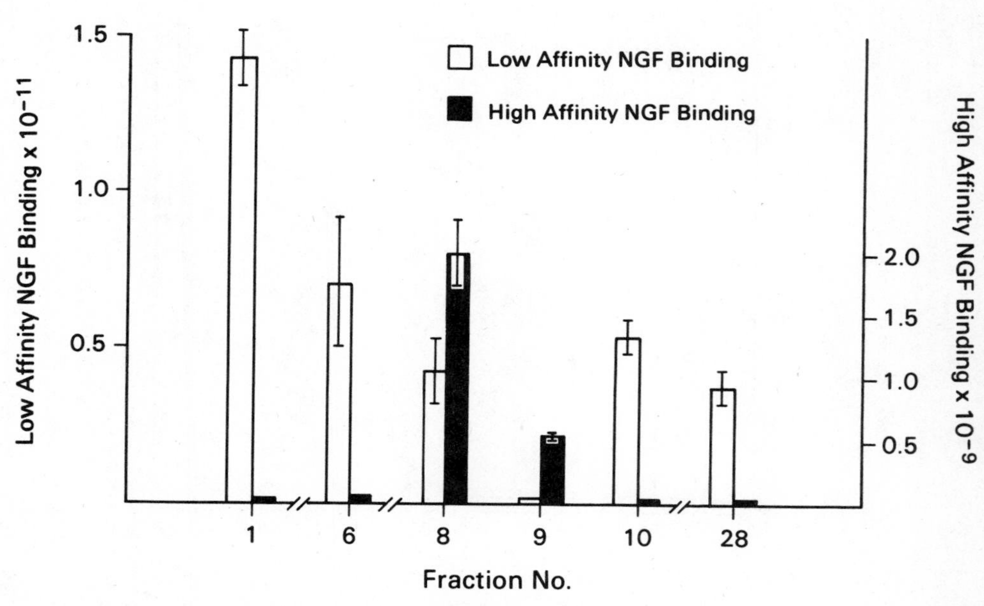

Fig. 9. Chromatograph of PEGG

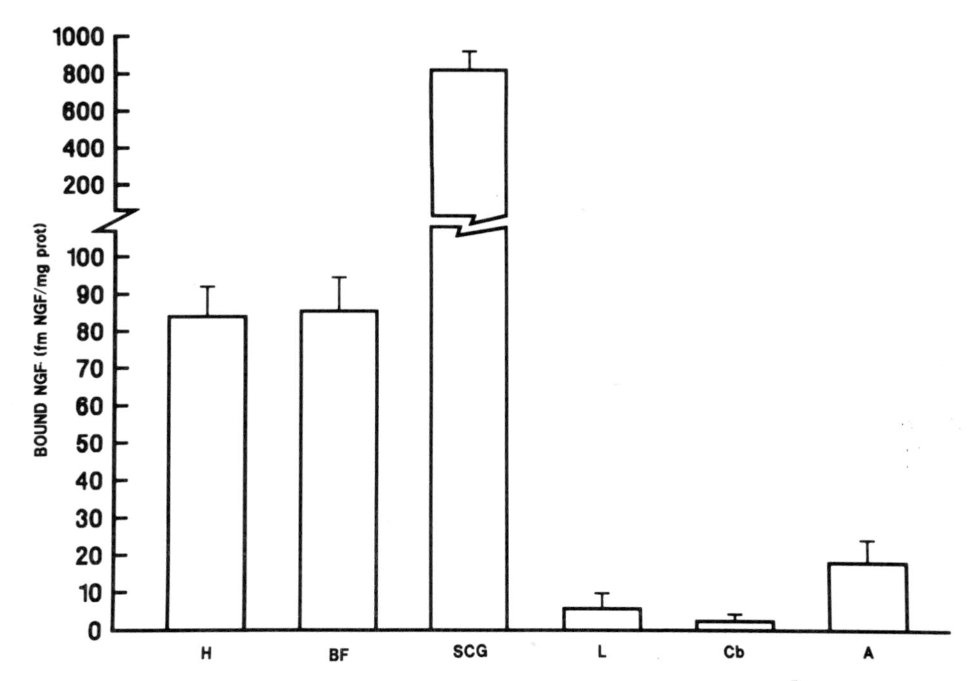

Fig. 10. Bmax obtained by Scatchard analysis of ^{1}25I- -NGF binding to NP-40-solubilized tissues of rat brain. H:hippocampus, BF:basal forebrain, SCG:superior cervical ganglion, L:liver, Cb:cerebellum, A:adrenal. (Angelucci et al, 1988b).

(Zimmermann and Sutter, 1983; Marchetti et al, 1987). When ^{125}I-NGF is infused into spinal cord there is demonstrable uptake and accumulation of the labeled NGF into the large sensory neurons of sensory ganglia but also into satellite cells (Khan et al., 1987). This is of interest since many of the determinations of levels of NGF binding, NGF binding proteins and NGF receptor mRNA do not differentiate between glial and neuronal compartments. These findings suggest that the strategy of using cell line models to better understand NGF-NGFR interactions prior to tackling in vivo situations is a prudent approach and that careful attention to culture conditions is necessary since steroids affect NGF and NGFR expression in vivo (Wright et al, 1987; Wright et al, 1988).

It is established that NGFR mRNA and NGFR protein are distributed throughout the CNS but less is known about the binding properties of NGFR in CNS. It has been shown that when levels of total NGF binding (fM NGF/mg protein) in 4 month old rat brains is compared to that in 26 month old rats, there is a reduction in NGF binding in the brains of the aged rats (Angelucci, et al., 1988). Specifically, there is a reduction in the levels of NGF binding present in cortex, basal forebrain and hippocampus of the aged rats but no detectable changes in cerebellum (Angelucci, et al., 1988b) (Fig. 10, 11). As part of these studies, it was also shown that acetyl-L-carnitine, shown to rescue cholinergic neurons in aged hippocampus (Angelucci et al, 1988a), appears to be rescuing NGF receptor bearing cells in these experiments (Fig. 11). Whether these effects in aged rodent brain are direct or via satellite cells has not been determined. This is an interesting finding because others have found equivalent reductions in NGF and NGF mRNA levels

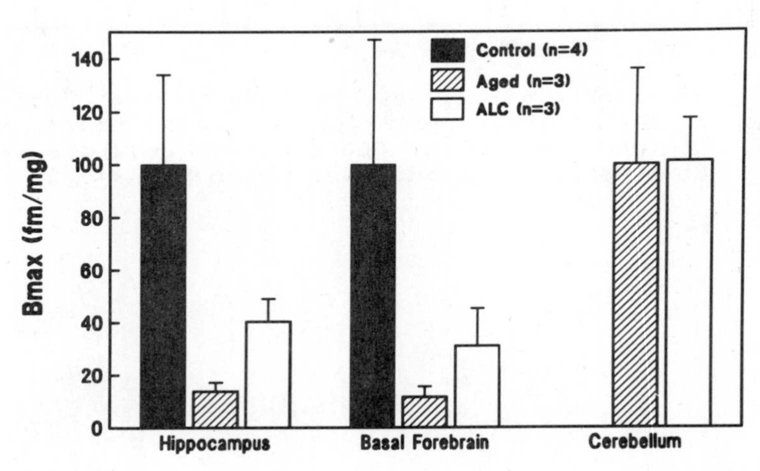

Fig. 11. Relative Bmax of NGF binding to solubilized tissues of young, aged or ALC-treated aged rats as % of values for 4 month old rats (Angelucci et al, 1988b).

in aged human CNS but no difference between aged matched controls and Alzheimer patients (Goedert et al., 1986). Since there are no significant changes in observed Kds, one must conclude that the brain samples from aged rats contain fewer receptor molecules. There may be fewer NGFR/cell or fewer cells containing the same number of NGFR/cell.

6. Oxidative Stress in Neurons

Analysis of the metabolic pathways involved in cellular oxidant-antioxidant balance would suggest that there are a limited number of antioxidant molecules and antioxidant enzymes whose measurement should provide a good index of oxidant-antioxidant balance and the extent of peroxidative damage. At the mitochondrial level, peroxidation will result in calcium release into the cytoplasm that in turn can activate calcium dependent proteases and lead to cell death. In addition, perturbation of the level of intracellular hydrogen peroxide can have toxic effects on glucose metabolism since energy metabolism in neurons is particularly dependent on glucose utilization via the pentose phosphate pathway and lipogenesis. Thus, neurons are particularly vulnerable to free radical induced imbalances in energy metabolism in addition to the effect of radicals on membrane lipids, nucleic acids and proteins. As a result, neuronal function is impaired and for some neurons the result is cell death. There are likely to be subsequent events that involve permanent changes at the blood brain barrier, astrocytic activation and immunogenic involvement that will not be addressed here, although they are important and free radicals may be involved.

Initial studies on the protection afforded to the SY5Y human neuroblastoma by NGF against 6-hydroxydopamine, 6-OHDA, a known generator of H_2O_2, showed that NGF protection was due to its stimulation of metabolic events and not due to morphological or extracellular substrata associated events or totally due to dopamine uptake (Perez-Polo et al., 1982; Perez-Polo and Werrbach-Perez, 1985; Perez-Polo et al., 1986; Perez-Polo and Werrbach-Perez, 1988) Fig.12. It has been shown that 6-OHDA generated free radicals are the principal toxic agent in this instance and that catalase, but not superoxide dismutase, SOD, provides protection from 6-OHDA (Perez-Polo and Werrbach-Perez, 1985; Perez-Polo et al., 1986; Perez-Polo and Werrbach-Perez, 1988). Dose response curves for H_2O_2 toxicity are similar for PC12 and SY5Y cells, as is NGF and catalase protection (Jackson et al, submitted). Also, treatment with NGF induced catalase and glutathione peroxidase but not SOD activity (Jackson et al, in preparation). This is of interest because both cell lines have very low endogenous levels of these antioxidant enzymes as compared to several non-neuronal cell lines tested. NGF protection was related directly to the induction of catalase since n the presence of aminotriazole, AZ, a small molecular weight inhibitor of catalase that can cross the cell membrane, NGF protection is abolished and AZ, by itself, renders PC12 and SY5Y cells more susceptible to H_2O_2 toxicity, supposedly by inhibiting endogenous unstimulated catalase (Jackson et al, submitted) Fig. 13,14,15.

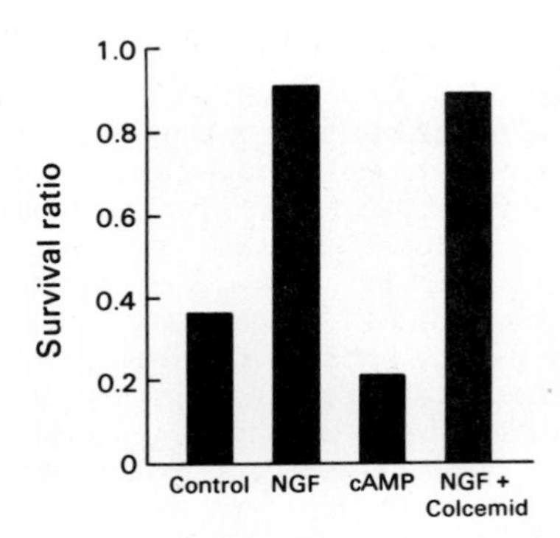

Fig. 12. Effect of colcemid (disrupts neurites) and cAMP (induces neurites) on NGF protection of 6-OHDA treated SY5Y.

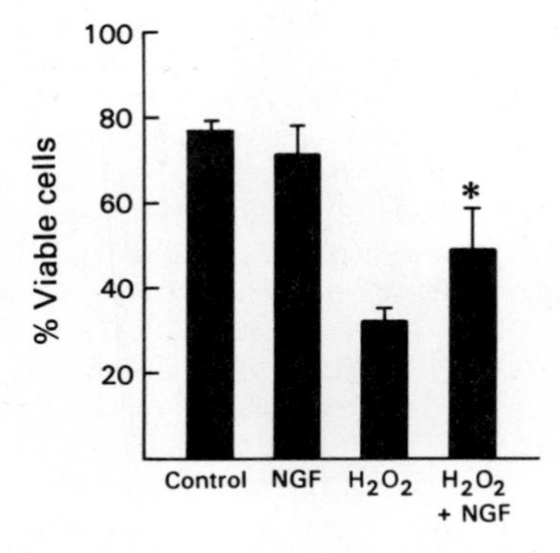

Fig. 13. NGF protection of SY5Y from H2O2.

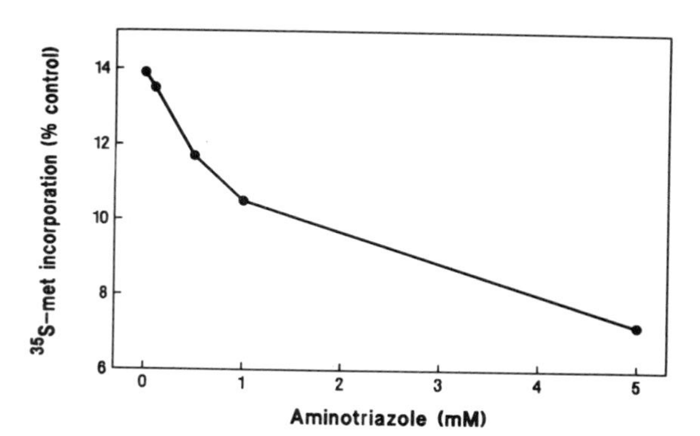

Fig. 14. SY5Y survival in H_2O_2 and aminotriazole.

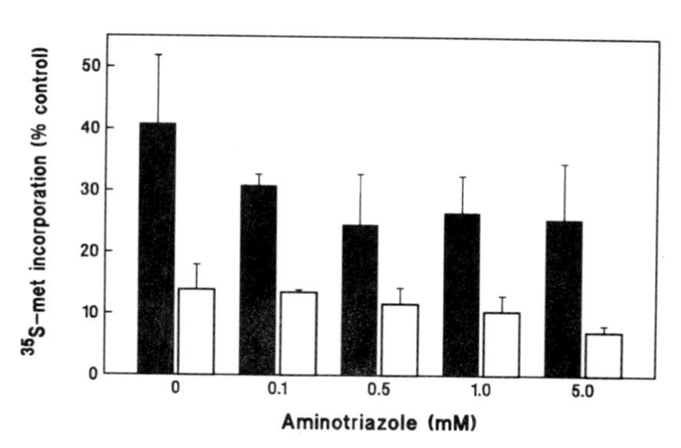

Fig. 15. Trophic effect on NGF (filled bars) vs. control (open bars) on PC12 cells in presence of H_2O_2.

CONCLUSION

In conclusion there is detailed, although not complete, information about the structure of NGF and NGFR. As to the molecular mechanism of action of NGF, most of the hypotheses elaborated are pertinent to neurite elongation and effects on neurotransmitter metabolism. Although cell rescue by NGF following lesions and aging in the CNS and the periphery is well documented, molecular rationale are not established at the present time.

ACKNOWLEDGEMENTS

It is a pleasure to acknowledge the collaborative efforts of R. Saneto, D. Marchetti, L. Angelucci, L. Apffel, G. Jackson, M.T. Ramacci, G. Taglialatela, E. Tiffany-Castiglioni, and K. Werrbach-Perez. Thanks to D. Masters for manuscript preparation. Supported in part from grants from NINCDS, CNR and the Sigma Tau Company.

REFERENCES

Agranoff, B.W., 1984, Lipid peroxidation and membrane aging. Neurobiol. of Aging, 5:337-338.
Angelucci, L., Ramacci, M.T., Taglialatela G., Hulsebosch, C., Morgan, B., Werrbach-Perez, K and Perez-Polo, R., 1988, Nerve growth factor binding in aged rat central nervous system: effect of acetyl-L-carnitine, JNR 20:491-496.
Angelucci, L., Ramacci, M.T., Amenta, F., Lorentz, G. and Maccari, F., 1988, Acetyl-L-carnitine in the rat's hippocampus aging: morphological, endocrine and behavioral correlates. In: Neural Development & Regeneration, Cellular & Molecular Aspects, A. Gorio, J.R. Perez-Polo, J. deVellis, B. Haber, eds., Springer-Verlag, Heidelberg, pp. 57-66.
Bostwick, J.R., Appel, S.H., Perez-Polo, J.R., 1987, Distinct influences of NGF and a CNS cholinergic factor on medial septal explants. Brain Res. 422:92-98.
Buck, C.R., Martinez, H.J., Black, I.B., Chao, M.V., 1987, Developmental regulated expression of the NGF Receptor gene in the PNS and brain. Proc. Natl. Acad. Sci. 84:3060-3063.
Chao, M.V., Bothwell, M.A., Ross, A.H., Koprowski, H., Lanahan, A.A., Buck, C.R., Sehgal, A., 1986, Gene transfer and mol. cloning of the human NGFR. Science 232:518-521.
Cutler, R.G., 1985, Peroxide-producing potential of tissues: inverse correlation with longevity of mammalian species. Proc. Natl. Acad. Sci. USA, 82:4798-4802.
Davies, K.J.A., Protein damage and degradation by oxygen radicals. J. Biol. Chem. 262:9895-9920.
Demopoulos, H.B., Flamm, E.S., Pietronigro, D.D., Seligman, M.L., 1980, The free radical pathology & the microcirculation in the major CNS disorders. Acta Physiol Scand. 492:(suppl):43-57.
Goedert, M., Fine, A., Hunt, S.P., Ullrich, A., 1986, NGF mRNA in PNS & CNS tissues & in the human CNS: lesion effects in the rat brain and levels in Alzheimer's Disease. Mol. Brain Res. 1:85-92. Greene, L.A., Shooter, E.M., 1980, The Nerve Growth Factor: Biochemistry, synthesis, and mechanism of action. Annu Rev. Neurosci. 3:353-402.
Grob, P.M., Ross, A.H., Koprowski, H., Bothwell, M., 1985,

Characterization of the human melanoma NGFR. J. Biol. Chem. 260:8044-8049.
Hatanaka, H., Nihonmatsu, I and Tsukui, H., 1988, Nerve growth factor promotes survival of cultured magnocellular cholinergic neurons from nucleus basalis of meynert in postnatal rats. Neurosci. Lett. 90:63-68.
Hosang, M., Shooter, E.M. 1987, The internalization of NGF by high affinity receptors on pheochromocytoma PC12 cells. EMBO J. 6:1197-1202.
Khan, T., Green, B., Perez-Polo, J.R., 1987, Effect of Injury on NGF uptake by sensory ganglia. J. Neurosci. Res. 18:562-567.
Levi-Montalcini, R, 1987, The NGF 35 Years Later, Science 237:1154-1162.
Lillien, L.E., Claude, P., 1985, NGF is a mitogen for cultured chromaffin cells. Nature 317:632-634.
Lyons, C.R., Stach, R.W., Perez-Polo, J.R., 1983, Binding constants of isolated NGFR from different species. Biochem. Biophys. Res. Comm. 115:368-374.
Marchetti, D., Perez-Polo, J.R., 1987, NGFR in human neuroblastoma cells. J. Neurochem. 49:475-486.
Marchetti, D., Stach, R.W., Saneto, R.P., deVellis, J., Perez-Polo, J.R., 1987, Binding constants of soluble NGFR in rat oligodendrocytes & astrocytes in culture. Biochem. Biophys. Res. Comm. 147:422-427.
Perez-Polo, J.R., Reynolds, C.P., Tiffany-Castiglioni, E., Ziegler, M., Schulze, I., Werrbach-Perez, K., 1982, NGF effects on human neuroblastoma lines: A model system. In: Proteins in the Nervous System: Structure and Functions. B. Haber, J.R. Perez-Polo, J.D. Coulter eds., Alan R. Liss, NY;285-299.
Perez-Polo, J.R., 1985, Neuronotrophic Factors. In: Cell Cultures in the Neurosciences, J.R. Bottenstein, G. Sato eds., Plenum Press, New York, 3:95-123.
Perez-Polo, J.R., Werrbach-Perez, K.,1985, Effects of NGF on the in vitro response of neurons to injury. In: Recent Achievements in Restorative Neurol Neuron Functions and Dysfunctions J. Eccles, J.R. Dimitrijevic, eds., Karger 30:321-377.
Perez-Polo, J.R., Apffel, L., Werrbach-Perez, K., 1986, Role of CNS and PNS trophic factors on free radical mediated aging events. Clin. Neuropharm. 9:98-100.
Perez-Polo, R., Werrbach-Perez, K., 1987, In vitro model of neuronal aging and develop. in the nervous system. In: Model Systems of Develop. and Aging of the Nerv. System. A. Vernadakis, ed., Martinus Nijhoff, Boston, pp.433-442.
Perez-Polo, J.R., 1987, Neuronal Factors. CRC, Boca Raton: 1-202.
Perez-Polo, J.R., Werrbach-Perez, K., 1988, Role of NGF in neuronal injury & survival. In: Neural Develop. & Regeneration, Cellular & Molecular Aspects, A. Gorio, J.R. Perez-Polo, J. de Vellis, B Haber, eds., Springer Verlag, Heidelberg, 399-410.
Radeke, M.J., Misko, T.P., Hasu, C., Herzenberg, L.A., Shooter, E.M., 1987, Gene transfer and molecular cloning of rat NGFR. Nature 325: 593-597.
Stach, R.W., Perez-Polo, J.R., 1987, Binding of NGF to its receptor. J. Neurosci. Res. 17:1-10.
Thoenen, H., Barde, Y.A., Davies, A.M., Johnson, J.E., 1981, Neuronotrophic factors & neuronal death. Ciba Found. Symp. 126:838-840.

Thorpe, L.W., Morgan, B., Beck, C., Werrbach-Perez, K., Perez-Polo, JR., 1988, NGF and the immune system. In: Neural Development & Regeneration Cellular & Molecular Aspects" A. Gorio, J.R. Perez-Polo, J. deVellis, B. Haber, eds., Springer Verlag, Heidelberg pp.583-594.
Uchida, Y, Tonionaga, M, 1987, Loss of NGFR in sympathetic ganglia form aged mice. Biochem. Biophys. Res. Comm. 146:797-801.
Whittemore, S.R., Seiger, A., 1987, The Expression, localization and functional significance of beta-NGF in the Central Nervous System, Brain Res. Rev., 12:439-464.
Whittemore, S.R., Person, H., Ebendal, T., Larkfors, L., Larhammar, D., Ericsson, A., 1988, Structure and Expression of beta-NGF in the rat Central Nervous System. In: Neural Development & Regeneration Cellular & Molecular Aspects, A. Gorio, J.R. Perez-Polo, J. deVellis, B. Haber eds., Springer-Verlag, Heidelberg, pp.245-256.
Wright, L.L., Beck, C., Perez-Polo, J.R. 1987, Sex differences in NGF levels in SCG and pineals. Int. J. Dev. Neurosci., 5:383-390.
Wright, L.L., Marchetti, D., Perez-Polo, J.R., 1988, Effects of gonadal steroids on NGF receptors in sympathetic and sensory ganglia of neonatal rats. Int. J. Dev. Neurosci., 6:3,217-222.
Yankner, B.S., Shooter, E.M. 1982, The biology and mechanism of action of NGF. Annu Rev. Biochem. 51:845-868.
Yau, Q., Johnson, Jr., E.M., 1987, A quantitative study of the developmental expression of the NGF receptor in rats. Dev. Biol. 121:139-148.
Yip, H.K., Johnson, R.M. 1984, Developing DRG neurons required trophic support form their central processes: Evidence for a role of retrogradely transported NGF from the CNS system to the PNS. Proc. Natl. Acad. Sci. 81:6245-6249.
Zimmermann, A, Sutter, A., 1983, NGFR on glial cells in sensory neurons. EMBO J. 2:879-885.

LOCALIZATION OF THE NORMAL CELLULAR SRC PROTEIN TO THE GROWTH CONE OF DIFFERENTIATING NEURONS IN BRAIN AND RETINA

Patricia F. Maness , Carol G. Shores
and Michael Ignelzi
Department of Biochemistry
University of North Carolina School of Medicine
Chapel Hill, NC 27515

ABSTRACT

The protooncogene c-src is implicated in the development of the vertebrate nervous system. Its product pp60^{c-src} is a tyrosine-specific protein kinase that is expressed in two phases of neural development. An activated form of the pp60^{c-src} is highly enriched in the membrane of nerve growth cones and in the proximal neuritic shaft of differentiating neurons, as shown in brain and retina. A possible role for pp60^{c-src} in neuronal process extension is suggested that may involve cell-substratum adhesion or motility.

INTRODUCTION

Proto-oncongenes are normal cellular genes that are believed to control cell growth and differentiation. Mutations in the coding or regulatory regions of these genes can induce malignant growth in appropriate target cells. The product of the c-src proto-oncogene (pp60^{c-src}) is a tyrosine-specific protein kinase that is homologous to the transforming protein (pp60^{v-src}) encoded by the v-src gene of Rous sarcoma virus (Collett et al., 1979; Opperman et al., 1979). Its biological function has not yet been elucidated, but studies concerning the expression of the c-src protein from our laboratory and others suggest that this tyrosine kinase plays a role in neuronal differentiation.

Two Phases of pp60^{c-src} Expression in the Neural Lineage

Two phases of pp60^{c-src} expression have been identified in the neural cell lineage of the vertebrate nervous system. For these studies immunoperoxidase staining of pp60^{c-src} in paraffin sections of developing chick embryos was carried out using a polyclonal antibody raised against pp60^{v-src} expressed in bacteria (Gilmer and Erikson, 1983). This antibody is specific for pp60^{c-src} in normal chicken tissues. In the first phase of pp60^{c-src} expression, pp60^{c-src} specific immunoreactivity (IR) appeared in proliferating neural ecto-

Molecular Aspects of Development and Aging of the Nervous System
Edited by J.M. Lauder
Plenum Press, New York, 1990

dermal cells of gastrula and neurula-stage chick embryos (stages 3-9 of Hamburger and Hamilton (1951)) at approximately the time when pluripotent stem cells become committed to pathways of neuronal or glial cell differentiation (Figure 1A; Maness et al., 1986). Immunoreactivity was restricted to the presumptive neural ectoderm as it was not observed in non-neural ectoderm, mesoderm, or endoderm. After the neural tube closed, pp60^{c-src} IR was no longer seen in the proliferating, committed neural precursors (Figure 1B).

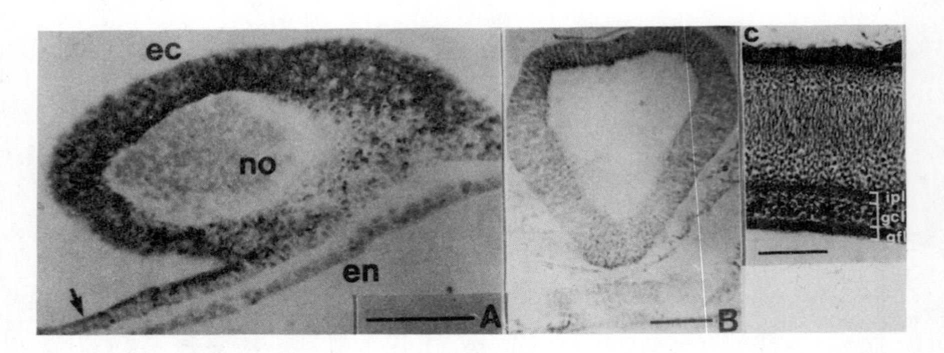

Fig. 1. Localization of pp60^{c-src} immunoreactivity in early chick embryos and neural retina by immunoperoxidase staining with antibodies raised against bacterially expressed pp60^{c-src}. (A) Transverse paraffin section through head process of a stage 6 chick embryo. Arrow indicates point of transition between stained neural ectoderm medially and unstained nonneural ectoderm laterally (ec, ecto-derm; en, endododerm; no, notochord). (B) Transverse section through metencephalon of stage 12 chick embryo. (C) Section through neural retina of stage 32 embryo. Immunoreactive staining was prominent in the ganglion fiber layer (gfl), ganglion cell layer (gcl), and inner plexiform layer (ipl). Bars in A, B = 100 µM. Bar in C = 3 µm.

Later, when neuroblasts underwent terminal differentiation and ceased replicating, a second phase of pp60^{c-src} expression was initiated in terminally differentiating neurons. As shown in the embryonic chick neural retina at stage 32 (Sorge et al., 1984), pp60^{c-src} IR was abundant in the inner plexiform layer, ganglion cell layer, and ganglion fiber layer (Figure 1C). The inner plexiform and ganglion fiber layers are rich in newly elaborated neuronal processes and growth cones of differentiated neurons, but contain few cell bodies. The ganglion fiber layer, which is also stained, contains the bodies of differentiated ganglion cell neurons. In contrast to ganglion cells, other neuronal cell bodies in the stage 32 retina did not contain detectable levels of pp60^{c-src}. The undifferentiated, dividing neuroepithelial precursors spanning the remainder of the retina were not immunoreactive. Similar studies in the developing chick cerebellum (Fults et al., 1985) and telencephalon (Sorge et al., 1984) showed pp60^{c-src} IR to be most prominent in the developing molecular layer, where processes of differentiated neurons are localized, but little staining was observed in proliferating neural precursors. All of the staining described was specific, as it could be effectively competed by preincubating the antibodies with purified antigen prior to application to the sections.

Most of the neurites located in the fiber layers at the stage and location where $pp60^{c-src}$ IR was observed are derived from neuronal cells. However, radially oriented Mueller glia are also present in these regions. We do not know if glia are also immunoreactive, but no obviously radial staining pattern was observed. $pp60^{c-src}$ is clearly a product of some neuronal cells, as $pp60^{c-src}$ immunofluorescence has been demonstrated in cultural dorsal root ganglion cell neurons (Maness, 1986) and retinal neurons (Maness et al., 1988), and $pp60^{c-src}$ has been detected biochemically in extracts of primary cortical neurons in culture (Brugge et al., 1985).

The onset of the second phase of $pp60^{c-src}$ expression in terminally differentiating neurons was unexpected in view of the ability of the closely related tyrosine protein kinase $pp60^{v-src}$ to stimulate cell growth in fibroblasts. These results suggest that $pp60^{c-src}$ may play a role in an aspect of neuronal differentiation such as process extension, rather than in proliferation. Consistent with such a possibility, rat PC12 pheochromocytoma cells can be induced to differentiate and extend neurites when $pp60^{v-src}$ is expressed in these cells during viral infection (Alema et al., 1985).

$pp60^{c-src}$ is Enriched in Nerve Growth Cone Membranes from Fetal Rat Brain

The high levels of $pp60^{c-src}$ IR in process-rich regions of the embryonic nervous system led us to investigate the hypothesis that $pp60^{c-src}$ is preferentially localized in the nerve growth cone (Maness et al., 1988). The growth cone represents the distal tip of the outgrowing process, and is important in controlling motility by dynamic changes in cytoskeletal architecture and adhesion to the substratum, through which it migrates in search of a synaptic partner. Subcellular fractionation of fetal rat brain (E18) by a procedure developed by Pfenninger et al., (1983) yields a fraction that is at least 70% enriched in nerve growth cone particles as assessed by electron microscopy. $pp60^{c-src}$ was analyzed in this and other subcellular fractions from fetal rat brain (E18) by quantitative immunoblotting with a monoclonal antibody that recognizes the mammalian form of $pp60^{c-src}$ (Lipsich et al., 1983), the generous gift of Joan Brugge (SUNY at Stony Brook). $pp60^{c-src}$ was found to be enriched approximately 9-fold in growth cone membranes compared to homogenate membranes, and 50-fold compared to total homogenate (Table 1). All of the $pp60^{c-src}$ detected was associated with membranes, in agreement with its localization in the membrane of nonneuronal cells, which has been shown to occur via a covalently bound N-terminal myristic acid residue (Courtneidge et al., 1989; Buss et al., 1984; Cross et al., 1984; Garber et al., 1985).

Table 1. Relative Levels of pp60^{c-src} in Fetal and Adult Rat Brain Fractions

	pp60^{c-src} per mg protein
Early Neural Tube (E11)	
Low-speed supernatant membranes	1.3
Fetal Brain (E18)	
Total homogenate	0.18
Homogenate membranes	1.0
Low-speed supernatant membranes	2.3
Growth cone membranes	9.0
Adult Brain	
Homogenate membranes	0.4
Low-speed supernatant membranes	0.3
Synaptosomal membranes	0.6

pp60^{c-src} was quantitated in fractions (100 µg) by scanning densitometrically the 60 kDa bands on autoradiograms obtained from immunoblots. Absorbance values were normalized to those measured in homogenate membranes, as described in Maness et al., (1988).

In synaptosomal membranes isolated from adult rat brain levels of pp60^{c-src} were approximately 15-fold lower than in fetal growth cone membranes (Table 1). Moreover, pp60^{c-src} was not substantially enriched in synaptosomal membranes compared to homogenate membranes or membranes from other subcellular fractions of adult brain, suggesting that pp60^{c-src} may not be as concentrated at synaptic terminals of mature neurons (Maness et al., 1988). Comparison of the levels of pp60^{c-src} in equivalent fractions from early neural tube (E11), fetal brain (E18), and adult rat brain (the low speed supernatant membrane fraction; Maness et al., 1988) revealed that pp60^{c-src} increased from embryonic day 11 to 18, concomitant with process extension, then decreased almost 8-fold in adult brain, further underlying a primary role in developing, rather than mature brain.

pp60^{c-src} in neurons of the central nervous system is a tyrosine-specific kinase with especially elevated specific activity (Brugge et al., 1985; Bolen et al., 1985). This form of pp60^{c-src} is the product of alternative splicing and contains six additional amino acids in its N-terminal region (Martinez et al., 1987; Levy et al., 1987). pp60^{c-src} in growth cone and synaptosomal membranes exhibits high specific activity toward the synthetic peptide substrate angiotensin I, and slightly retarded electrophoretic mobility than the nonneuronal form of pp60^{c-src}, expressed in fibroblasts, suggesting that at least in part, the growth cone-associated pp60^{c-src} represents, the activated, alternatively spliced form of the enzyme (Maness et al., 1988).

$pp60^{c-src}$ is Localized in the Nerve Growth Cone and Proximal Neuritic Shaft of Retinal Neurons

To confirm the growth cone enrichment of $pp60^{c-src}$ observed by biochemical fractionation, indirect immunofluorescence staining of primary cultures of chick retinal neurons was carried out. Neurofilament-positive neuronal cells with long processes exhibited more intense fluorescence in the growth cone and proximal region of the neurite than in the distal, older region of the neurite (Figure 2A), consistent with the subcellular fractionation results (Maness et al., 1988). The cell body was also stained, but at least some of the staining was nonspecific, as indicated by failure of the fluroescence to compete completely when the antibody was preincubated with purified antigen (Figure 2B).

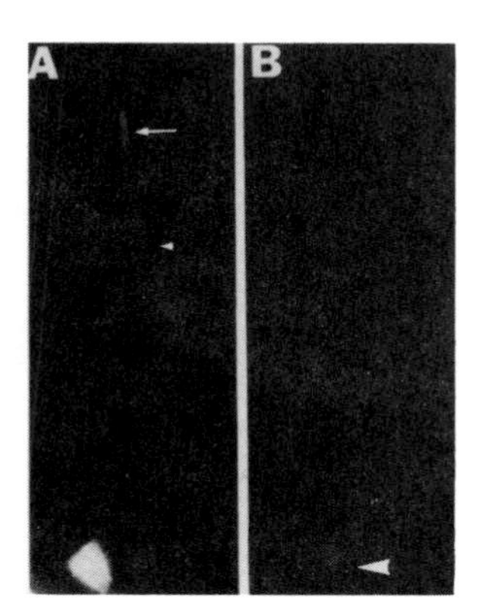

Fig. 2. Indirect immunofluorescence staining of retinal neurons with anti-$pp60^{c-src}$ antibodies. (a) Growth cone, contiguous region of neurite, and cell body are labeled with anti-$pp60^{c-src}$ antibodies. (b) Competition control photographed under conditions identical to a, showing a low level of staining only in the cell body when the primary antibody was preincubated with purified antigen. Arrow, growth cone; small arrowhead, neurite; large arrowhead, cell body. (X 860).

In the chick embryo, growth cones of retinal ganglion cell axons migrate within the optic tract to the optic tectum, where they form synapses with target neurons. To investigate whether $pp60^{c-src}$ exhibits a distribution _in vivo_ consistent with an enrichment in nerve growth cones, immunoperoxidase staining for $pp60^{c-src}$ was carried out in paraffin sections of the stage 36 chick embryo in which the optic tract could be seen in its entirety, including the neural retina, optic nerve, optic chiasm, and optic tectum. $pp60^{c-src}$ IR was observed at all sites within the optic tract and at the surface of the optic tectum, where growth cones of arriving retinal ganglion cell axons accumulate, unobscured by processes of tectal neurons (Figure 3A). $pp60^{c-src}$ IR was especially prominent in the ganglion fiber layer of the neural retina and at the surface of the optic tectum (Figure 3C). This pattern was similar to the distribution of neurofilament protein revealed in an adjacent section by staining with a mouse monoclonal antibody directed against the 200 kDal subunit of neurofilament protein (Wood and Anderton, 1981) provided by John Wood (Sandoz Institute for Medical Research, London) (Figure 3B). $pp60^{c-src}$ staining was effectively competed by preincubation of the antiserum with purified antigen (not shown). Although identification of single growth cones was not possible at the level of resolution afforded in these studies, the results are consistent with a localization of $pp60^{c-src}$ in growth cones and proximal neurites of ganglion cells _in vivo_.

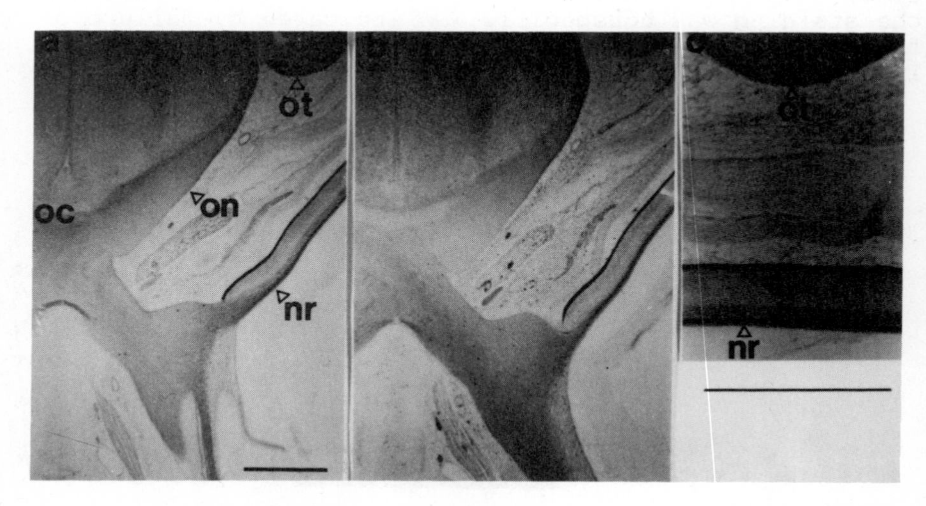

Fig. 3. Localization of $pp60^{c-src}$ in the retinotectal projection of the chick embryo at stage 36. Paraffin sections (5 µm) were subjected to immuno-peroxidase staining as described previously (Sorge et al., 1984) using as primary antibodies: (A,C) a rabbit polyclonal antibody raised against $pp60^{v-src}$ (Gilmer and Erikson, 1983), and (B) a mouse monoclonal antibody against the 200 kDal subunit of neurofilament protein (Wood and Anderton, 1981). oc, optic chiasm; on, optic nerve; nr, neural retina; ot, optic tectum. Bar in A, and below C = 25 µM.

DISCUSSION

Collectively, studies on $pp60^{c-src}$ expression demonstrate that $pp60^{c-src}$ is expressed at early and late phases of neural differentiation. Moreover, $pp60^{c-src}$ persists, albeit at lower levels, in the mature central nervous system. The temporal and spatial pattern of $pp60^{c-src}$ expression indicates that $pp60^{c-src}$ may be more functionally important in an aspect of neural differentiation than in proliferation of precursors.

In differentiating neuronal cells, $pp60^{c-src}$ is highly enriched in the membrane of nerve growth cones and the proximal region of the axon shaft. It is possible to speculate that the activated form of pp60^{c-src} in the growth cone may be a component of signal transduction used to convey dynamic changes in cell-substratum adhesion or motility, which are undoubtedly crucial for neurite extension. In mature neurons, pp60^{c-src} may have an analogous role in plasticity. Its retroviral homologue, pp60^{v-src}, is known to be localized in adhesion plaques of transformed cells, further suggesting that these closely related tyrosine-kinases may influence adhesion (Rohrschneider, 1979; Willingham et al., 1979; Shriver an Rohrschneider, 1981). Disruption of cell-cell or cell-substratum contact by pp60^{v-src} action could conceivably achieve anchorage-dependent cell growth, a well known property of transformed cells that is poorly understood at the molecular level. The function of pp60^{c-src} in early gastrulating embryos coinciding with committment of neuroectodermal cells, at a time when there are no growth cones, could similarly be related to adhesion or cell-cell contact. Such cell-cell interactions may be crucial in conveying positional information to the embryo in a manner analogous to that of the Drosophia _sevenless_ gene product, a tyrosine-specific protein kinase that regulates photoreceptor cell differentiation (Banerjee et al., 1987; Hafen et al., 1987; Basler and Hafen, 1988).

ACKNOWLEDGEMENTS

This work was supported by National Institutes of Health Grant NS26620. P.F.M. is a recipient of a National Institutes of Health Research Career Development Award. Colleagues who were involved in previously published parts of the work presented are referenced in the text.

REFERENCES

Alema, S., Casalbore, P., Agostini, E., and Tato, F., 1985, Differentiation of PC12 phaeochromocytoma cells induced by _v-src_ oncogene, _Nature_, 316, 557-559.

Banerjee, U., Renfranz, P.J., Pollock, J.A. and Benzer, S., 1987, Molecular characterization and expression of _sevenless_, a gene involved in neuronal pattern formation in the Drosophila eye, _Cell_, 49, 281-291.

Basler, K. and Hafen, E., 1988, Control of photoreceptor cell fate by the _sevenless_ protein requires a functional tyrosine kinase domain, _Cell_, 54, 299-311.

Bolen, J.B., Rosen, N., and Israel, M.A., 1985, Increased pp60^{c-src} tyrosyl kinase activity in human neuroblastomas is associated with amino-terminal tyrosine phosphorylation of the src gene product, _Proc. Natl. Acad. Sci. USA_, 82, 7275-7279.

Brugge, J.S., Cotton, P.C., Queral, A.E., Barett, J.N., Nonner, D., and Keane R.W., 1985, Neurones express high levels of a structurally modified, activated form of pp60^{c-src}, _Nature_,316, 554-557.

Buss, J.E., Kamps, M.P., and Sefton, B.M., 1984, Myristic acid is attached to the transforming protein of Rous sarcoma virus during or immediately after its synthesis and is present in both soluble and membrane-bound forms of the protein, _Mol. Cell. Biol_., 4, 2697-2704.

Collett, M.D., Erikson, E., Purchio, A.F., Brugge, J.S., and Erikson, R.L., 1979, A normal cell protein similar in structure and function to the avian sarcoma virus transforming gene product, Proc. Natl. Acad. Sci., 76, 3159-3163.

Courtneidge, S.A., Levinson, A.D., and Bishop, J.M., 1980, The protein encoded by the transforming gene of avian sarcoma virus (pp60^{c-src}) and a homologous protein in normal cells (pp60$^{proto-src}$) are associated with the plasma membrane, Proc. Natl. Acad. Sci. USA, 77, 3783-3787.

Cross, F.R., Garber, E.A., Pellman, D., and Hanafusa, H., 1984, A short sequence in the p60^{src} N-terminus is required for p60^{src} myristylation and membrane association and for cell transformation, Mol. Cell. Biol., 4, 1834-1842.

Fults, D.W., Towle, A.C., Lauder, J.M., and Maness, P.F., 1985, pp60^{c-src} in the developing cerebellum, Mol. Cell Biol., 5, 27-32.

Garber, E.A., Cross, F.R., and Hanafusa, H., 1985, Processing of pp60^{v-src} to its myristylated membrane-bound form, Mol. Cell. Biol., 5, 2781-2788.

Gilmer, T.M. and Erikson, R.L., 1983, Development of anti-pp60^{src} serum antigen produced in Escherichia coli, J. Virol., 45,462-465.

Hafen, E., Basler, K., Edstoem, J.-E. and Robin, G.M., 1987, Sevenless, a cell-specific homeotic gene of Drosophila encodes a putative transmembrane receptor with a tyrosine kinase domain, Science, 236, 55-63.

Hamburger, V. and Hamilton, H.L., 1951, A series of normal stages in the development of the chick embryo, J. Morphol., 88, 49-92.

Levy, J.B., Dorai, T., Wang, L.-H., and Brugge, J.S., 1987, The structurally distinct form of pp60^{c-src} detected in neuronal cells is encoded by a unique c-src mRNA, Mol. Cell. Biol., 7, 4141-4145.

Lipsich, L.A., Lewis, A.J., and Brugge, J.S., 1983, Isolation of monoclonal antibodies that recognize the transforming proteins of avian sarcoma virus, J. Virol., 48, 352-360.

Maness, P.F., Aubry, M., Shores, C.G., Frame, L., and Pfenninger, K.H., 1988, The c-src gene product in developing brain is enriched in nerve growth cone membranes, Proc. Natl. Acad. Sci., 85, 5001-5005.

Maness, P.F., Sorge, L.K., and Fults, D.W., 1986, An early developmental phase of pp60^{c-src} expression in the neural ectoderm, Dev. Biol., 117, 83-89.

Martinez, R., Mathey-Prevot, B., Bernards, A., and Baltimore, D., 1987, Neuronal pp60^{c-src} contains a six-amino acid insertion relative to its non-neuronal counterpart, Science, 237, 411-415.

Opperman, H., Levinson, A.D., Varmus, H.E., Levintow, L., and Bishop, J.M., 1979, Uninfected vertebrate cells contain a protein that is closely related to the product of the avian sarcoma virus transforming gene src, Proc. Natl. Acad. Sci. USA, 76, 1804-1808.

Pfenninger, K.H., Ellis, L., Johnson, M.P., Friedman, L.B., and Somlo, S., 1983, Nerve growth cones isolated from fetal rat brain: Subcellular fractionation and characterization, Cell, 35, 573-584.

Rohrschneider, L.R., 1979, Immunofluorescence on avian sarcoma virus-transformed cells: localization of the src gene product, Cell, 16, 11-24.
Shriver, K., and Rohrschneider, L., 1981, Organization of pp60^{c-src} and selected cytoskeletal proteins within adhesion plaques and junctions of Rous sarcoma virus-transformed rat cells, J. Cell Biol., 89, 525-535.
Sorge, L.K,. Levy, B.T., and Maness, P.F., 1984, pp60^{c-src} is developmentally regulated in the neural retina, Cell, 36, 249-257.
Willingham, M.C., Jay, G., and Pasta, I., 1979, Localization of the avian sarcoma virus src gene product to the plasma membrane of transformed cells by electron microscopic immunocytochemistry, Cell, 18, 125-134.
Wood, J.N., and Anderton, B.H., 1981, Monoclonal antibodies to mammalian neurofilaments, Biosci. Rep., 1, 263-268.

A ROLE OF RAS ONCOGENES IN CARCINOGENESIS AND DIFFERENTIATION

Demetrios A. Spandidos[1,2] and Margaret L.M. Anderson[3]

[1]The Beatson Institute for Cancer Research, Garscube Estate, Bearsden, Glasgow G61 1BD, Scotland, UK
[2]Biological Research Center, The National Hellenic Research Foundation, 48 vas Constantinou Avenue Athens, Greece
[3]Department of Biochemistry, University of Glasgow Glasgow, G12 8QQ, Scotland, UK

INTRODUCTION

As we learn more about the process of carcinogenesis it is becoming clear that several oncogenes are involved in growth and differentiation of cells. There is also growing evidence for a role of proto-oncogene products in differentiation for example in the neuronal and hemopoietic lineages (for a review see ref.1). Proliferation and differentiation appear to be interdependent processes, in that a cycle of cell division often seems to be a prerequisite for differentiation. Further growth factors are frequently also differentiation factors.

One of the most intensively studied family of oncogenes is the ras family. The p21 protein encoded by the ras gene has been shown to be essential for control of both cellular proliferation (2,3) and differentiation (4,5). Ras genes are the most frequently activated genes in human cancer. Quantitative or qualitative changes in ras expression can bring about one step towards the cancer phenotype (for a review see ref.6).

Here we discuss further experimental evidence suggesting a role of ras oncogenes in carcinogenesis and differentiation.

MATERIALS AND METHODS

Rat pheochromocytoma PC12 cells (7) were grown in RPMI medium containing 15% fetal bovine serum (FBS). The vector (Homer 6) and recombinant plasmids carrying the mutant T24 (pH06T1) and the normal (pH06N1) H-ras1 genes (8) as well as plasmid pMCGM1 carrying the normal human myc gene (9) were employed. Plasmids were introduced into cells by

Molecular Aspects of Development and Aging of the Nervous System
Edited by J.M. Lauder
Plenum Press, New York, 1990

electroporation as previously described (10). Briefly, electroporation was carried out using a BioRad gene pulser at 2 kv/cm and 10 μg DNA/5x10^6 recipient cells. After electroporation the cells were plated in RPMI medium containing 15% FBS. The medium was changed every 2 days.

RESULTS AND DISCUSSION

Action of Ras Oncogenes in Multi-stage Carcinogenesis

Carcinogenesis is a multi-step process. Although the number of intermediate steps are not completely defined some of the stages e.g. immortalization, morphological transformation, anchorage independence, tumorigenicity and metastasis have been studied extensively. Oncogenes have been shown to be involved in all these stages. *Ras* oncogenes in particular can either immortalize, cause morphological transformation, anchorage independence, tumorigenicity or metastasis depending on the cell type (for a review see ref.11). They can also trigger malignant conversion of "normal" (early passage) rodent cells (8). Although it was earlier thought that *ras* genes act at a particular stage in transformation, it is now clear that they can be involved in any of the recognizable stages of the transformation process (for a discussion see ref.11).

Ras Oncogenes Trigger the Release of Growth Factors in Transfected Cells

Both fibroblasts (9,12) and epithelial (11) cells transfected with *ras* oncogenes secrete transforming growth factors. Interleukin-3 (IL-3) like activity is also produced by *ras* transformed fibroblasts (13).

The mechanism by which transforming growth factors act to stimulate proliferation of cells is not known. However, some tumors, but not normal counterparts produce growth factors which stimulate their own growth i.e. autocrine stimulation (14). In several cases alterations in receptor structure, causes permanent stimulation of growth even in the absence of the cognate ligand (15). A third possibility is that proteins which act intracellularly to transduce the proliferation signal from the cell surface to the nucleus are altered in such a way that the signal is transmitted continuously. Thus mutations in the *ras* protein or proteins which interact with it may result in a constitutive signal transduction (for a review see ref.16).

Ras Oncogenes and Differentiation

The *src*, *fos*, *fms*, *myc* and *myb* proto-oncogenes are known to play a role in cellular differentiation (for a review see ref.1). Evidence that the *ras* proto-oncogene also affects differentiation comes from a variety of systems. Thus, the rat pheochromocytoma cells PC12 which can be induced to differentiate by nerve growth factor, also differentiate on infection with v-H-*ras* or v-K-*ras* viruses (6) or on microinjection of oncogenic H-ras p21 protein but not the normal p21 (5).

We have electroporated a subclone (PC12-A) of the rat pheochromocytoma cell line PC12 with a plasmid vector alone (Homer 6) or with recombinants carrying the mutant T24 or the normal H-ras1 gene (17). Six days after electroporation with the T24 H-ras1 gene transfectants had undergone neuronal differentiation similar to that induced by NGF. Neither the Homer 6 vector alone nor the normal H-ras1 gene elicited this effect. Introduction of the human c-myc oncogene into PC12-A cells) blocked NGF-induced differentiation into neuronal type of cells (17).

These results on the effects of ras and myc genes on PC12 rat pheochromocytoma differentiation are similar to recent observations in the same system with oncogene N-ras (18) and c-myc (19) genes. However, in mouse embryonal carcinoma cells expression of the exogenous human T24 H-ras1 gene did not interfere with differentiation but cells developing along a certain lineage(s) became malignantly transformed (20).

Block of differentiation by activated ras genes has been observed in a number of systems e.g. introduction of the T24 H-ras1 gene into F4-12B2TK- mouse erythroleukemic cells interferes with differentiation as shown by the failure of about 80% of transfectants to differentiate when induced by hexamethylene bis-acetamide (HMBA) (11). The same mutant ras gene also inhibited adipocytic cell differentiation (21). Activated H-ras1 gene prevents mouse muscle myoblast differentiation (22) and inhibits myogenesis of the muscle cell line BC3H1 (23). On the other hand it was found that the Harvey and Kirsten sarcoma viruses promote the growth and differentiation of erythroid precursor cells in vitro (24). Therefore, depending on the assay system the same ras gene can promote, have no effect or block differentiation.

CONCLUDING REMARKS

Controlled transfer and expression of oncogenes into normal cells and analysis of their products by the use of monoclonal antibodies to the ras proteins has led to a better understanding of several steps in the multistage process of carcinogenesis. Because of the involvement of activated ras oncogenes in human cancer, it is important to determine the normal function of proto-oncogene and how this is affected by the alterations which occur during oncogenic conversion.

Rat pheochromocytoma PC12 cells are a good model system for studying differentiation since they can be induced to differentiate into neuron like cells by nerve growth factor. In the present study we have examined the effects on differentiation of ras and myc genes. We have found that whereas the human mutant T24 H-ras1 gene can induce differentiation the human myc gene blocks differentiation induced by nerve growth factor. Further studies are being undertaken to investigate the mechanisms of these effects.

ACKNOWLEDGEMENTS

The Beatson Institute for Cancer Research is supported by the Cancer Research Campaign of Great Britain and the National Hellenic Research Foundation by the Secretariat for Research and Technology of Greece. M.L.M.A. is supported by the Medical Research Council of Great Britain.

REFERENCES

1. R. Muller. Proto-oncogenes and differentiation. TIBS 11: 129 (1986).
2. L.S. Mulcahy, M.R. Smith and D.W. Stacy. Requirement for ras proto-oncogene function during serum-stimulated growth of NIH 3T3 cells. Nature 318: 73 (1985).
3. J.R. Feramisco, R. Clarc, G. Wong, N. Arheim, R. Milley and F. McCormick. Transient reversion of ras oncogene-induced cell transformation by antibodies specific for amino acid 12 of ras protein. Nature 314: 639 (1985.
4. D. Bar-Sagi and J.R. Feramisco. Microinjection of the ras proteins into PC12 cells induces morphological differentiation. Cell, 42: 841 (1985).
5. M. Noda, M. Ko, A. Ogura, D.-G. Liu, T. Amano and Y. Ikawa. Sarcoma viruses carrying ras oncogenes induce differentiation-associated properties in a neuronal cell line. Nature 318: 73 (1985).
6. D.A. Spandidos. A unified theory for the development of cancer. Bioscience Rep. 6: 691 (1986).
7. L.A. Greene and A.S. Tischler. Establishment of a noradrenergic clonal line of rat adrenal pheochromocytoma cells which respond to nerve growth factor. Proc. Natl. Acad. Sci. USA 73: 2424 (1976).
8. D.A. Spandidos and N.M. Wilkie. Malignant transformation of early passage rodent cells by a single mutated human oncogene. Nature 310: 469 (1984).
9. D.A. Spandidos. Mechanisms of carcinogenesis. The role of oncogenes, transcriptional enhancers and growth factors. Anticancer Res. 5: 485 (1985).
10. D.A. Spandidos. Electric field mediated-gene transfer (electro- poration) into mouse Friend and human K562 eryuthroleukemic cells. Gene Anal. Tech. 40: 50 (1987).
11. D.A. Spandidos and M.L.M. Anderson. A study of mechanisms of carcinogernesis by gene transfer of oncogenes into mammalian cells. Mutat. Res. 185: 271 (1987).
12. I.B. Pragnell, D.A. Spandidos and N.M. Wilkie. Consequences of altered oncogene expression in rodent cells. Proc. Roy. Soc. London, U.K. B226: 107 (1985).
13. M. Yiagnisis and D.A. Spandidos. Interleukin 3 like activity secreted from human ras or myc transfected rodent cells. Anticancer Res. 7, 1293 (1987).
14. M.B. Sporn and G.J. Todaro. Autocrine secretion and malignant transformation of cells. New Engl. J. Med. 303, 878 (1980).
15. R.M. Kris, I. Lax, W. Gullick, M.S. Waterfield, A. Ullrich, M. Fridkin and J. Schlessinger. Antibodies against a synthetic peptide as a probe for the kinase activity of the avian EGF receptor and v-erbB protein. Cell 40: 619 (1985).

16. D.A. Spandidos. Ras oncogenes in cell transformation. ISI Atlas of Science. Immunology. 1:1 (1988).
17. D.A. Spandidos. The effect of exogenous ras and myc oncogenes in morphological differentiation of the rat pheochromocytoma PC12 cells. Int. J. Dev. Neurosci. In Press (1989).
18. I. Guerrero, A. Pellicer and D.E. Burstein. Dissociation of c-fos from ODC expression and neuronal differentiation in a PC12 subline stably transfected with an inducible N-ras oncogene. Biochem. Biophys. Res. Comm. 150: 1185 (1988).
19. K. Maruyama, S.C. Schiavi, W. Huse, G.L. Johnson and H.E. Ruley. Myc and E1A oncogenes alter the responses of PC12 cells to nerve growth factor and block differentiation. Oncogene 1: 361 (1987).
20. J.C. Bell, K. Jardine and M.W. McBurney. Lineage-specific transformation after differentiation of multipotential murine stem cells containing a human oncogene. Mol. Cell. Biol. 6: 617 (1986).
21. R. Gambari and D. Spandidos. Chinese hamster lung cells transformed with the human HA-RAS-1 oncogene: 5 azacytidine mediated induction of adipogenic conversion. Cell Biol. Int. Rep. 10: 173 (1986).
22. E.N. Olson, G. Spizz and M.A. Tainsky. The oncogenic forms of N-ras or H-ras prevent skeletal myoblast differentiation. Mol. Cell. Biol. 7: 2104 (1987).
23. J.M. Caffrey, A.M. Brown and M.D. Schneider. Mitogens and oncogenes can block the induction of specific voltage-gated ion channels. Science 236: 570 (1987).
24. W.D. Hankins and E.M. Scolnick. Harvey and Kirsten sarcoma viruses promote the growth and differentiation of erythroid precursor cells in vitro. Cell 26:91 (1981).

NEURONAL LOCALIZATION OF THE NERVE GROWTH FACTOR PRECURSOR-LIKE IMMUNOREACTIVITY IN THE CORTEX AND HIPPOCAMPUS OF THE RAT BRAIN

Marie-Claude Senut[1], Yvon Lamour[1], Philippe Brachet[2] and Eleni Dicou[2]*

[1]INSERM U 161, 2, rue d'Alésia, Paris, France
[2] INSERM U 298, Centre Hospitalier Régional, 49033 Angers, France

INTRODUCTION

Several lines of evidence have indicated that the nerve growth factor (NGF) is active not only in the peripheral nervous system but also in the brain. Exogenously administered NGF induced choline acetyltransferase in neonatal rat cortex, hippocampus, septum[1] striatum[2] and in lesioned mature septum[3]. NGF receptors were also found in the brain[4,5], while NGF mRNA and protein were detected in brain regions which are the targets of the NGF-responsive cholinergic neurons of the basal forebrain nuclei [6-10]. NGF injected into cortical and hippocampal regions is transported retrogradely to the nucleus basalis and the medial septum-diagonal band of Broca region respectively[11].

In order to identify NGF-synthesizing cells in the peripheral and central nervous system we have raised antibodies to two synthetic peptides that reproduce sequences of the NGF precursor molecule. These antisera cross-react with the NGF precursor as verified by ELISA techniques using a chimeric prepro NGF protein synthesized in Escherichia Coli[12]. These antisera were also shown to stain the basal portion of the cells forming the granular convoluted tubules of the submandibular gland of the mouse[13,14] (MSG) which are known to be the site of synthesis of this factor as well as some parafollicular cells of the thyroid and cells in the parathyroid gland of the rat[13].

Evidence that these sera cross-react with the precursor protein was further provided by in situ hybridization experiments which indicated colocalization of the NGF mRNA and proNGF-like immunoreactivity in the MSG[14].

We present results of proNGF-like immunoreactivity detected with these affinity-purified sera in the cortex and hippo-

*to whom correspondence should be addressed.

Molecular Aspects of Development and Aging of the Nervous System
Edited by J.M. Lauder
Plenum Press, New York, 1990

campus of the adult rat. Retrograde tracing experiments using wheat germ agglutinin apohorseradish peroxidase conjugate coupled to colloidal gold (WGA-apoHRP-gold) suggest that the immunoreactive material is localized in neurons.

MATERIAL AND METHODS

Sprague-Dawley rats (250-300 g) were anesthetized with ketamine and were sacrificed by perfusion through the ascending aorta, with 100 ml of 0.1M phosphate buffer saline (PBS), followed by 400 ml of 4% paraformaldehyde with 0.1M lysine HCl and 0.01M sodium metaperiodate in 0.1M phosphate buffer (PB) [15]. After perfusion, brains were dissected out, postfixed at 4°C for an additional 3 to 4 hours and cryoprotected in a 30% sucrose solution in PB for at least 24 hours. Coronal 40 um thick sections were cut on a freezing microtome and collected in PB.

Antisera

Polyclonal rabbit sera were raised against a chemically synthesized 26-amino acid peptide, N4, that corresponds to the sequence (-71 to -46) and a 10-amino acid peptide, N3, that corresponds to the sequence (-49 to -40) of the NGF precursor protein[16]. Rabbits were immunized with peptides coupled to keyhole limpet hemocyanin as previously described[14]. Antibodies to synthetic peptides were purified by affinity chromatography over a column of peptide coupled to AH-Sepharose (Pharmacia). Adsorbed immunoglobulins were eluted with 0.1M glycine-HCl, pH 2.3 and quickly neutralized with 1M K_2HPO_4. The eluate and the unadsorbed fractions were tested by an enzyme-linked immunoassay for the presence or absence of specific antibodies[12,14].

Immunohistochemical procedure

Antisera were diluted in PBS containing 1% normal goat serum and 0.3% Triton X100 at a protein concentration of about 10-50 µg/ml and applied on free floating sections for 16 hours to 3 days at 4°C. The avidin-biotin-peroxidase (ABC) technique described by Hsu et al.[17] was employed. The tissue bound peroxidase was visualized by using 3-3' diaminobenzidine (DAB; 10mg/50ml) and 0.03% hydrogen peroxide in PB. The DAB reaction was monitored under the microscope in order to reduce the background staining. Sections were rinsed in buffer, collected on slides, air-dried, embedded in Eukitt and examined with a light microscope.

Retrograde tracing combined with immunohistochemistry

Seven animals received stereotaxic injections of a wheat germ agglutinin apohorseradish peroxidase conjugate coupled to 10 nm colloidal gold particles[18,19] into the lateral septum, ventro-basal complex of the thalamus or the cervical level of the spinal cord. Pressure injections were performed through a glass micropipette (tip diameter : 40 µm) attached to a 1 µl Hamilton syringe. Animals were allowed to survive from 2 to 7 days; brains were perfused, processed and sectioned as above. In order to visualize the retrogradely labeled neurons containing colloidal gold, sections were submitted to a silver intensification protocol immediately before immunohistochemistry.

RESULTS

Topographical distributions of anti-peptide immunoreactivity in the rat cortex and hippocampus

The cellular distrubution patterns obtained in the rat cortex and hippocampus with anti-peptide N_3 and N_4 sera were rather similar. The majority of the immunolabeling was in the form of a diffuse reaction product accumulated in the cell somata but a punctate labeling circling cell bodies was also observed. Strong immunoreactivity was consistently observed in the fronto-parietal and cingular cortices in cell bodies almost exclusively located in layer V and VIb (Fig. A). In layer V large pyramidal cell somata and their apical dendrites coursing perpendicularly to the pial surface were immunoreactive while in layer VIb immunolabeled cell somata appeared smaller and were oriented tangentially to the cortical surface. Scattered stained cell somata were also present in layer VIa and, in some animals, in layer IV. A significant number of N_3 positive cells were also observed in layer II although their number was smaller to that observed in layer V. Some immunolabeling, unlike with anti-peptide N_4 serum, was also present in the pyriform cortex.

In the hippocampal formation strong immunoreactivity was observed within the pyramidal cells and was characterized by a gradient of intensity with the strongest immunolabeling observed in the CA3-CA4 subfields and decreasing to a much weaker staining in subfield CA1 (Fig. B). Some scattered positive cells were present in the stratum oriens, the stratum radiatum and the stratum lacunosum-moleculare. The granular layer of the dentate gyrus was also immunoreactive but to a much lesser extent than hippocampal subfields CA3-CA4 and the hilar polymorphic layers.

Fiber immunoreactivity

The majority of the immunolabeling was found in cellular profiles, however a consistent staining of fibers was also noticed with both anti-peptide sera especially within the corpus callosum.

Specificity of the immunoreactivity

Several control experiments were performed to verify the specificity of the sera. 1) Affinity-purified anti-peptide immunoglobulins were routinely tested by ELISA for their specificity to each peptide. 2) Unretained serum fractions after passage through peptide-Sepharose columns were tested by ELISA for the absence of any anti-peptide immunoglobulins and sections tested with such fractions were negative. 3) Immunoglobulins were purified from sera of non-immunized rabbits by protein-A sepharose chromatography and when tested at the same protein concentration as the anti-peptide sera showed no immunoreactivity. 4) Each of the immunoreagents in the immunohistochemical sequence were omitted and staining was abolished. 5) The intensity of the histochemical staining was diminished after preadsorption of the affinity-purified immunoglobulins with their corresponding peptide.

Neuronal localization of the proNGF immunoreactivity

In order to assess the type of cells that are immunoreactive with the anti-peptide sera WGA-apoHRP-gold was

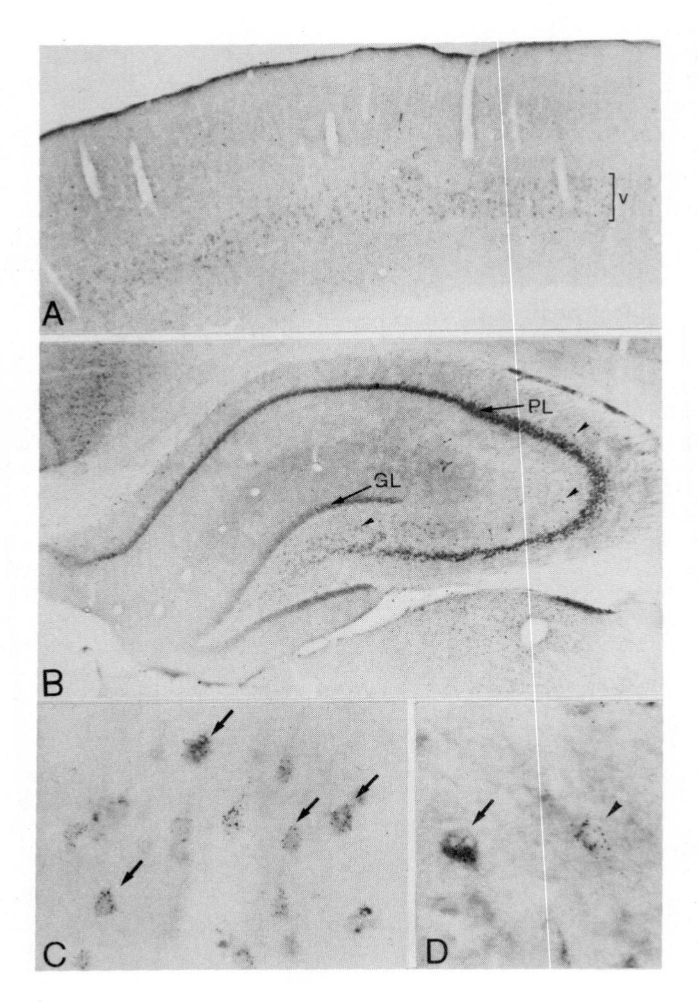

Fig. 1. ProNGF-like immunoreactivity in the cortex and hippocampus of the rat brain. A : Low-power view of a section of the cerebral cortex stained with anti-peptide N4 serum. Immunoreactive cell bodies are concentrated particularly in layer V, x 36. B : Low-power view of a section through the hippocampal formation stained with the same antiserum. Immunoreactive cell somata are present in all subfields of the hippocampal pyramidal layer (PL) and in the granular layer (GL) of the dentate gyrus. Some scattered positive cells are present in other regions of the hippocampal formation (arrowheads), x 40. C : High magnification of double-labeled neurons (arrows) in the cortical layer V after a tracer injection into the cervical spinal cord. The immunolabeling is diffusely ditributed within the cell bodies whereas the punctate retrograde labeling is due to the silver intensification procedure and is clearly visible on the background as black dots, x 214. D : High magnification of double-labeled neurons in the CA3 subfields of the hippocampal pyramidal layer after a tracer injection in the lateral septal nucleus. The black arrow indicates a double-labeled neuron; the arrowhead shows a retrogradely labeled neuron with no immunoreactivity, x 300.

injected in various sites of the brain and the distribution pattern of the retrogradely labeled neurons was compared to the immunoreactive cells with the antipeptide sera in the same section (double labeling experiments). Tracer injections in the ventrolateral and the ventroposterior lateral thalamic nuclei resulted in retrograde labeling of numerous neurons in layer VIa and to a lesser extent in layer V of the neocortex. Some double-labeled neurons were observed, especially, in layer V and more occasionally in layer VI. A tracer injection was performed in the cervical spinal cord, at the C4-C5 level, including mostly the dorsal horn but also more ventral regions including lamina VII and resulted in the retrograde labeling of numerous neurons in layer V. Almost all of these neurons were also immunoreactive (Fig. C).

A large tracer injection was performed in the dorsal and intermediate parts of the lateral septal nucleus in order to retrogradely label neurons within the hippocampal formation. Numerous double-labeled neurons were observed in the hippocampus and dentate gyrus (Fig. D). In the hippocampus, they were mainly localized in the pyramidal layer; however the retrograde labeling is so dense in this layer that it is not certain that every retrogradely labeled neuron is also immunoreactive. In the other hippocampal layers, as well as in the dentate gyrus layers, scattered double-labeled cells mixed with populations of singly immunoreactive or retrogradely labeled neurons.

CONCLUSION

In the present study we used antisera to synthetic peptides that reproduce sequences of the proNGF protein in an attempt to localize NGF precursor-like immunoreactivity in the cortex and hippocampus of the adult rat and to identify cell types which synthesize it. Both affinity-purified antisera gave relatively comparable distributions of immunoreactive material in these brain regions, as it would be expected if they cross-react with a commonly recognized antigen. ProNGF-like immunoreactivity was detected in the cortex, known to contain high amounts of NGF mRNA[6-10], and was essentially localized in layers V and VIb. ProNGF-like immunoreactivity was also detected in the pyramidal layer and the dentate gyrus of the hippocampus. In the pyramidal layer, subfields CA3-CA4 displayed more intense immunoreactivity as compared to CA1-CA2. This is in agreement with previous NGF mRNA measurements which indicated high NGF mRNA levels in the hippocampus, especially in subfields CA3-CA4[6-10] and with *in situ* hybridization studies which identified NGF producing cells in neurons of the pyramidal layer and the dentate gyrus[20-22]. Furthermore, retrograde tracing using WGA-apoHRP-gold combined with immunohistochemistry demonstrated localization of the immunoreactive material in retrogradely labeled neurons. Thus, another conclusion of this study is the localization of the proNGF-like immunoreactivity in neurons in the cortex and hippocampal formation.

REFERENCES

1. H. Gnahn, F. Hefti, R. Heumann, M. Schwab and H. Thoenen. NGF-mediated increase of choline acetyltransferase (ChAT) in the neonatal forebrain : evidence for a physiological role of NGF in the brain ? *Dev. Brain Res.* 9:45-52 (1981).

2. W.C. Mobley, J.L. Rutkowski, G.I. Tennekoon, K. Buchanan and M. V. Johnston. Choline acetyltransferase activity in striatum of neonatal rats increased by nerve growth factor. Science 229:284-287 (1985).

3. F. Hefti, A. Dravid and J.Hartikka. Chronic intraventricular injections of nerve growth factor elevate hippocampal choline acetyltransferase activity in adult rats with partial septo-hippocampal lesions. Brain Res. 293:305-309 (1984).

4. P.M. Richardson, V.M.K. Verge Issa and R.J. Riopelle. Distribution of neuronal receptors for nerve growth factor in the rat. J. Neuroscience 6(8):2312-2321 (1986).

5. M. Taniuchi, J.B. Schweitzer and E.M. Johnson. Nerve growth factor receptor molecules in the rat brain. Proc. Natl. Acad. Sci. USA 83:1950-1954 (1986).

6. S. Korsching, G. Auburger, R. Heumann, J. Scott and H. Thoenen. Levels of nerve growth factor and its mRNA in the central nervous system of the rat correlate with cholinergic innervation. EMBO J. 4(6):1389-1393 (1985).

7. D.L. Shelton and L.F. Reichardt. Studies on the expression of the nerve growth factor (NGF) gene in the central nervous system : Level and regional distribution of NGF mRNA suggest that NGF functions as a trophic factor for several distinct populations of neurons. Proc. Natl. Acad. Sci. USA 83:2714-2718 (1986).

8. S.R. Whittemore, T. Ebendal, L. Lärkfors, L. Olson, A. Seiger, I. Strömberg and H. Persson. Developmental and regional expression of nerve growth factor messenger RNA and protein in the rat central nervous system. Proc. Natl. Acad. Sci. USA 83:817-821 (1986).

9. T.H. Large, S.C. Bodary, D.O. Clegg, G. Weskamp, U. Otten and L.F. Reichardt. Nerve growth factor gene expression in the developing rat brain. Science 234:352-355 (1986).

10. M. Goedert, A. Fine, S.P. Hunt and A. Ullrich. Nerve growth factor mRNA in peripheral and central rat tissues and in the human central nervous system : Lesion effects in the rat brain and levels in Alzheimer's disease. Molec. Brain Res. 1:85-92 (1986).

11. M.E. Schwab, U. Otten, Y. Agid and H. Thoenen. Nerve growth factor (NGF) in the rat CNS : absence of specific retrograde axonal transport and tyrosine hydroxylase induction in locus coeruleus and substantia nigra. Brain Res. 168:473-483 (1979).

12. E. Dicou, R. Houlgatte, J. Lee and B. von Wilcken-Bergmann. Synthesis of chimeric mouse nerve growth factor precursor and human β-nerve growth factor in Escherichia Coli : Immunological properties. J. Neurosc. Res., in press (1988).

13. E. Dicou, J. Lee and P. Brachet. Synthesis of nerve growth factor mRNA and precursor protein in the thyroid and parathyroid glands of the rat. Proc. Natl. Acad. Sci. USA 83:7084-7088 (1986).

14. E. Dicou, J. Lee and P. Brachet. Co-localization of the nerve growth factor precursor protein and mRNA in the mouse submandibular gland. Neuroscience Lett. 85:19-23 (1988).

15. I.W. Mc Lean and P.K. Nakane. Periodate-lysine paraformaldehyde fixative. A new fixative for immunoelectron microscopy. J. Histochem. Cytochem. 22:1077-1083 (1974).

16. A. Ullrich, A. Gray, C. Berman and T.J. Dull. Human β-nerve growth factor gene sequence highly homologous to that of mouse. Nature 303:821-825 (1983).

17. S. Hsu, L. Raine and H. Fanger. A comparative study of the peroxydase-antiperoxydase method and an avidin-biotin complex method for studying polypeptide hormones with radioimmunoassay antibodies. Am. J. Clin. Pathol. 75:734-738 (1981).

18. A.I. Basbaum and D. Menetrey. Wheat germ agglutinin-apoHRP-gold : a new retrograde tracer for light- and electron-microscopic single and double-label studies. J. Comp. Neurol. 261:306-318 (1987).

19. D. Menetrey. Retrograde tracing of neural pathway with a protein-gold complex. I. Light microscopic detection after silver intensification. Histochemistry 83:391-395 (1985).

20. P.D. Rennert and G. Heinrich. Nerve growth factor mRNA in brain : Localization by in situ hybridization. Biochem. Biophys. Res. Commun. 138:813-818 (1986).

21. C. Ayer-LeLièvre, L. Olson, T. Ebendal, A. Seiger and H. Persson. Expression of the β-nerve growth factor gene in hippocampal neurons. Science 240:1339-1341 (1988).

22. S.R. Whittemore, P.L. Friedman, D. Larhammar, H. Persson, M. Gonzalez-Carvajal and V.R. Holets. Rat β-nerve growth factor sequence and site of synthesis in the adult hippocampus. J. Neur. Res. In press (1988).

DEVELOPMENT OF NILE GLYCOPROTEIN IN CHICK BRAIN

Anna Batistatou and Elias D. Kouvelas

Department of Physiology, School of Medicine
Uviversity of Patras
Patras, Greece

INTRODUCTION

The nerve growth factor (NGF)-inducible large external (NILE) glycoprotein is a cell surface component and it was first identified in the NGF-induced neuronal differentiation of the PC12 pheochromocytoma cell line[1,2]. It has been shown to be widely distributed in the mammalian central and peripheral nervous system, but not in nonneural tissues[3]. Recent reports have indicated a role for this molecule in fasciculation of neurites in culture, and its appearance in the developing mammalian central nervous system appears to be correlated with genesis of fiber tracts[4,5,6,7]. Furthermore reports from several groups[7,8,9,10] have indicated that this molecule is immunologically indistiguishable from the high-molecular weight component of the brain L1 antigen and the neuron-glia cell adhesion molecule (Ng-CAM). Both of these antigens are also believed to have a role in regulating cell-cell adhesion and fasciculation in the developing nervous system[10,11,12]. With these considerations in mind we became interested in the understanding of the mechanisms which regulate NILE development in the central nervous system. A first step towards the understanding of this kind of regulatory mechanism, is the study of the spatiotemporal distribution of the protein under investigation in a specific tissue. For such developmental studies chick brain appears to provide certain advantages. Chick brain develops mostly during the in ovo period of life and is therefore a closed system, which offers also the opportunity for making experimental manipulations during embryogenesis. Furthermore certain areas of chick brain have been well characterized developmentally and anatomically. The results which we report here describe the spatiotemporal distribution of NILE in chick cerebellum and optic lobes.

MATERIALS AND METHODS

Brains from chicks at ages E9-P25 were fixed for 24 hrs in Karnovsky fixative in cold and paraffin horizontal sections (5μm) of cerebellum and optic lobes were cut. Sections on glass

Molecular Aspects of Development and Aging of the Nervous System
Edited by J.M. Lauder
Plenum Press, New York, 1990

slides were heat-fixed for 2 hrs at 58° and rehydrated using xylene 5min, followed by 100% ethyl alcohol 2X3 min, 95% ethyl alcohol 2X3 min, and PBS buffer (pH 7,4) 5 min. Sections were then incubated in 3% hydrogen peroxide for 5 min and washed 2X10 min with PBS; then incubated with PBS containing 5% normal serum for 20 min and washed 2X5 min with PBS. Sections were incubated overnight in the cold with rabbit anti-NILE raised against protein purified from NGF-treated PC12 cells on SDS gels[3,4], kindly offered by Drs M. Shelanski and L.A. Greene; then washed 2X10 min with PBS: incubated with anti-rabbit immunoglobulin (linking reagent) for 20 min and washed with PBS. Sections were finally incubated with peroxidase-anti-peroxidase complex for 20 min and washed 2X10 min with PBS. Color was developed with 3-amino-9-ethyl carbazole. All immunostaining reagents were purchassed from Ortho, New Jersey.

RESULTS

In the cerebellum staining is evident in the white matter and within the external part of the molecular layer by embryonic day 15. At embryonic day 18 staining becomes intense in the developing molecular layer. At embryonic day 21 staining is very intense in the molecular layer and the white matter. Similar staining is detected at post-hatching day 1 and it persists until post-hatching day 5. After this stage staining becomes more diffuse, it declines in the molecular layer but it persists in the white matter. No staining was detected in the layer of the Purkinje cell bodies (Fig. 1A-I).

In the optic lobes of embryonic day 9 several cellular elements in the periphery (probably post-mitotic neurons which have finished their migration) and in the periventricular layers of the developing optic lobes showed a weak staining. By E12 a very weak staining for NILE GP was evident in two rather broad layers of the superficial part of the optic tectum which were separated by a broad layer which was not stained and which at this age probably corresponds to the immature cellular layer g[13]. At this age staining was more evident in the fibrous area of the tectal ventricle which corresponds to the immature stratum album centrale. Similar pattern was observed at embryonic days15 and 18 with increasing amounts of staining in the fibrous deep layers of the developing optic tectum. By hatching when the avian optic tectum is fully developed one could observe a very distinct pattern of staining. Staining was very intense in the superficial plexiform and cellular layers of the optic tectum up to cellular layer g which was not stained. Staining was also intense in layer h, which contains the neurites of cellular layer g, it decreased in the cellular layer i and it was also very intense in the fibers of stratum album centrale. No staining for NILE GP could be detected in the ependymal layer around the ventricle. This pattern of staining remained until post-hatching day 5. After this age an increased diffuse staining was observed which was more intense in the stratum opticum and in the fibers of stratum album centrale and was not extended in the periventricular ependymal layers (Fig. 1J-O).

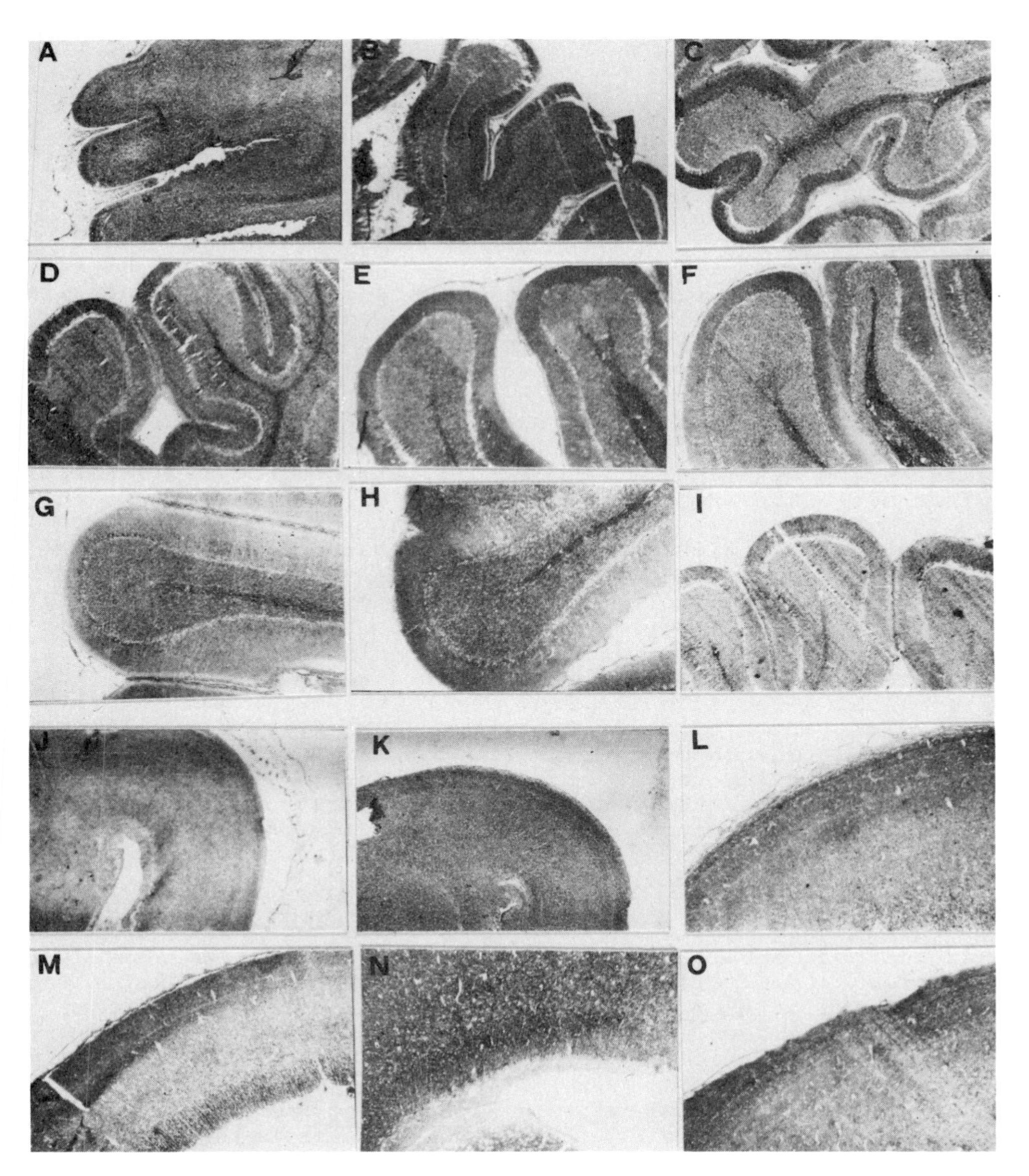

Fig. 1. Staining with anti-NILE antibody of chick cerebellum (A-I) and optic tectum (J-O). A:E15 cerebellum staining is seen in the white matter and the external part of the molecular layer. B-F: E18, E21, P1, P3, P5 cerebellum. Staining is intense in the white matter and the molecular layer. G-H: P10, P25 cerebellum. Staining is seen in the white matter. I:P1 cerebellum stained with control rabbit preimmune serum at 1:100 dilution. J-K: E15-E18 optic tectum. Staining is seen in two layers of the external part and in the fibers of SAC. L:P1 external part of optic tectum. Staining covers SO, and the layers of SGFS until layer h.A thin layer (g) is not stained. M:P5 optic tectum. Staining is intense in the fibers of SAC and the layers of SGFS as until layer h. Layer g is not stained. N:P25 deep layers of the optic tectum. O: Superficial layers of the optic tectum. Staining persists in the fibers of SAC and in SO. Magnification 20X.

DISCUSSION

The development of chick cerebellar cortex has been studied by several investigators[14,15] and their results can be briefly described as follows: The differentiation of granule cells begins at 11 to 13 days of incubation when Purkinje cells are disposed in groups just under the external granular layer. Bundles of parallel fibers are seen between Purkinje elements and external granules at the 11 to 13 days in ovo. The number of parallel fibers rises then until some days after hatching with a most marked increase from day 15 to 20 of incubation. This is approximately the period when the external granular layer most rapidly decreases in thickness and the molecular and internal granular layers show a rapid devepolment. Furthermore by hatching, a significant number of synaptic contacts can be detected between parallel fibers and the dentritic spines of Purkinje cells.

The avian optic lobe is itself composed of the optic tectum and a region subjacent to the tectal ventricle which includes among others three nuclei. The development of the chick optic tectum may be divided broadly into three, somewhat overlapping, phases. An initial phase extending from day E3 to E6, characterized by the rapid proliferation of cells in the neural epithelium. A second phase from day E6 to E12 is marked by the outward migration of cells from the neural epithelium. The third phase in the development of the tectum extends from embryonic day 12 to maturity and is marked by the rapid growth of the cells and fiber systems and the establishment of final synaptic contacts. By embryonic day 12 retinal fibers have begun to invade the outer layers of the optic tectum and retinal tectal synapses begin to appear[13,16,17].

With these considerations in mind our results have shown that antibodies raised against NILE-GP purified from PC12 pheochromocytoma cells can stain the developing cerebellum and optic tectum of the chick in a specific manner. Our study of spatiotemporal development of this protein has shown that it is expressed on postmitotic neurons and is subsequently associated with the propagation of neuronal axons, formation of neuronal fiber bundles and development of synaptic connections. These results are similar to those previously described for the expression of NILE/L1 antigens in mouse cerebellum[6] and in embryonic spinal neurons[7]. Our results have shown also that the pattern of development of NILE glycoprotein in chick cerebellum differs to that of neurofascin and N-CAM but it is very similar with the pattern of spatiotemporal development of G4 antigen[18,19]. This result suggests that antibodies raised against NILE glycoprotein of PC12 cells recognize the G4 antigen of chick brain, a finding which could be expected given the molecular similarities of G4 and L1 antigens[20]. Finally our results have shown the pattern of spatiotemporal development of NILE is similar but not identical with the pattern of staining previously described for Ng-CAM[18]. The major differences are that the anti Ng-CAM antibody staining of SO is very intense by embryonic day 8 and is absent in the abult brain, whereas with the anti-NILE staining of SO appears during the late stages of embryonic development and it persists until maturity. Immunoblotting studies have shown that the anti-NILE and anti-Ng-CAM antibodies recognize different polypeptide chains from chick brain. Thus the anti-NILE antibody which was used in our studies was raised against PC12 NILE glycoprotein isolated as single 230KD band on SDS gels and recognized two major bands of 180KD and 140KD on immunoblots of P2 optic tectum (J. Zarkadis and E.D. Kouvelas unpublished results), whereas the anti-Ng-CAM antibody recognized on immunoblots of E18 chick optic tectum a major band of 135KD and a minor band of 200KD[18]. Thus it seems possible that the differences which were detected on the developmental pattern of the two antigens reflect a differential temporal expression of the different polypeptide forms of the axonal surface glycoproteins. The isolation of cDNA clones encoding these proteins and the use of these clones for the study of the expression of the related messengers should resolve the relationship between the different polypeptide forms and their function as regulators of the different developmental processes. Using a NILE related cDNA probe isolated from a PC12 cDNA library[21] we have recently detected a cDNA probe from a λgt11 cDNA library of two days old chick brain (J. Zarkadis, A. Athanassiadou, D. Thanos, J. Papamatheakis and E.D. Kouvelas in preparation). Sequencing of this probe demonstrated distinct similarities with the recently reported nucleotide sequence of L1 antigen[22] and therefore we believe that it will be a very useful tool for our future studies.

ACKNOWLEDGEMENTS

Supported by a grant (E.D.K) of the Secretariat of Research and Technology of Greece.

REFERENCES

1. J. C. McGuire, L.A. Greene and A.V. Furano, NGF stimulates incorporation of fucose or glucosamine into an external glycoprotein in cultured rat PC12 pheochromocytoma cells. Cell 15:357 (1978).
2. S. R. J. Salton, M.L. Shelanski and L.A. Greene, Biochemical properties of the nerve growth factor-inducible large external (NILE) glycoprotein J. Neurosci. 3:2420 (1983).
3. S. R. J. Salton, C. Richter-Landsberg, L.A. Greene and M. L. Shelanski, Nerve growth factor inducible large external (NILE) glycoprotein: studies of a central and peripheral neuronal marker, J. Neurosci. 3:441 (1983).
4. W. B. Stallcup and L. Beasley, Involvement of the NGF-inducible large external glycoprotein (NILE) in neurite fasciculation in primary cultures of rat brain, PNAS 82:1276 (1985).
5. W. B. Stallcup, L. Beasley and J.M. Levine, Antibody against NGF-inducible large external (NILE) glycoprotein labels nerve fiber tracts in developing rat nervous system, J. Neurosci. 5:1090 (1985).
6. P. Sajovic, E. Kouvelas and E. Trenkner, Probable identity of NILE glycoprotein and the high-molecular-weight component of L1 antigen, J. Neurochem. 47:541 (1986).
7. J. Dodd, S.B. Morton, D. Karagogeos, M. Yamamoto and T.M. Jessell, Spatial regulation of axonal glycoprotein expression on subsets of embryonic spinal neurons, Neuron 1:105 (1988).
8. E. Bock, C. Richter-Landsberg, A. Faissner and M. Schachner, Demonstration of immunochemical identity between the

NGF-inducible large external (NILE) glycoprotein and the cell adhesion molecule L1, EMBO J. 4:2765 (1985).

9. D. R. Friedlander, M. Grumet and G.M. Edelman. NGF induces expression of Ng-CAM in PC12 cells, J. Cell Biol. 102: 413 (1986).

10. M. Schachner, A. Faissner, J. Kruse, J. Lindner, D.H. Meier, F.G. Rathjen and H. Weinecke, Cell type specificity and developmental expression of neuronal cell surface components involved in cell interactions and of structually related molecules, Cold Spring Harbor Symp. Quant. Biol. 48:557 (1983).

11. S. Hoffman, D.R. Friedlander, C-M. Chuong, M. Grumet and G. M. Edelman, Differential contributions of Ng-CAM and N-CAM to cell adhesion in different neural regions, J. Cell Biol. 103:145 (1986).

12. C-M. Chuong, K.L. Crossin and G.M. Edelman, Sequential expression and differential function of multiple adhesion molecules during the formation of cerebellar cortical layers, J. Cell Biol. 104:331 (1987).

13. J. H. LaVail and W.M. Cowan. The development of the chick optic tectum. I. Normal morphology and cytoarchitectonic development, Brain Res. 28:391 (1971).

14. R. Foelix and R. Oppenheim, The development of synapses in the cerebellar cortex of chick embryo. J. Neurocyt. 3:277 (1974).

15. E. Mugnaini and P.J. Forstronen, Ultrastructual studies on the cerebellar histogenesis I. Differentiation of granule cells and development of gromeruli in the chick embryo, Z. Zellforsch. Mikrosk. Anat. 77:115 (1967).

16. D. Cantino and L.S. Daneo, Synaptic junctions in the developing chick optic tectum, Experientia 29:85 (1973).

17. W. J. Crossland, W.M. Cowan and L.A. Rogers, Studies on the development of chick optic tectum. IV An autoradiographic study on the development of retino-tectal connections, Brain Res. 91:1 (1975).

18. J. K. Daniloff, C-M. Chuong, G. Levi and G.M. Edelman, Differential distribution of cell adhesion molecules during histogenesis of the chick nervous system, J. Neurosci.6:739 (1986).

19. F. G.Rathjen, J.M. Wolff, S. Chang, F. Bonhoeffer and J.A. Raper, Neurofascin: A novel chick cell-surface glycoprotein involved in neurite-neurite interactions, Cell 51:841 (1987).

20. F. G. Rathjen, J.M. Wolff, R. Frank, F. Bonhoeffer and U. Rutishauser, Membrane glycoprotein involved in neurite fasciculation, J. Cell Biol. 104:343 (1987).

21. P. Sajovic, D.J. Ennulat, M.L. Shelanski and L.A. Greene, Isolation of NILE glycoprotein-related cDNA probes, J. Neurochem. 49: 756 (1987).

22. M. Moos, R. Tacke, H. Scherer, D. Teplow, K. Fruh and M. Schachner, Neural adhesion molecule L1 as a member of the immunoglobulin superfamily with binding domains similar to fibronectin, Nature 334:701 (1988).

STRUCTURAL AND MOLECULAR POLARITY IN RETINAL PHOTORECEPTOR NEURONS: ROLES FOR THE CYTOSKELETON

Ruben Adler and Steven A. Madreperla

The Johns Hopkins University
600 North Wolfe Street
Baltimore, MD 21205

I. INTRODUCTION

The differentiation of neuronal cells involves the synthesis of cell-specific macromolecules, as well as the assembly of the latter into supramolecular entities responsible for complex patterns of cell organization. For example, axons are distinguished from dendrites by the presence or absence of ribosomes and by the specific distribution of molecules such as cytoskeletal proteins, α-bungarotoxin receptors and GAP-43, among others (Banker and Waxman, 1988; Goslin et al., 1988). Asymmetries in molecular distribution also occur within individual neuronal processes, as illustrated by the concentration of voltage-sensitive sodium channels at the nodes of Ranvier (Ritchie, 1977; Angelides, et al., 1988). Although all neurons seem to share a similar polarized pattern of organization, different types of neurons can be distinguished by unique variations of this common pattern. Cell-specific morphology and branching patterns of axons and dendrites provide clear examples of this type of diversity (i.e., Wuerker and Kirkpatrick, 1972; Peters et al., 1976; Bartlett and Banker, 1984).

The contributions of intracellular determinants and micro-environmental influences to the generation and maintenance of neuronal form and polarity are not fully understood. In vitro studies have shown that neuroblastoma cells (Solomon and Zurn, 1980) and isolated neurons (eg, Scott et al., 1969; Fischbach, 1970; Dichter, 1978; Bartlett, et al., 1984; among others) can express cell type-specific neurite patterns even when cultured in the absence of intercellular contacts, suggesting a protagonistic role for intracellular determinants. On the other hand, glial cells and extracellular matrix molecules have been suggested not only to affect the capacity of some types of neurons to develop and maintain specific patterns of morphological organization, but also to be involved in the generation of molecular asymmetries such as Na^+ channel accumulations at nodes of Ranvier (Waxman, 1986).

II. DEVELOPMENT AND MAINTENANCE OF STRUCTURAL POLARITY BY CULTURED PHOTORECEPTOR PRECURSOR CELLS

We will summarize below studies from our laboratory that

Molecular Aspects of Development and Aging of the Nervous System
Edited by J.M. Lauder
Plenum Press, New York, 1990

have characterized the development and maintenance of structural and molecular polarity by isolated retinal photoreceptor cells grown in vitro in the absence of intercellular contacts. A few comments on the structure and development of these cells will be presented to provide the necessary background.

A. Properties of photoreceptor cells in vivo

Photoreceptor cells can be readily distinguished from other retinal neurons in vivo based on their elongated shape and high degree of polarity. As illustrated in Fig. 1, photoreceptors are subdivided along their longitudinal axis into a series of compartments, which include a short axon, the cell body (occupied almost exclusively by the nucleus), an inner segment (which contains the energy-producing and synthetic organelles of the cell as well as a lipid droplet in some cones), and a unique outer segment that contains parallel stacks of membranous disks and is connected to the inner segment by a cilium. Inner and outer segment differ not only in morphology, but also in the constituents of their plasma membranes. For example, the visual pigment opsin is concentrated in the outer segment, while the enzyme Na^{+},K^{+}-ATPase is much more abundant in the inner segment plasma membrane (Bok, 1985; Stirling and Sarthy, 1985; Besharse, 1986; Papermaster et al, 1986; Yazulla and Studholme, 1987, Spencer, et al., 1988; among others). These structural and molecular differences between inner and outer segments are likely to have functional importance.

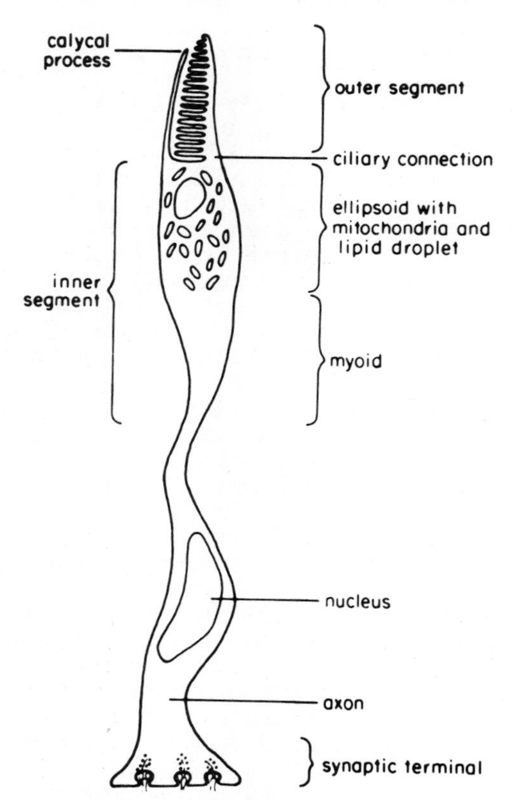

Figure 1. Diagram of a chick cone photoreceptor as seen in vivo. From Madreperla, S. A. and R. Adler (1989), Dev. Biol. 131:149-160. With permission.

B. Neural retina development in vivo

The different cell types present in the adult retina derive from an apparently homogeneous population of neuroepithelial, matrix cells. The timetable of events such as terminal mitosis, migration and differentiation of various cell types during retinal development has been studied in many species, including the chick embryo (rev: Grun, 1982). In the 8-day chick embryo, the stage used in our studies, most retinal cells are already postmitotic, but the expression of differentiated properties is seen only in some neuronal cells in the inner retina and no signs of photoreceptor differentiation can be detected.

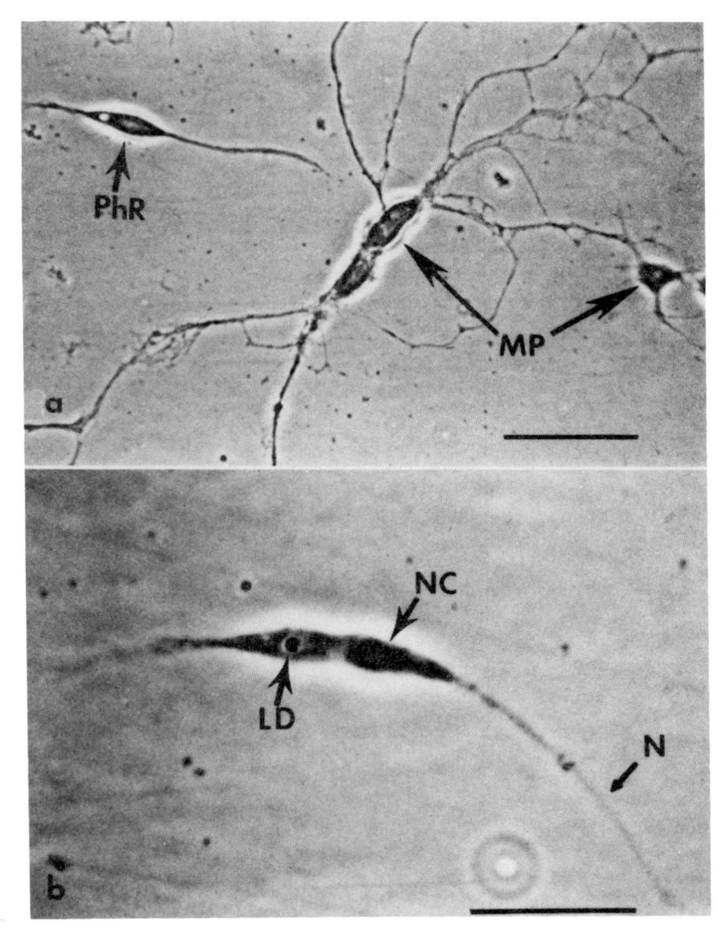

Figure 2. **a)** Culture of dissociated cells from 8-day chick embryo retina, grown in vitro for 6 days. Multipolar neurons (MP) and a cone photoreceptor (PhR) are seen. Note that the photoreceptor is free of contacts with other cells. **b)** Cultured chicken cone photoreceptor showing an elongated, polarized, compartmentalized organization with a short neurite (N), a nuclear compartment (NC), and an inner segment containing a lipid droplet (LD). Previous ultrastructural and immunocytochemical studies have also shown the presence of a small outer-segment-like process. Magnification bar: 30 um. From Madreperla, S. A. and R. Adler (1989), Dev. Biol. 131:149-160. With permission.

C. General description of the Retinal Cultures

The 8-day chick embryo retina can be readily dissected from other intraocular tissues, including the pigment epithelium, and dissociated into a suspension of single cells after mild trypsinization. When the cells are seeded on highly adhesive, polyornithine-coated substrata, a culture initially appears as a morphologically homogenous population of process-free, round cells. Subsequently, some of these round cells differentiate as multipolar neurons, while others develop as cone photoreceptors (Fig. 2). Even though they develop in the absence of intercellular contacts, photoreceptor precursor cells acquire many of the properties that are characteristic of photoreceptors in vivo (Adler, et al., 1984; Adler 1986, 1987). Particularly relevant is that they become highly elongated and polarized, with subcellular compartments similar to those found in vivo (compare Figs. 1 and 2). They also show molecular polarity as evidenced by opsin immunoreactive materials that accumulate in their apical, outer segment-like process (Adler, 1986) and the enzyme Na^{+} ,K^{+}-ATPase that is concentrated in the inner segment region (Madreperla, et al., 1989 a, b).

D. Role of the cytoskeleton in the development and maintenance of structural polarity

Investigations of the role of the cytoskeleton in photoreceptor morphogenesis were carried out with cultures grown on glass coverslips containing an etched grid, that allowed individual cells to be repeatedly identified based on their position relative to grid landmarks (Madreperla and Adler, 1989). In these experiments the cytoskeletal inhibitors nocodazole, which depolymerizes microtubules, and cytochalasin D (CCD), which inhibits actin filament formation, were used. Individual photoreceptors were identified and their responses to treatments with these drugs were determined by serial photomicrography and computer-assisted image analysis. In some cases the drugs were removed by washing to investigate the reversibility of their effects. A model based on the results of these studies is shown in Fig. 3.

Elongated photoreceptor cells showed opposite responses to nocodazole and CCD. Exposure to nocodazole resulted in decreases in cell body length, and with prolonged treatment the photoreceptors reverted to a circular configuration (Fig 4). Conversely, CCD treatment caused cell body elongation (Fig. 5). The length changes induced by nocodazole and CCD were concentration-dependent and fully reversible, with the cells recovering a normal pattern of organization when returned to drug-free medium. These observations suggested that the elongated, polarized photoreceptor phenotype is maintained by an equilibrium between two sets of constantly active, opposing cytoskeletal forces: one set is microtubule-dependent and tends to cause cell elongation, while the other is actin-dependent and causes cell shortening (Fig. 3). The results are consistent with cytochemical information about the distribution of microtubules and filamentous actin in these cultured cells (Madreperla and Adler, 1989), as well as with electron microscopical observations in vivo (Burnside and Dearry, 1986). Interactions between microtubules and microfilaments have been suggested by studies with other cell types (cf, Solomon and Zurn, 1981).

Similar inhibitor studies were also performed at earlier in vitro times, when photoreceptor precursors are undergoing morphogenesis. In untreated cultures, round isolated photoreceptor

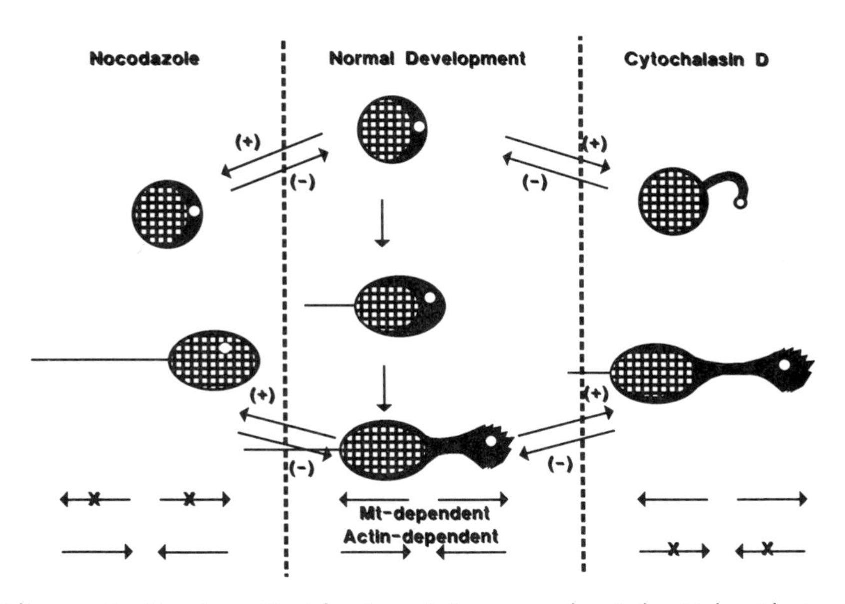

Figure 3. The hypothetical model summarized in this figure proposes that the development and maintenance of the elongated configuration of photoreceptor cells results from a balance between two sets of continuously active, opposing forces (central panel, bottom arrows). In the schematic representation of photoreceptor cells the cross-hatched, large circles represent the nucleus, and the small, empty circles represent the lipid droplet. Microtubule-dependent forces tend to push the nuclear compartment away from the lipid droplet-containing apical region of the cells, while actin-dependent forces tend to pull these two regions together. The left panel illustrates the effects of nocodazole. When this inhibitor is added to round precursors (upper figure; "+" diagonal arrows), their elongation is prevented. When nocodazole is added to elongated photoreceptors (lower figure) they revert to a round configuration. Both effects are reversible ("-" diagonal arrows). The model proposes that these behaviors are the result of inhibition of microtubule-dependent forces in the presence of active actin-dependent contractile forces (arrows, bottom left). When CCD is added to cultures (right panel, "+" diagonal arrows) round precursors fail to elongate, and form a long, abnormal process which contains the lipid droplet at its distal end (upper figure). Although not illustrated in the figure, this process is rich in microtubules, and its formation is blocked by nocodazole. On the other hand, elongated photoreceptors respond to CCD treatment by becoming longer (right panel, lower figure). All CCD effects are reversible ("-" diagonal arrows). As shown by the arrows at the bottom of the right panel, photoreceptor responses to CCD are interpreted to reflect the inhibition of actin-dependent contractile forces, in the presence of constantly active microtubule-dependent forces. From Madreperla, S. A. and R. Adler (1989), Dev. Biol. 131:149-160. With permission.

precursors were found to undergo a stereotyped and predictable sequence of morphogenetic transformations leading to an elongated, polarized pattern of organization (Madreperla and Adler, 1989). Both nocodazole and CCD blocked these morphogenetic changes reversibly. However, while nocodazole simply kept photoreceptor precursors "frozen" in their original round shape, CCD caused the extension of an abnormal cell process, which developed through a microtubule-dependent mechanism. As summarized in Fig 3, these observations suggested that opposing cytoskeletal forces similar to those involved in the maintenance of the elongated photoreceptor shape are also involved in its development (Madreperla and Adler, 1989).

III. DEVELOPMENT AND MAINTENANCE OF MOLECULAR POLARITY IN CULTURED PHOTORECEPTORS

As mentioned in the Introduction, Na^+,K^+-ATPase (ATPase) is concentrated in the plasma membrane of the inner segment region of photoreceptor cells in vivo. In our studies, ATPase anti-sera developed by McGrail and Sweadner (1986) and Siegel, et al., (1988) were used to investigate the mechanisms for generating the asymme-

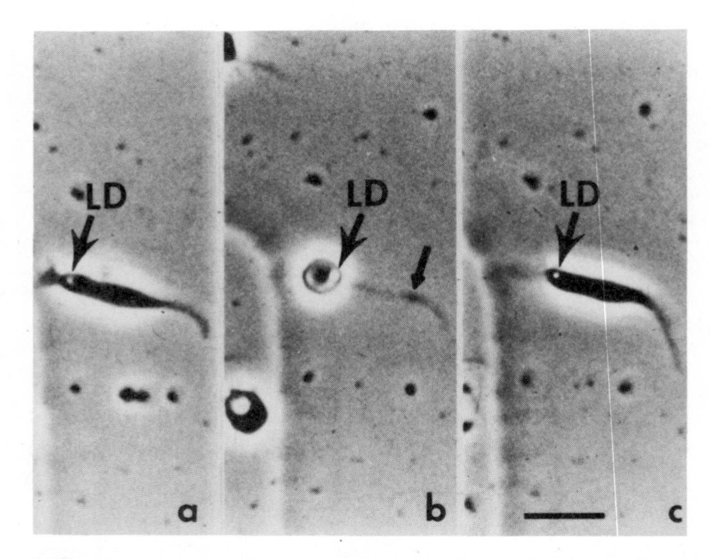

Figure 4. Effect of nocodazole on an elongated photoreceptor and reversal after nocodazole removal. An elongated photoreceptor is shown before treatment in panel **(a)**. Twelve hours after addition of 2 uM nocodazole **(b)** the cell has lost its elongated configuration, and become round. The nucleus has apparently moved to a position more apical than that of the lipid droplet (LD). Some materials appear to remain attached to the substratum in the area previously occupied by the photoreceptor cell body [arrow in **(b)**]. Panel **(c)** shows the same cell 24 hr after the nocodazole was removed by three 5-min washes with fresh medium; the photoreceptor appeared to have regained an elongated, polarized morphology very similar to that before treatment. Magnification bar: 15 um. From Madreperla, S. A. and R. Adler (1989), Dev. Biol. 131:149-160. With permission.

tric distribution of this enzyme in developing photoreceptors in culture (Madreperla, et al., 1989 a, b). Immunocytochemical studies showed that the enzyme is uniformly distributed in the round photoreceptor precursor cells in early cultures. However, ATPase immunoreactivity becomes progressively restricted to the developing inner segment region as the photoreceptors elongate. A series of experiments indicated that cytoskeletal attachments, perhaps involving an analogue of the erythrocyte protein spectrin, are responsible for creating ATPase asymmetry. Studies using fluorescence photobleaching and recovery (FPR) demonstrated that the fraction of mobile ATPase molecules was 50% less in elongated photoreceptors than in precursor cells, suggesting that decreasing ATPase mobility was associated with increasing ATPase polarity. Parallel experiments showed that ATPase is resistent to extraction with detergents such as Triton X-100. Moreover, there was good quantitative correlation between the immobile and detergent-resistent ATPase fractions in round precursors and in elongated photoreceptors. Immunocytochemical studies showed co-localization of ATPase and spectrin immunoreactivities throughout photoreceptor morphogenesis in vitro, and quantitative image analysis confirmed that their distributions were similar. The resistance of both ATPase and spectrin to detergent extraction was similarly affected by variations in pH and KCl concentration in the extraction medium (Madreperla, et al., 1989 a, b).

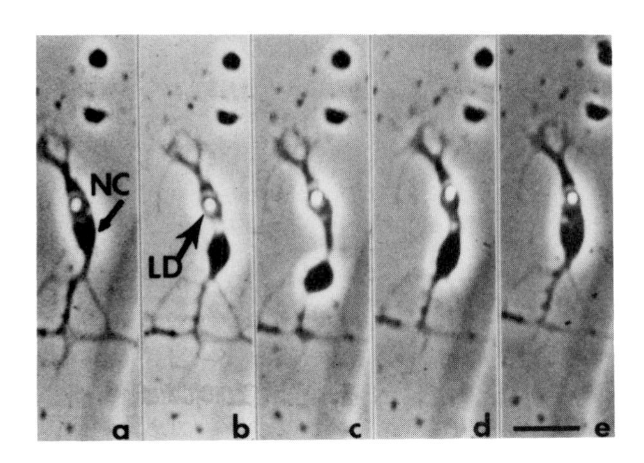

Figure 5. Reversible effects of CCD on photoreceptor cell length. An elongated photoreceptor is shown before treatment **(a)**, 1 hr **(b)** and 2 hr **(c)** after exposure to 5 uM CCD, and 1 hr **(d)** and 2 hr **(e)** after the CCD was removed by three 5-minute washes with fresh medium. A constriction in the inner segment, between the lipid droplet (LD) and the nucleus, can be seen after 1 hr exposure to CCD **(b)**, as the nuclear compartment moves away from the lipid droplet. By 2 hrs **(c)** the nuclear compartment has moved further basally. Thus, the distance between the nuclear compartment and the lipid droplet is markedly increased. One hr after the CCD was removed the cell begins to return towards its original length **(d)** and by 2 hr **(e)** it shows a configuration similar to that seen before treatment onset. Magnification bar: 15 um. From Madreperla, S. A. and R. Adler (1989), Dev. Biol. 131:149-160. With permission.

These results are consistent with the concept that interactions between plasma membrane proteins and the cytoskeleton are important for the establishment of specialized plasma membrane domains, as proposed for other cell systems (eg, Axelrod, 1983; Almers and Stirling, 1984; Wolf, 1987). For example, the restriction of Na^+,K^+-ATPase to the basolateral domains of epithelial cells involves interactions with a spectrin-containing subcortical cytoskeleton (eg, Nelson and Veshnock, 1986). It is noteworthy, however, that the presence of tight junctions is also necessary for ATPase polarization in these epithelial cells. Our results, on the other hand, show that photoreceptor cells can develop and maintain ATPase polarity in vitro through a cytoskeleton-dependent mechanism that does not require tight junctions or other contact-mediated intercellular interactions.

IV. CONCLUSIONS

The availability of an in vitro system in which isolated precursor cells differentiate as photoreceptors in the absence of intercellular contacts has been advantageous for the investigation of mechanisms controlling structural and molecular cell polarity. One of the useful features of the system is that, unlike epithelia, photoreceptors become polarized along a plane parallel to the culture substratum. Therefore, their morphogenesis and their responses to treatment with cytoskeletal inhibitors can be dynamically monitored, and the cells are also directly amenable to analysis with immunocytochemical techniques or with biophysical methods such as the quantitation of lateral mobility of plasma membrane molecules by fluorescences photobleaching and recovery.

Several conclusions can be derived from these studies. One is that, at the time of their isolation from the retina, some precursor cells appear to be committed to a photoreceptor developmental program which includes not only the expression of cell-specific genes such as opsin (Adler, 1986), but also the acquisition of a complex and polarized pattern of structural and molecular organization. The observation that this developmental program can be expressed by cells grown in vitro in the absence of intercellular contacts suggests that intracellular mechanisms play an important role in these phenomena. The data reviewed here emphasize that cytoskeleton-dependent forces are critical for photoreceptor morphogenesis, and indicate that the cytoskeleton is involved in the development and maintenance of molecular polarity as well. This experimental system should allow further investigation of these mechanisms.

Acknowledgements

Original research reported in this article was supported by USPHS grant EY 04859 (to Ruben Adler) and AI4584 (to Michael Edidin). SM was supported by NIH-MSTP GM 07309. Many thanks are due to Doris Golembieski for secretarial assistance and to M. Lehar for photographic work.

References

Adler, R., J. D. Lindsey, and C. L. Elsner. 1984. Expression of cone-like properties by chick embryo neural retinal cells in glial-free monolayer cultures. J Cell Biol. 99:1173-1178.

Adler, R. 1986. Developmental predetermination of the structural and molecular polarization of photoreceptor cells. Dev. Biol. 117:520-527.

Adler, R. 1987. The differentiation of retinal photoreceptors and neurons in vitro. In: Progress in Retinal Research. Vol. 6. N. Osborne and G. Chader, editors. Pergamon Press, London. 1-27.

Almers, W. and C. Stirling. 1984. Distribution of transport proteins over animal cell membranes. J. Membrane Biol. 77:169-186.

Angelides, K. J., L. W. Elmer, D. Loftus and E. Elson. 1988. Distribution and lateral mobility of voltage-dependent sodium channels in neurons. J. Cell Biol. 106:1911-1925.

Axelrod, D. 1983. Lateral motion of membrane proteins and biological function. J. Memb. Biol. 75:1-10.

Banker, G. A., and A. B. Waxman. 1988. Hippocampal neurons generate natural shapes in cell culture. In Intrinsic Determinants of Neuronal Form and Function. R. J. Lasek and M. M. Black, editors. Alan R. Liss, Inc., New York, pp. 61-82.

Bartlett, W. P., Banker, G. A. 1984. An electron microscopic study of the development of axons and dendrites by hippocampal neurons in culture. I. Cells which develop without intercellular contacts. J. Neurosci. 4(8):1944-1953.

Besharse J. C. 1986. Photosensitive membrane turover: differentiated domains and cell-cell interaction. In The Retina: A Model for Cell Biology Studies, Part I, R. Adler and D. Farber, editors. Academic Press, New York, 297-339.

Bok D. 1985. Retinal photoreceptor-pigment epithelial interactions. Friedenwald Lecture. Invest. Ophth. and Vis. Sci. 26:1659-1694.

Burnside, B., and A. Dearry. 1986. Cell motility in the retina. In The Retina: A Model for Cell Biology Studies. R. Adler and D. Farber, editors. Academic Press, Orlando, 152-206.

Dichter, M. A. 1978. Rat cortical neurons in cell culture: Culture methods, cell morphology, electrophysiology, and synapse formation. Brain Res. 149:279-293.

Fischbach, G. D. 1970. Synaptic potentials recorded in cell cultures of nerve and muscle. Science 169:1331-1333.

Goslin, V., D. J. Schreyer, J. H. P. Skene, and G. Banker. 1988. Development of neuronal polarity: GAP-43 distinguishes axonal from dendritic growth cones. Nature 336:672-674.

Grun, G. 1982. The development of the vertebrate retina: a comparative survey. Adv. Anat. Embryol. Cell Biol. 78:1-85.

Madreperla, S. A., and R. Adler. 1989. Opposing microtubule- and actin-dependent forces in the development and maintenance of structural polarity in retinal photoreceptors. Dev. Biol. 131:149-160.

Madreperla, S. A., M. Edidin and R. Adler (1989a) Role of the cytoskeleton in the polarized distribution of the Na^{+},K^{+}-ATPase in isolated photoreceptors. J. Cell Biol. 107:783a.

Madreperla, S. A., M. Edidin and R. Adler (1989b) Na^+,K^+-ATPase in retinal photoreceptors: a role of cytoskeletal attachments. Submitted.

McGrail, K. M., and K. J. Sweadner. 1986. Immunofluorescent localization of two different Na,K-ATPases in the rat retina and in identified dissociated retinal cells. J. Neurosci. 6(5): 1272-1283.

Nelson, W. J., and P. J. Veshnock. 1986. Dynamics of membrane-skeleton (fodrin) organization during development of polarity in Madin-Darby canine kidney epithelial cells. J. Cell Biol. 103:1751-1765.

Papermaster, D. S., B. G. Schneider, D. Defoe, and J. C. Besharse. 1986. Biosynthesis and vectorial transport of opsin on vesicles in retinal rod photoreceptors. J. Histochem. Cytochem. 34:5-16.

Peters, A., Palay, S. L. Webster, H. 1976. The fine structure of the nervous system. In: The neurons and the Supporting Cells. Philadelphia: W. B. Saunders. 1-406.

Ritchie, J. M., Rogart, R. B. 1977. Density of sodium channels in mammalian myelinated nerve fibers and nature of the axonal membrane under the myelin sheath. Proc. Natl. Acad. Sci. USA 74(1):211-5.

Scott, B. E. Engelbert, V. E., Fisher, K. C. 1969. Morphological and electrophysiological characteristics of dissociated chick embryonic spinal ganglion cells in culture. Exp. Neurol. 23:230-248.

Siegel G. J., S. A. Ernst, Lin, and T. J. Desmond. 1988. Heterogeneity of the α^+-catalytic subunit in mouse brain Na^+,K^+-ATPase. Trans. Am. Soc. Neurochem. 19:252.

Solomon, F., Zurn, A. 1981. The cytoskeleton and specification of neuronal morphology. Neurosci. Res. Prog. Bull. 19(1):100-135.

Stirling, C. E., and P. V. Sarthy. 1985. Localization of the Na-K pump in turtle retina. J. Neurocytol. 14:33-47.

Waxman, S. G. 1986. The astrocyte as a component of the mode of Ranvier. Trends in Neurosci., 9:250-253.

Wolf, D. E. 1987. Diffusion and the control of membrane regionalization. Ann. NY Acad. Aci. 247-261.

Wuerker, R. B., Kirkpatrick, J. B. 1972. Neuronal microtubules, neurofilaments, and microfilaments. Int. Rev. Cytol. 33:45-75.

Yazulla, S., and K. M. Studholme. 1987. Ultracytochemical distribution of ouabain-sensitive, K^+-dependent, p-nitrophenylphosphatase in the synaptic layers of goldfish retina. J. Comp. Neurol. 261(1):74-84.

MATRIX INTERACTIONS REGULATING MYELINOGENESIS IN CULTURED OLIGODENDROCYTES

Leonard H. Rome, Michael C. Cardwell, Phyllis N. Bullock and Steven P. Hamilton

Department of Biological Chemistry, and the Mental Retardation Research Center, UCLA School of Medicine, Los Angeles, California 90024

INTRODUCTION

Myelin is a membrane unique to the nervous system that is deposited in segments along selected nerve fibers. Myelin functions as an insulator to increase the velocity of impulses transmitted between the cell body of a nerve and its target. In the central nervous system (CNS) myelin is produced by oligodendroglial cells. Each cell extends numerous processes that ensheathe segments of several different axons simultaneously. A different mechanism of myelination appears to occur in the peripheral nervous system where Schwann cells myelinate only single segments of single axons. Other differences are seen between Schwann cells and oligodendroglia such as morphology, growth factor requirements, extracellular matrix involvement and composition of the myelin produced.

Myelination In Vitro - We have focused our efforts on examining the process of myelination in the CNS, utilizing primary cultures of highly purified oligodendrocytes. Unlike Schwann, cells which require neurons to produce myelin in culture, we and others have shown that oligodendrocytes will produce a myelin-like membrane in the absence of neurons that is strikingly similar in morphology and biochemistry to that seen *in vivo* (1-3). Therefore, these cells are ideally suited for experiments on the control of myelinogenesis.

As is observed *in vivo*, the synthesis of myelin membranes *in vitro* is a developmentally regulated process. Purified oligodendrocytes show a maximal rate of cerebroside and sulfatide synthesis between 15 and 20 days "equivalent brain age" with a peak at 17 days for both glycolipids (Figure 1).

The myelin-like membrane produced in culture is present as large sheets as well as vesicular swirls. These swirls vary in their size and complexity, and many resemble myelin figures. At times the membranes are observed wrapped around extracellular debris. Upon close examination (see Figure 2 top), the myelin-like membrane sheets were composed of two to ten or more layers,

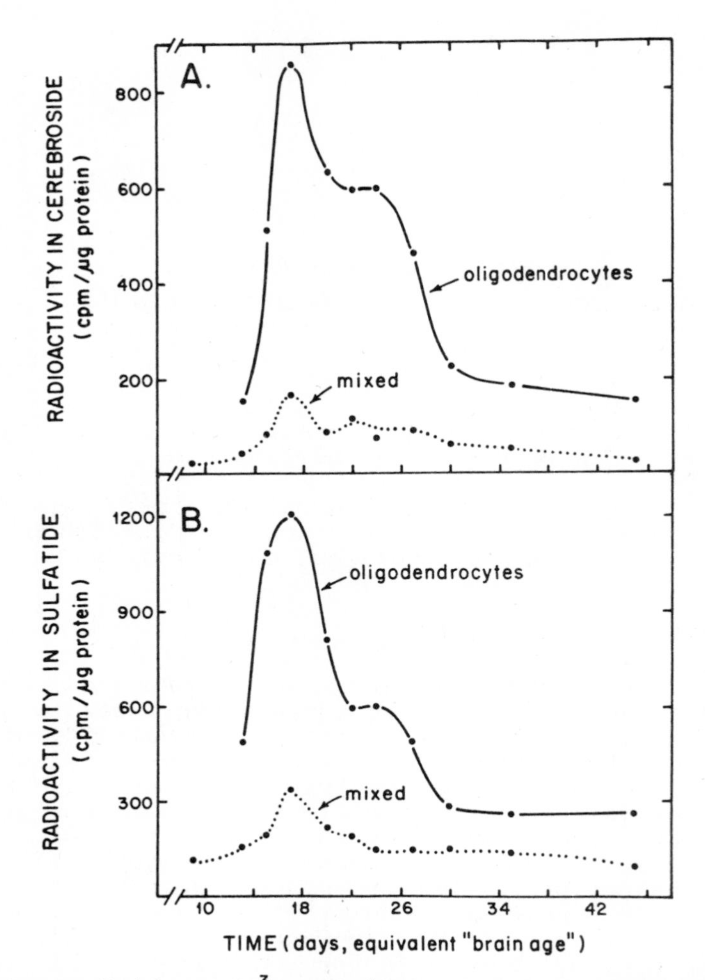

Fig. 1. Accumulation of[^{3}H]palmitate-labeled cerebrosides (A) and sulfatide (B) in mixed and pure oligodendrocyte cultures. Data from reference 2.

which exhibited alternating major and minor dense lines. The periodicity of the most compact regions was approximately 11 nm between dense lines. This is slightly less compact than the periodicity of the myelin sheath of a CNS axon, which is shown in Figure 2 (bottom) for comparison.

The Role of an Extracellular Matrix - The surface interactions of cells with neighboring cells and with the extracellular matrix are among the most critical determinants of form and function in a developing tissue. Many cells in primary culture require the use of a specific substratum in order to express those differentiated properties which typify the cells *in vivo* (4-7). Schwann cells, when co-cultured with neurons, produce an extracellular matrix that is similar to the matrix formed in peripheral nerve endoneurium (8,9). The presence of a basal lamina has been found to be an absolute requirement for normal Schwann cell differentiation and for the myelination of neuronal axons (10-12). A matrix requirement has yet to be determined for oligodendroglia, however it has been shown with mature ovine oligodendrocytes that an adhesive interaction with substrate is

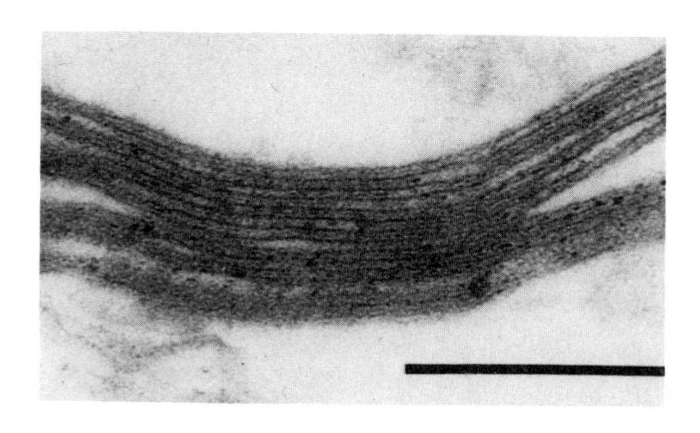

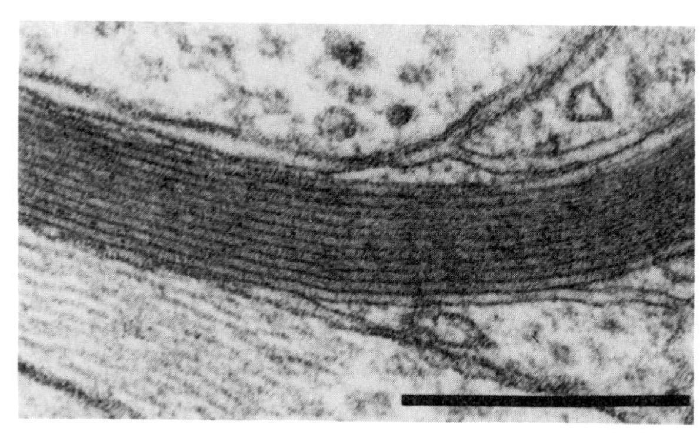

Fig. 2. Electron micrographs of purified myelin-like membrane and myelin. Top: an example of a myelin-like membrane figure produced by purified oligodendrocytes at 23 days equivalent "brain age". Bottom: myelin sheath surrounding an axon within the endopeduncular nucleus. Bars = 0.25 μm. From reference 2.

necessary in order for production of a myelin-like membrane to occur (13). We have examined oligodendrocyte adhesion to various surfaces with the long term aim of assessing the contribution of these interactions to oligodendrocyte maturation and to the development of the cell's myelin-like properties. Our initial interest has been in a substratum produced by mixed glial cultures, the astroglial matrix (AGM). This interest in the AGM stems from the observations that oligodendrocytes plated on AGM, rather than plastic or poly-L-lysine, tend to more rapidly extend processes and develop morphologically. In addition, cells plated on AGM show a two-fold higher rate of incorporation of [^{3}H]thymidine into trichloroacetic acid precipitable material than cells grown either on plastic or poly-L-lysine (P.N. Bullock and L.H. Rome, unpublished observations).

<u>RGD-dependent Cell Adhesion</u> - The most thoroughly studied example of cellular adhesion to an extracellular protein is the interaction of fibronectin with its cell surface receptor (for review 14-16). The study of fibronectin has, to some extent, guided the direction of progress in this field. Fibronectin is a large (220 kD) glycoprotein with the ability to independently bind to several extracellular matrix constituents in addition to its cell binding activity. The region within the fibronectin molecule that contains cell binding activity has been localized

to a tripeptide of arginine-glycine-aspartic acid (RGD[1] in single letter code) (17-19). Synthetic peptides that contain this sequence inhibit fibronectin's interaction with its receptors and, when immobilized, mimic fibronectin's cell binding activity.

The mechanism defined for binding of cells to fibronectin appears to be of general significance. The RGD sequence has been found in the primary sequence of at least five other extracellular proteins with cell binding functions, suggesting the existence of a family of proteins that bind cells via a common mechanism (6,20). Cell surface receptors for many of these RGD-containing extracellular ligands have been purified and characterized (15,20-22). It has been suggested that these receptors are related members of a group of proteins with roles in cell adhesion and have been termed "integrins" (23).

The role that these RGD interactions play in cell function has been examined largely in cell culture studies where general roles in anchorage, maintenance of cell shape, and cell migration have been characterized. Cellular functions that are responsive to occupancy of these RGD-dependent receptors remain to be clarified, however recent evidence suggests that occupancy of an RGD-dependent receptor is coupled to the differentiation of myoblasts. Immature cells cultured with antibodies to the ß subunit of Integrin, or with RGD-containing peptides, are prevented from differentiating into myotubes (24).

We have examined the possibility that an RGD-dependent interaction might be an early step in the sequence of events leading to formation of the myelin sheath. Utilizing a simple adhesion assay (25) we have been able to measure the interaction between oligodendrocytes and various substrata, including AGM. Oligodendrocytes bound to surfaces coated with fibronectin, vitronectin and to AGM. The binding of cells to all of these substrates required divalent cations and was inhibited by a synthetic peptide (GRGDSP) modeled after the cell binding domain of fibronectin (Figure 3). The specificity of this inhibition was determined by screening a group of peptides, both related and unrelated to GRGDSP. These peptides were unable to inhibit adhesion of oligodendrocytes to either AGM or to an untreated plastic surface. One of these peptides differed from the active peptide by the single substitution of alanine for glycine within the RGD sequence (GRADSP). Another inactive peptide, a tetramer with the sequence SDGR, represents the inverse sequence of the active region within the GRGDSP peptide (25).

The component of the glial matrix responsible for oligodendrocyte adhesion appears to be a protein which is secreted by the glial cells. Although we have been unable to identify the extracellular matrix protein that gives the adhesive property to AGM, we have been able to demonstrate that

---

[1]The abbreviations used in this paper: RGD, arginine-glycine-aspartic acid; GRGDSP, glycine-arginine-glycine-aspartic acid-serine-proline; GRADSP glycine-arginine-alanine-aspartic acid-serine-proline; VYPNGA, valine-tyrosine-proline-asparagine-glycine-alanine; SDGR, serine-aspartic acid-glycine-arginine; AGM, astroglial matrix,

several known RGD-containing extracellular proteins are not responsible, these including fibronectin, vitronectin, thrombospondin, type-I and type-IV collagen, and tenascin.

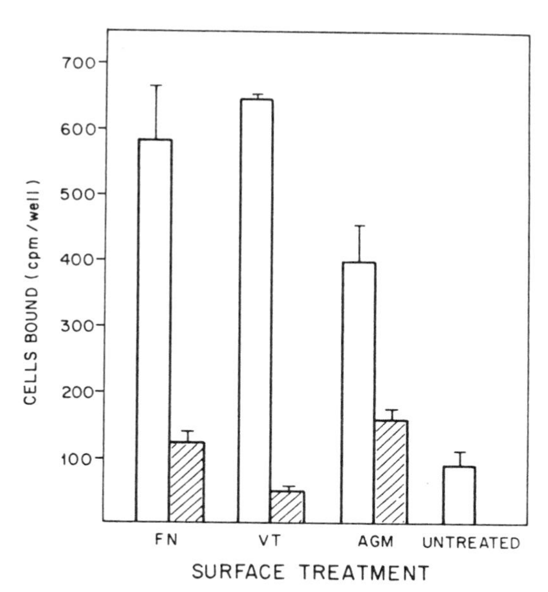

Fig. 3. Inhibition of oligodendrocyte adhesion by GRGDSP. Cells were incubated on bovine fibronectin (B-FN), vitronectin (VT), AGM, or untreated plastic. Hatched bars indicate the addition of 0.1 mg/ml of GRGDSP in the incubation media. Values given are the average of 3 wells and the error bars indicate standard deviation. From reference 25.

<u>Regulation of Gene Expression; Inhibition of sulfolipid synthesis by GRGDSP</u> - As described above, oligodendrocytes adhere to a number of different substrates in an RGD-dependent manner. However, these cells can be maintained in culture in the presence of a concentration of GRGDSP which does not cause a significant amount of cell loss. Cells grown with 0.1 mg/ml GRGDSP show only slight differences in overall morphology. Notable among these changes is a reduced tendency of cells to aggregate into the multi-cellular clusters which have often been observed with oligodendrocytes in culture (26,27).

In addition to these differences in cell morphology, peptide treated cultures showed marked difference in their ability to produce myelin specific components (28). As shown in Figure 4, pure oligodendrocytes maintained in culture progress through a characteristic developmental change in the ability to produce the myelin enriched galactolipids sulfatide and cerebroside. In the experiment shown in Fig. 4, the biosynthetic peak for [^{35}S]sulfolipid in untreated cultures occurred on day 16. When cells were grown in the presence 0.1 mg/ml GRGDSP from the time of plating as pure oligodendrocytes and labeled after various times in culture, the normal developmental pattern of sulfatide synthesis was completely blocked. Sulfolipid synthetic levels for peptide treated cultures never exceeded the baseline level of the untreated controls.This inhibition of sulfolipid synthesis is specific to GRGDSP. Other peptides, GRADSP and VYPNGA, were unable to inhibit sulfolipid synthesis.

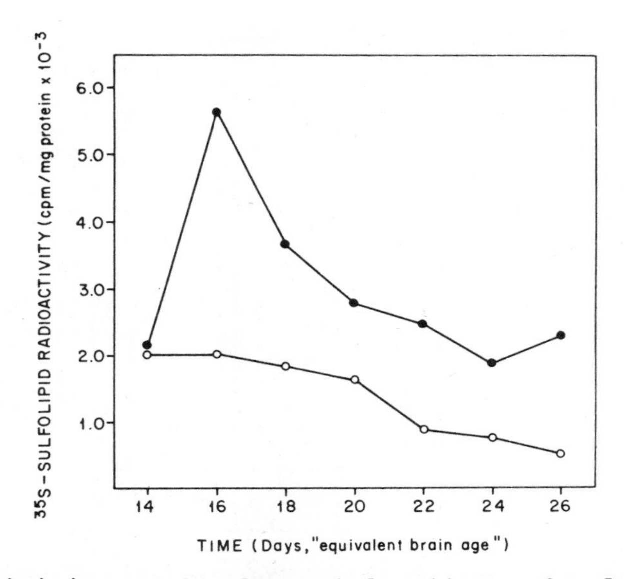

Fig. 4. Inhibition of developmental pattern of sulfolipid synthesis by GRGDSP. Oligodendrocytes obtained from mixed glial cells were plated on AGM. The cultures were supplemented with 0.1 ml of 1.0 mg/ml GRGDSP in DMEM/F12 with 5% calf serum (open circles) or with 0.1 ml DMEM/F12 with 5% calf serum (closed circles). The initial concentration of peptide was maintained in subsequent feedings. Cultures were labeled for 16 hours in low sulfate DMEM containing 1% calf serum and 4.0 μCi $[^{35}S]H_2SO_4$/ml beginning on the day shown. The cell layer was washed, solubilized and lipid extracted and counted as described (28). Each bar represents the average of 3 separate determinations and the variation between samples was less than 10%. From reference 28.

<u>Timing of GRGDSP Effects</u> - To determine the length of exposure to GRGDSP required for maximal inhibition of sulfolipid synthesis, Cultures of pure oligodendrocytes were treated for up to 7 days in media containing the GRGDSP peptide. Cells were labeled with $[^{35}S]H_2SO_4$ on day 17 for 16 hours in media that did not contain the peptide. Sulfolipid synthesis was reduced to 7-15% of control values with 3-5 days of pretreatment, although even a one day pretreatment had a significant inhibitory effect.

Oligodendrocytes were viable in culture for several months and continued to synthesize the myelin enriched galactolipids throughout this time. The cells were variably inhibited by peptide treatment depending on the age of the cultures. While measurable levels of sulfolipid were produced in newly plated cells, this basal level of synthesis did not appear to be inhibited by GRGDSP. Inhibition was not apparent until days 14-16. While cells were impaired by peptide treatment at any age in culture after this time, the extent of inhibition was greatest when the cells were labeled during the developmental peak in sulfolipid synthesis.

While we have shown that the oligodendrocytes adhered in an RGD-dependent manner to AGM and not to plastic, cells plated on either plastic or AGM showed a decrease in sulfolipid synthesis when cultured in the presence of the RGD-containing hexapeptide. This result might suggest that it is not the interaction of oligodendrocytes with a component of the AGM which is inhibited by culture with peptide, but perhaps, the interaction of oligodendrocytes with each other.

In order to further measure the rapidity of the sulfolipid inhibition, cells were cultured without peptide until the time that lipid synthesis was maximal. The cells were then labeled in media supplemented with 0.1 mg/ml GRGDSP. Cultures were labeled for up to 24 hours and compared to control cultures which contained no added peptide. Inhibition of sulfolipid synthesis was evident after only 2-4 hours of exposure to peptide (28).

Reversibility of Inhibition of Sulfolipid Synthesis - The inhibition of [^{35}S]sulfolipid synthesis was also found to be reversible. Oligodendrocytes were cultured with and without added GRGDSP from the time of plating on day 9. Peptide treated cultures were then handled in three different ways: 1) cells were kept in the presence of GRGDSP for the duration of the experiment (including the labeling period); 2) peptide was removed on day 14 for up to six days; or 3) peptide was removed on day 16 and cultured for an additional four days.

The synthesis of [^{35}S]sulfolipid in cells grown in the presence of peptide was depressed relative to controls by 50% in the day 14-16 labeling period, by 70% in days 16-18, and by 69% in days 18-20. When the peptide was removed on day 14, the synthesis of [^{35}S]sulfolipid was essentially at the level of untreated controls. When the peptide was removed from cultures on day 16, the synthesis of sulfolipid remained depressed to the level of cultures continually treated with peptide (28). Since a maximal inhibition of [^{35}S]sulfolipid synthesis was seen after only 3-5 days of treatment with peptide, it appears that this finding is not simply the result of longer exposure to the peptide, but of exposure at some critical period of the cells *in vitro* development. Cells treated in such a way never returned to control levels of synthesis even if allowed up to 10 days to recover. Further, if the peptide was removed at any point after day 16, the oligodendrocyte was also permanently impaired in its ability to synthesize sulfolipid *in vitro*.

Inhibition of Myelin-Like Membrane and Myelin Basic Protein Synthesis by GRGDSP - It is conceivable that the effect of the GRGDSP peptide on sulfolipid synthesis was due to a direct inhibition of the enzymes of sulfatide biosynthesis or on altered sulfate uptake or metabolism, and not a specific effect related to the cells ability to produce a myelin-like membrane in culture. We therefore examined more directly the production of myelin-like membranes in cultures treated with GRGDSP. Cultured oligodendrocytes were grown either in media supplemented with 0.1 mg/ml GRGDSP from day 14 to day 18, in media supplemented with 0.1 mg/ml GRADSP from day 14 to day 18, or without added peptide. All cultures were maintained in media without peptide from day 18 to day 24 at which time they were harvested and used to prepare a crude myelin-like membrane fraction. There was a small degree of cell loss in GRGDSP treated cultures which was minimized by limiting the length of exposure to peptide to only four days. The protein content of the myelin-like membrane fraction isolated from GRGDSP-treated cultures was 27% of that derived from untreated cultures. For GRADSP-treated cultures this value was 75%. The content of myelin basic protein in the myelin-like membrane fractions of these cultures was greatly reduced relative to either GRADSP-treated or untreated cultures.

Cultures of pure oligodendrocytes not only produce the major myelin proteins and lipids, they also assemble them into a readily identifiable multi-lamellar membrane. In cultures treated with GRADSP for 4 days between days 14 and 18, membranes were very prominent and commonly displayed as many as 10 semi-compacted lamellae. In contrast, in cultures treated with GRGDSP for this same length of time, multi-lamellar membrane was very infrequently observed. Less than 20% of the treated oligodendrocytes displayed any multi-lamellar membrane in their extracellular surroundings and, when it was observed, the lamellae never exceeded 2-3 layers (28).

<u>A Model for the Design of Future Experiments</u> - We have devised a simple working model to explain the adhesive interactions of the astroglial matrix and the RGD inhibitory effects observed. Fig. 5 shows a schematic representation of this model where the earliest adhesive interactions between the oligodendrocyte and the astroglial matrix is shown in the center and in a stylized close-up version on the right. The mature oligodendroglial cell <u>in vivo</u> is illustrated on the left.

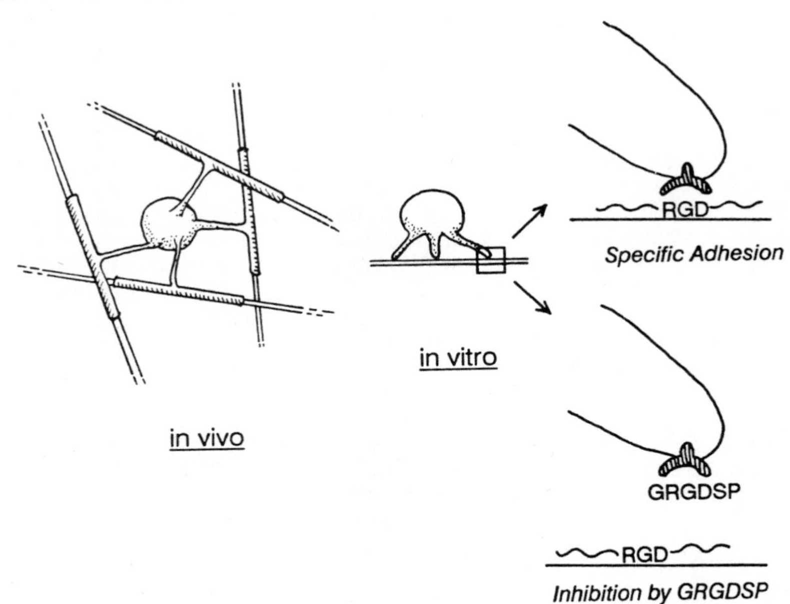

Fig. 5. A model that describes the inhibition of adhesion by RGD-containing peptides. This model assumes that early adhesive contacts like those shown to be RGD-dependent <u>in vitro</u> (25), could signal the directed synthesis of myelin <u>in vivo</u>.

Adhesion *in vitro* is presumed to occur between specialized regions of the oligodendrocyte plasma membrane containing an integrin-like receptor and the substratum which includes a matrix protein with an active RGD sequence able to mediate the process. Addition of the GRGDSP peptide blocks this early adhesive interaction. These interactions may be similar to the early recognition events that occur between oligodendroglia and neuronal axons *in vivo*, shown on the left. Productive contacts (perhaps via integrin receptors) may either directly or indirectly signal the oligodendrocyte to induce myelin membrane synthesis and in addition, direct the newly produced membrane to the areas of contact. Thus, the addition of GRGDSP peptides could not only interrupt adhesion but in addition, interfere with the signal necessary for myelin production. We plan to challenge this model by a further characterization of the oligodendrocyte integrin and identification of its *in vivo* ligand(s). Furthermore a connection between RGD dependent adhesive interactions and the regulation of myelin synthesis will need to be elucidated.

One obvious difference between the *in vitro* and *in vivo* systems is that *in vitro*, contacts between oligodendrocytes and their ligands can only occur in a single plane (the surface of the tissue culture plate). Perhaps additional "productive" contacts between oligodendrocytes and RGD-containing ligands would lead to an increased signal resulting in an increased production of myelin membranes. We have recently begun to modify our culture system to enable oligodendrocytes to form adhesive contacts in 3 dimensions. This has been achieved by the inclusion of glass microfibers into the cultures to act as "surrogate" axons able to be contacted by oligodendrocytes in all directions. Preliminary studies suggest that oligodendrocytes do interact with AGM-coated glass fibers but not native uncoated fibers. Furthermore, the presence of AGM-coated fibers appears to stimulate the synthesis of myelin-specific sulfolipids (Bullock, Hamilton & Rome, *In Preparation*).

ACKNOWLEDGEMENTS

This work was supported by USPHS Grant HD-06576. L.H.R. is the recipient of an American Cancer Society Faculty Research Award. M.C.C was supported as a predoctoral trainee on USPHS Grant HD-07032.

REFERENCES

1. Bradel E.J., and F.P. Prince. 1983. Cultured neonatal rat oligodendrocytes elaborate myelin membrane in the absence of neurons. *J. Neurosci. Res.* 9:381-392.

2. Rome L.H., P.N. Bullock, F. Chiappelli, M.C. Cardwell, A.M. Adinolfi, and D. Swanson. 1986. Synthesis of a myelin-like membrane by oligodendrocytes in culture. *J. Neurosci. Res.* 15:49-65.

3. Szuchet, S., S.H. Yim, and S. Monsma. 1983. Lipid metabolism of isolated oligodendrocytes maintained in long-term culture mimics events associated with myelinogenesis. *Proc. Natl. Acad. Sci. USA* 80:7019-7023.

4. Enders, G.C., J.H. Henson and C.F. Millette. 1986. Sertoli cell binding to isolated testicular basement membrane. J. Cell Biol. 103:1109-1119.

5. Li, M.L., J. Aggeler, D.A. Farson, C. Hatier, J. Hassel, and M.J. Bissel. 1987. Influence of a reconstituted basement membrane and its components on casein gene expression and secretion in mouse mammary epithelial cells. Proc. Natl. Acad. Sci. USA 84:136-140.

6. Ruoslahti, E., E.G. Hayman, and M.D. Pierschbacher. 1985. Extracellular matrices and cell adhesion. Arteriosclerosis 5: 581-594.

7. Sanes, J.R. 1983. Roles of extracellular matrix in neural development. Ann. Rev. Physiol. 45:581-600.

8. Carey D.J., M.S. Todd, and C.M. Rafferty. 1986. Schwann cell myelination: induction by exogenous basement membrane-like extracellular matrix. J. Cell Biol. 102:2254-2263.

9. McGarvey, M.L., A. Baron-Van Evercooren, H.K. Kleinman, and M. Dubois-Dalcq. 1984. Synthesis and effects of basement membrane components in cultured rat Schwann cells. Dev. Biol. 105:18-28.

10. Baron-Van Evercooren, A., A. Gansmuller, M. Gumpel, N. Baumann and N.K. Kleinman. 1986. Schwann cell differentiation in vitro: extracellular matrix deposition and interaction. Dev. Neurosci. 8:182-196.

11. Bunge, R.P., M.B. Bunge, and C.F. Eldridge. 1986. Linkage between axonal ensheathment and basal lamina production by Schwann cells. Ann. Rev. Neurosci. 9:305-328.

12. Carey D.J., M.S. Todd, and C.M. Rafferty. 1986. Schwann cell myelination: induction by exogenous basement membrane-like extracellular matrix. J. Cell Biol. 102:2254-2263.

13. Yim, S.H., S. Szuchet, and P.E. Polak. 1986. Cultured oligodendrocytes: a role for cell-substratum interaction in phenotypic expression. J. Biol. Chem. 261:11808-11815.

14. Akiyama, S.K., S.S. Yamada, and K.M. Yamada. 1986. Characterization of a 140-kD avian cell surface antigen as a fibronectin-binding molecule. J. Cell Biol. 102:442-448.

15. Hynes, R.O. 1981. Fibronectin and its relation to cellular structure and behavior. In Cell Biology of Extracellular Matrix. E.D. Hay, editor. Plenum Press, New York. 295-333.

16. Ruoslahti, E., E.G. Hayman, and M.D. Pierschbacher. 1985. Extracellular matrices and cell adhesion. Arteriosclerosis 5:581-594.

17. Pierschbacher, M., E.G. Hayman, and E. Ruoslahti. 1983. Synthetic peptide with cell attachment activity of fibronectin. Proc. Natl. Acad. Sci. USA 80:1224-1227.

18. Pierschbacher, M.D., and E. Ruoslahti. 1984. Cell attachment activity of fibronectin can be duplicated by small synthetic fragments of the molecule. Nature 309:30-33.

19. Yamada, K.M., and D.W. Kennedy. 1984. Dualistic nature of adhesive protein function: fibronectin and its biologically active peptide fragments can autoinhibit fibronectin function. J. Cell Biol. 99:29-36.

20. Ruoslahti, E., and M.D. Pierschbacher. 1986. Arg-Gly-Asp: a versatile cell recognition signal. Cell 44:517-518.

21. Pytela, R., M.D. Pierschbacher, S. Argraves, S. Suzuki, and E. Rouslahti. 1987. Arginine-glycine-aspartic acid adhesion receptors. Meth. Enzymol. 144:475-489.

22. Ruoslahti, E., and M.D. Pierschbacher. 1987. New perspectives in cell adhesion: RGD and Integrins. Science 238:491-497.

23. Hynes, R.O. 1987. Integrins: a family of cell surface receptors. Cell 48:549-554.

24. Menko, A.S., and D. Boettinger. 1987. Occupation of the extracellular matrix receptor, Integrin, is a control point for myogenic differentiation. Cell 51:51-57.

25. Cardwell, M.C., and L.H. Rome. 1988. Evidence that an RGD-dependent receptor mediates the binding of oligodendrocytes to a novel ligand in a glial-derived matrix. J. Cell Biol. 107:1541-1549.

26. Holmes, E., Hermanson, R. Cole, and J. deVellis. 1988. Developmental expression of glial-specific mRNAs in primary cultures of rat brain visualized by in situ hybridization. J. Neurosci. Res. 19:389-396.

27. Walker, A.G., J.A. Chapman, and M.G. Rumsby. 1985. Immunocytochemical demonstration of glial-neuronal interactions and myelinogenesis in subcultures of rat brain cells. J. Neuroimmunol. 9:159-177.

28. Cardwell, M. C. and L.H. Rome. 1988. RGD-containing peptides inhibit the synthesis of myelin-like membranes by cultured oligodendrocytes. J. Cell Biol. 107:1551-1559.

CEREBELLAR GRANULE CELL MIGRATION INVOLVES PROTEOLYSIS

Nicholas W. Seeds, Susan Haffke, Kathleen Christensen and Judith Schoonmaker

Department of Biochemistry/Biophysics/Genetics
University of Colorado Health Science Center
Denver, Colorado

INTRODUCTION

Histogenesis in the central nervous system requires extensive cell migration. The migration of granule neurons in the neonatal mouse cerebellum has been a focus of study. Although granule cell migration occurs in close association with Bergmann glial fibers[1] and Purkinje cell dendrites[2], the actual mechanisms of movement that probably involve cytoskeletal systems as well as the interaction of cell surface and extracellular matrix components remain to be established.

The possibility that these morphogenetic cell movements may be facilitated by proteolytic mechanisms has only recently been examined. The protease plasminogen activator (PA), which has been implicated in tumor cell invasiveness and metastatic behavior[3], is also synthesized and secreted by certain normal cell types[4,5] which are migratory or engaged in tissue remodeling. Plasminogen activator levels were shown to be elevated at a time of active cell proliferation and migration in the developing cerebellum[6,7]; subsequently the secreted PA activity was localized to a subpopulation of granule neurons in dissociated cerebellar cell cultures[8]. Other studies with cells of the peripheral nervous system have demonstrated the secretion of PA at neuronal growth cones in neuroblastoma cells[9], dissociated sensory[10] and sympathetic neurons[11], as well as PA secretion by Schwann cells[10,12]; all are elements involved in active movement.

The aim of this study is to examine the role of proteolysis in granule cell migration *in vitro* and *in vivo*, as indicated by the sensitivity of migrating cells to several different protease inhibitors. A preliminary report of these findings has appeared[13].

IN VITRO GRANULE CELL MIGRATION

The *in vitro* culture system for studying granule cell migration is similar to that described by Trenkner and Sidman[14], where cell migration can be viewed directly. However, rather than using polylysine coated microwells with $5x10^4$ cells, the cultures described here contain $1.5x10^6$ cells dissociated from cerebella of 6 to 8-day postnatal C57Bl/6 mice as previously described[8] plated on polylysine or rat tail collagen coated

Molecular Aspects of Development and Aging of the Nervous System
Edited by J.M. Lauder
Plenum Press, New York, 1990

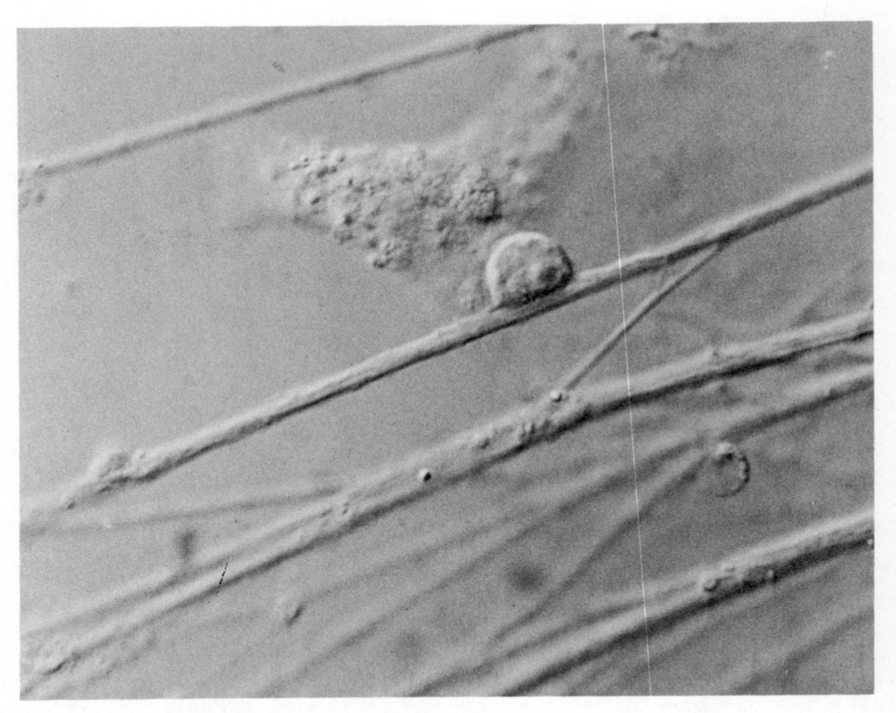

Fig. 1. Differential interference contrast photomicrograph of a migratory granule neuron on a neural fascicle. Leading process extends to the left and a small trailing process is seen to the right of the cell.

25 mm glass coverslips. The cells are cultured in basal Eagle's medium with 10% acid-treated fetal calf serum[15]. Cerebellar neurons readily adhere to the polylysine surface and extend neurites which often fasciculate, and by 3 days of culture granule neurons, as identified by their small size and characteristic nuclear morphology with differential interference contrast microscopy, are observed on these fascicles as seen in Figure 1. In contrast, the cerebellar neurons do not readily adhere to collagen and prefer to reaggregate. After a day or so these aggregates attach to the collagen substratum and by 3 days of culture are interconnected by cellular cables, containing numerous attached cells as seen in Figure 5.

Many of the cells attached to the cables bind the neuronal specific antibody Cbl-1[16]. These cells can be birthdated, as to having undergone their last mitosis, with [^{3}H]-thymidine to the day prior to removal of the cerebellum (Figure 2), thus identifying these cells as migratory (or possibly mitotic) granule neurons.

The cellular composition of these cables was investigated by double label immunofluorescence using tetanus toxin binding as an indicator of neuronal processess and antibody to glial fibrillary acidic protein as a marker for Bergmann fibers or astrocytic processes (Figure 3). In all cases these cables contained fasiculated neuronal processes (a & c) and often contained only neuronal processess (a & b), about 30% of the cables also possessed glial processes (d) when grown on collagen, or 60% when grown on polylysine substrata (Table 1).

Video time-lapse recording of the fascicle associated cells, shows that many of these cells migrate along the fascicle. The more closely associated (as seen in Fig. 1) or stretched-out cells are migratory while the round or loosely associated cells often fail to move. This finding is in agreement with those of Trenkner and coworkers[17]. Cell movement is not continuous nor unidirectional with time, rates of migration from 70 up to 410 um/day are seen in a given time period; therefore, average rates of migration calculated over several hours give a

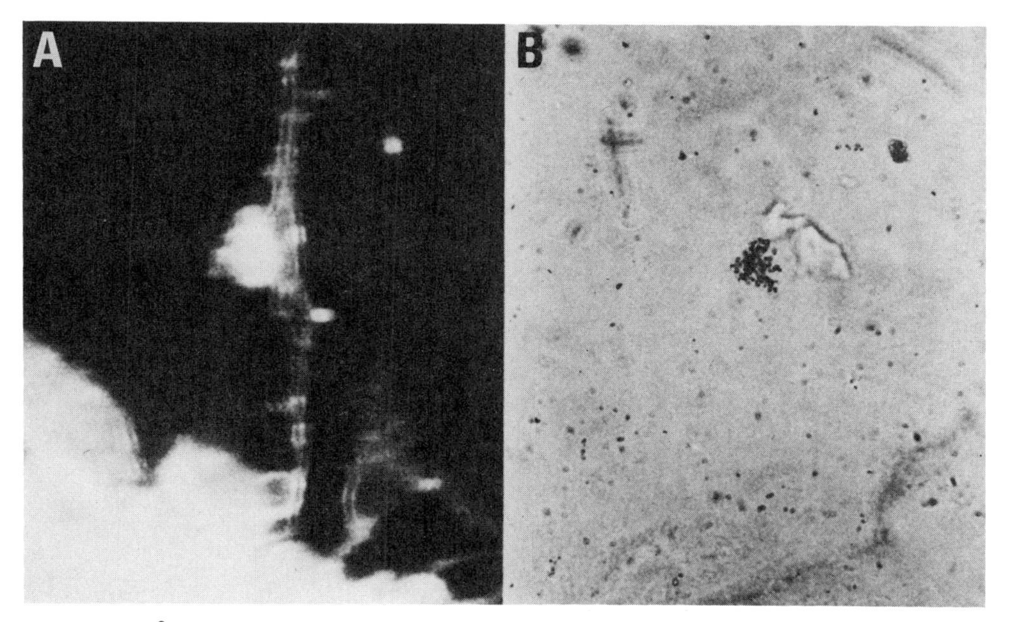

Fig. 2. [^{3}H]-thymidine birthdating of cells on fascicles. A neurite fascicle after 2 days in culture spans two cerebellar reaggregates. Both the associated cell and cable possess the neuronal specific cell surface antigen Cbl-1 as demonstrated by anti Cbl-1 immunofluorescence in panel A. This neuron (granule cell) was "birthdated" with [^{3}H]-thymidine the day before the mouse was sacrificed, and the abundant silver grains (panel B) and small size indicate this cell would probably be migratory if it was still *in vivo*.

more typical profile of a particular cell's migratory behavior. Over one hundred cells have been monitored during the past six years, and in general, average rates of 220 um/day or about 10 um/hr. are common. Similar rates were found when cells were grown on polylysine or collagen substrata. These migration rates are in reasonably good agreement with the rates calculated for granule cell migration *in vivo* of 70 um/day[18] and 100 um/day[19], which are based on a static analysis of birthdated cells in neonatal cerebellar tissue sections. However, these rates are slower than the 33 + 20 um/hr. found for granule cell migration on glial fibers *in vitro*[20].

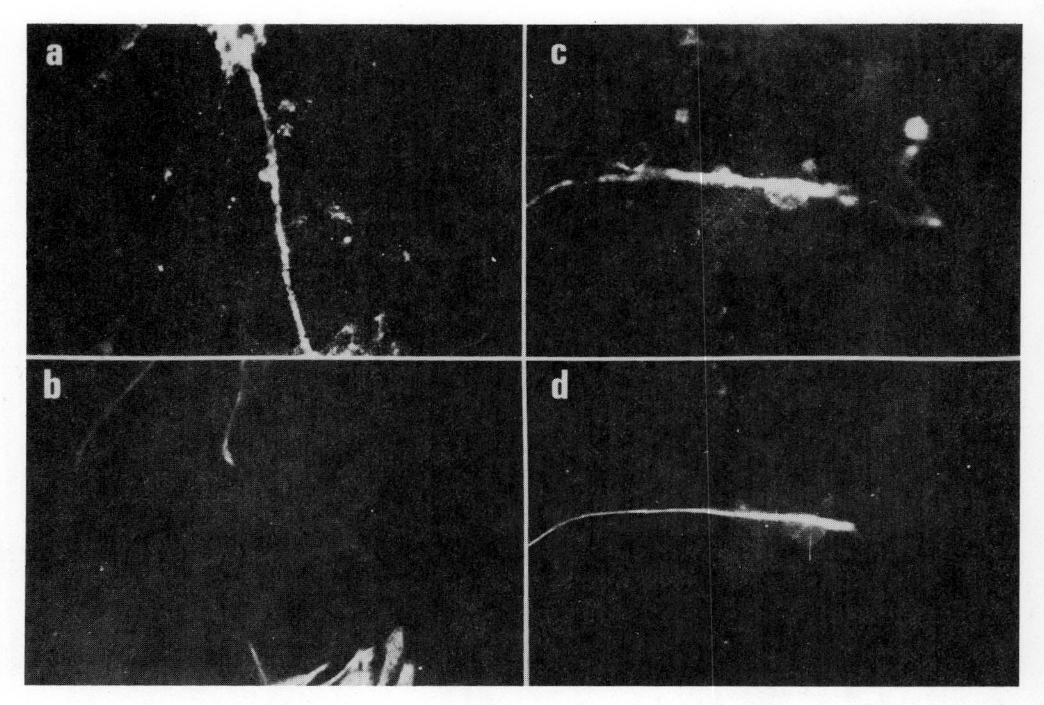

Fig. 3. Cellular composition of fascicles and cables. Panels a and c are tetanus toxin binding to granule neurons and the fascicle to which they are associated, demonstrating the neuronal composition of the cable. Comparison figures b and d represent the glial fibrillary acidic protein associated with the respective fascicle, as visualized with fluorescent 2nd antibodies. Fascicle in a and b is solely neuronal while the cable in c and d contains both neuronal and glial elements.

Table 1. Granule Neurons on Neuronal or Glial Cables

Cable Composition	Substratum Collagen	Polylysine
Tetanus toxin $^{+}$ GFAP $^{-}$ cable	69%	38%
Tetanus toxin $^{+}$ GFAP $^{+}$ cable	31%	62%
Tetanus toxin $^{-}$ GFAP $^{+}$ cable	0%	0%

Cerebellar cells cultured 4 days on the indicated substrata were examined for their ability to bind the neuronal marker tetanus toxin, and following methanol fixation to bind antibody to glial fibrillary acidic protein(GFAP), as seen with fluorescent 2nd antibodies.

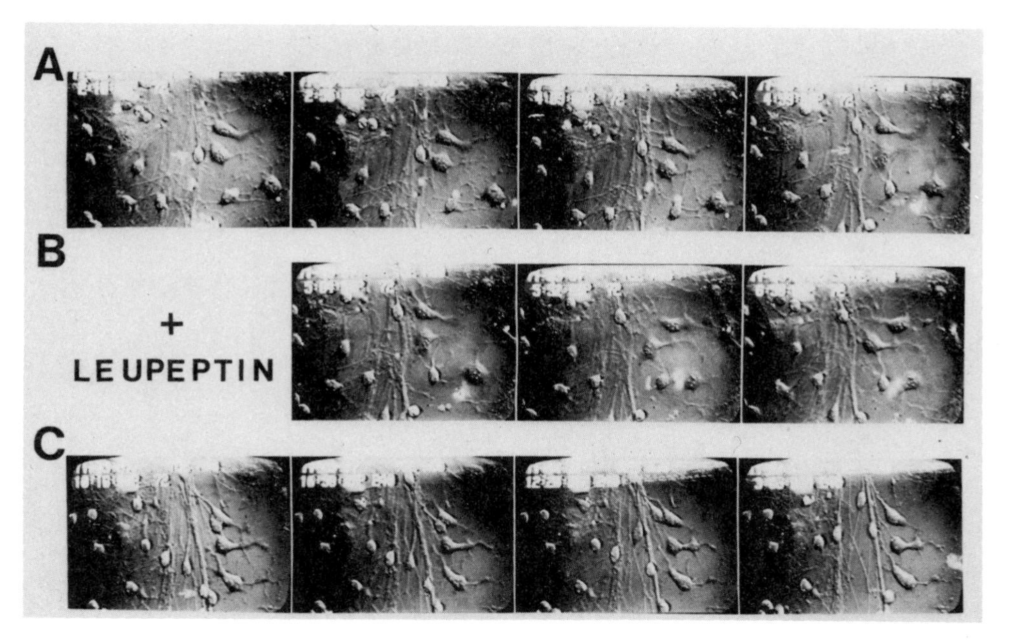

Fig. 4. Video time-lapse of granule cell migration and sensitivity to leupeptin. Dissociated cerebellar cells from 6 day old mice were cultured for 3 days, then observed by differential interference contrast enhanced-video time-lapse. In panel A a granule cell (arrow) and its leading process migrate (at an average rate of 90 um/d) for several hours upward along a neural fascicle, as indicated by the reference arrow in the last photo. The addition of 10^{-4}M leupeptin (panel B) causes a dramatic decrease in migration of this cell (avg. rate <25um/d) over the next two hours. In panel C the leupeptin has been removed by two media changes over 3 hrs., by which time cell migration has reinitiated and the original cell from panel A is moving out of the field of view, while another granule cell (arrow) is rapidly migrating up the same fascicle at an average rate of 230um/d.

The secretion of the protease plasminogen activator by a subpopulation of granule neurons in the developing cerebellum, suggested that PA secretion may be related to granule cell migration[8]. One approach to the question of whether protease activity is necessary for granule cell migration, is to assess the effect of proteolytic inhibitors of varying specificity on migration. Cells were examined for several hours to document their migratory nature, then the culture medium was changed to include a protease inhibitor and the same cells monitored for several additional hours. Figure 4A shows a migratory granule cell moving along a fasiculated neural cable at an average rate of 90um/day. The addition of 10^{-4}M leupeptin (panel B) strongly inhibited the migration. Similar results were seen with the synthetic tripeptide, D-phe-pro-arg-CH_2Cl at a concentration of 10^{-5}M. Both leupeptin and D-phe-pro-arg-CH_2Cl are inhibitors of plasminogen activator. The possibility that leupeptin was toxic to the granule cells was ruled-out by adding fluorescein diacetate to some cultures at the end of the study and monitoring cell viability

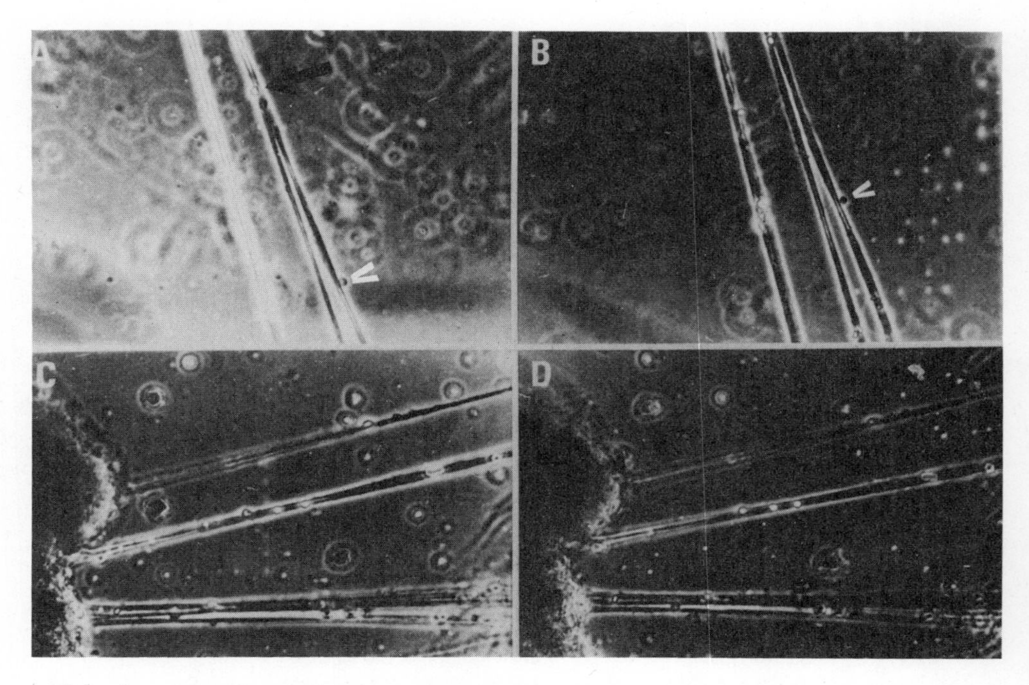

Fig. 5. Granule cell migration along neurite fascicles is insensitive to aprotinin but sensitive to leupeptin. Several granule cells on cables connecting cerebellar reaggregates, which form when the cerebellar cells are cultured on a collagen substratum, were examined at 20 min. intervals by microphotography using 25x phase contrast optics. After the cells were followed for 1.5 hrs. to establish their migratory behavior, protease inhibitor aprotinin (200KIU/ml) was added to A and observed over the next 2 hrs. (B), while the phase bright granule cell (arrow) moved uninterrupted toward a phase dark cell ghost (V). The field of view shifted upward during the study. Similarly, cells on cables (C) in the presence of aprotinin were demonstrated to move versus reference points; however, they abruptly stopped their migration upon the addition of 10^{-4}M leupeptin, as seen 2 hrs later in panel D. Removal of the leupeptin after two hrs. led to a resumption of cell migration.

by fluorescence. Usually the inhibited cultures were rinsed over two hours with several chamber volumes of media and re-incubated in fresh media at which time the cells re-initiated migration as seen in panel 4C. Following the removal of protease inhibitors, migration rates varied from 80 - 410 um/d, with an average of rate of 230 um/d in the study in 4C; thus demonstrating the viability of these cells and the reversibility of the leupeptin inhibition.

The possibility that PA was activating plasminogen to the broad spectrum protease plasmin, was examined with the protease inhibitor aprotinin which completely blocks plasmin activity at about 50 KIU/ml but does not affect PA activity. In Figure 5 even a high concentration of aprotinin does not affect migration; this is a very reproducible

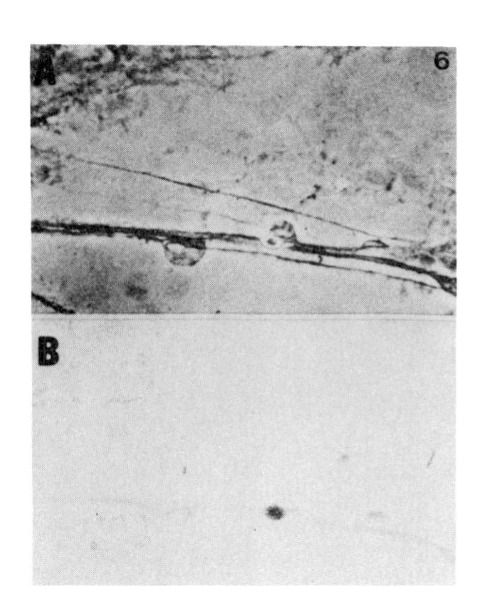

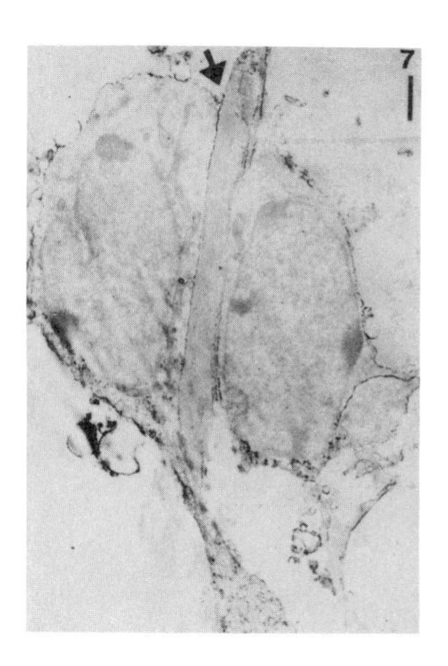

Fig. 6. Reactivity of cable associated granule cells to anti-tPA. Cultures of dissociated cerebellar cells used to assay cell migration were incubated with rabbit antibody to tPA for 20 min., then rinsed, fixed and processed with ABC peroxidase. Panel A is a phase contrast image of a granule cell on a neurite fascicle; B is the bright field image showing the peroxidase reaction product associated with this cell and a weaker but suggestive activity along the fascicle.

Fig. 7. EM localization of tPA in cerebellar cell cultures. Cerebellar cultures incubated with antibody to tPA in Fig. 6 were embedded in epon and fascicle associated granule cells selected for EM observation. Two granule cells are associated with opposite sides of an unindentified cellular process. The cell on the left has a leading process containing numerous vescicles and ribosomes. Although tPA is present over most of the cell surface, it appears to be enriched at the trailing edge of the cell (arrow). Magnification bar is 1.0 um.

finding in six different cultures and more than two dozen cells. However, addition of leupeptin to these same cultures causes a marked decrease or cessation of migration (panel D). Similar studies with the protease inhibitor pepstatin, which does not inhibit PA activity, showed no inhibition of cell migration even at 10^{-3}M.

Localization of PA release and proteolytic activity around these migrating granule cells, similar to our previous cell localization studies[8,9,10], is not possible since PA release from the large reaggregates nearby generates lysis zones that obscure release around single cells. However, rabbit antisera against the tissue type plasminogen activator (tPA) can be used to show that these live granule cells on neurite fascicles secrete or have bound to their surface tPA (Figure 6). Electron microscopic observations (Figure 7) of these same cultures show that tPA activity is associated with the entire granule cell surface. However, tPA appears to be concentrated at the trailing edge of the cell where it may facilitate granule cell migration by promoting detachment of the trailing edge as the cell moves forward.

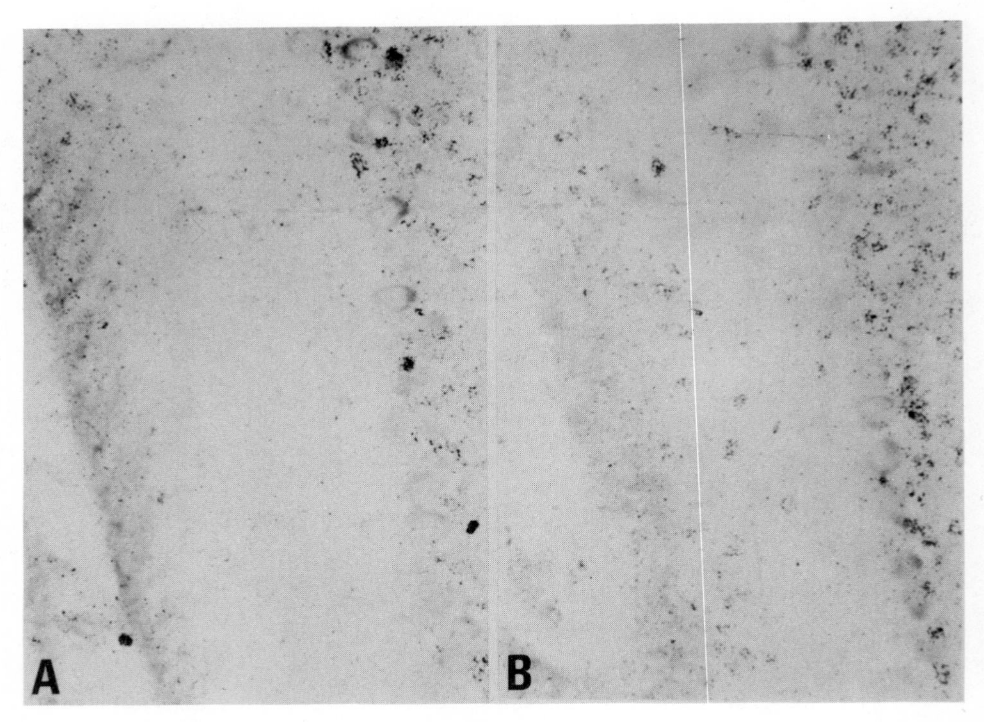

Fig. 8. Protease inhibitors retard granule cell migration in vivo. Autoradiography of 12 day mouse cerebella whose EGL granule cells were birthdated with ^{3}H-thymidine on day 6, and received intracerebellar injections of SBTI (3ug/ml) on day 7. Panel A shows a section through a folia at an "away" site(>0.4mm) from the injection, where granule cell migration is unaffected. In panel B near(<0.4mm) the injection site radiolabeled granule cell migration is markedly slowed with many labeled cells appearing the the molecular layer.

IN VIVO GRANULE CELL MIGRATION

The role of proteolytic activity in granule cell migration has also been examined in vivo. Migratory granule cells were birthdated by intraperitoneal injection of ^{3}H-thymidine at 6 days of age. On the next day(7) the mice were anesthetized and given 1 ul injections of soybean trypsin inhibitor(SBTI) (3ug/ml), D-Phe-Pro-Arg-CH_2-Cl (DPPA) (10^{-4}M) or phosphate buffered saline. The injection was made into the external granule cell layer immediately beneath the pial membrane and the site was marked with charcoal or ink. The animal's skull was covered with gelfoam, scalp sutured closed and pups were returned to their mother where they nursed well and developed normally compared to uninjected littermates. After five days(day 12) the mice were sacrificed and the cerebella removed, fixed in Cajal's formalin, dehydrated, embedded in polyester wax, sectioned and processed for autoradiography with NTB2 emulsion. The sections were examined and heavily labeled cells(ie., >18 silver grains/10um, compared to background <2 grains/10um) counted "near"(<0.4mm) the injection site as seen in Figure 8B or at distant sites ("away", >0.4mm) as shown in panel A, or other folia in the same animal. These findings are quantified in Table 2. Both SBTI and DPPA retarded granule cell migration as indicated by the increased number

Table 2. Protease Inhibitors and *in vivo* granule cell migration

Intracerebellar injection	Cells in EGL and Mol. layer near-site (<0.4mm)	away-site (>0.4mm)
Phosphate-buffered Saline	106 cells/mm^2	127 cells/mm^2
Soybean Trypsin Inhibitor	406 "	116 "
D-Phe-Pro-Arg-CH_2Cl	428 "	102 "

Mice were injected as described in Fig. 8 with 1 ul of SBTI(3ug/ml) or DPPA (10^{-4}M) or PBS. Labeled cell counts were made within 0.4mm of the injection site (near), or at distant sites or other folia in the same cerebellum.

of labeled cells in the molecular layer as compared to the controls, where migration of these labeled granule cells into the internal granule cell layer was virtually complete by day 12. The values for "away" sites are internal controls and agree well with the saline injected control group. Statistical analysis of the two experimental groups showed that both differed from their controls("away" values) at $p < 0.001$. The morphology of the counterstained labeled cells retarded in the molecular layer looks normal and does not suggest that they are dying cells; furthermore, most dead cells undergo fairly rapid lysis and degradation of the DNA *in situ*, such that they do not contribute to the autoradiography after 5 days. Both of these protease inhibitors have been shown to inhibit mouse uPA and tPA activity at the injected concentrations.

CONCLUSIONS

Migrating cells and migratory cell appendages, such as nerve growth cones, must make their way through a tissue matrix to their appropriate targets or destination. During this process they migrate over specific cell surfaces, extracellular matrix components and move between contiguous cells and cellular processes. Electron micrographs show migrating cells to be in close association with certain other cells and extracellular matrix components *in vivo*, while firm attachments to neural fascicles and cellular processes are seen *in vitro*. Thus, a role for extracellular proteolysis in locally digesting cell-matrix or cell-cell attachments, as well as mediating cell-substratum detachment events seem probable in facilitating cell migration. In the first reports of the protease PA in developing cerebellum, the authors[6,7,8] suggested its possible role in granule cell migration. More recent studies[21] have shown that PA is deposited on the substratum by growing neurites where it could act to promote detachment events. Furthermore, we[22] have shown the local digestion of extracellular matrix molecules (fibronectin) directly under growing neurites from sensory neurons.

Earlier studies by Moonen and coworkers[23] using peptide chloromethylketone inhibitors, implicated both PA activity and plasmin to be necessary for cell migration in cerebellar explant cultures. Although our studies with DPPA inhibition of granule cell migration are in good agreement with the earlier studies of Moonen suggesting that PA or another protease may be involved in granule cell migration, they question any role for plasmin, since even high concentrations of the plasmin inhibitor, aprotinin, do not inhibit cell migration. Although plasminogen is the primary and only well documented substrate for PA, our findings suggest that PA may act directly on some other substrate. Studies by Quigley and coworkers[24] with transformed fibroblasts have shown that plasminogen activators can hydrolyze fibronectin in the absence of a plasminogen-plasmin system.

The studies presented here have demonstrated that granule neuron migration both _in vivo_ and _in vitro_ is sensitive to protease inhibitors that can act on plasminogen activators; however, direct demonstration of a role for PA in this process will require more specific inhibitors for plasminogen activator than are currently available.

Acknowledgements

The authors express their gratitude to Drs. D. Dahl, W. Habig, D. Collen and F. Castellino for their generous gifts of anti-GFAP serum, tetanus toxin, and anti-tPA serum. This study was supported in part by grants from the National Institutes of Health NS-09818 and the National Science Foundation BNS-86-07719.

REFERENCES

1. P. Rakic and R. L. Sidman, _Proc. Nat. Acad. Sci. USA._ 70:240 (1973).
2. C. Sotelo and J.-P. Changuex, _Brain Res._ 77:484 (1974).
3. K. Dano, P. Andreasen, J. Grondahl-Hansen, P. Kristensen, L. Nielsen and L. Skiver, _Advances in Cancer Res._ 44:139 (1985).
4. J. Unkeless, J. Gordon and E. Reich, _J. Exp. Med._ 139:834 (1974).
5. D. Beers, S. Strickland and E. Reich, _Cell_ 6:387 (1975).
6. A. Krystosek and N. W. Seeds, _Fed. Proc._ 37:1702a (1978).
7. H. Soreq and R. Miskin, _Brain Res._ 216:361 (1981).
8. A. Krystosek and N. W. Seeds, _Proc. Nat. Acad. Sci. USA._ 78:7810 (1981).
9. A. Krystosek and N. W. Seeds, _Science_ 213:1532 (1981).
10. A. Krystosek and N. W. Seeds, _J. Cell Biol._ 98:773 (1984).
11. R. Pittman, _Develop. Biol._ 110:91 (1985).
12. A. Alvarez-Buylla and J. Valinsky, _Proc. Nat. Acad. Sci. USA._ 82: 3519 (1985).
13. N. W. Seeds, R. Hawkins, S. Verrall, S. Haffke, J. Schoonmaker and A. Krystosek, _J. Cell Biol._ 103:439a (1986).
14. E. Trenkner and R. L. Sidman, _J. Cell Biol._ 75:915 (1977).
15. P. Jones, W. Benedict, S. Strickland and E. Reich, _Cell_ 5:323 (1975).
16. N. W. Seeds, _Proc. Nat. Acad. Sci. USA._ 72:4110 (1975).
17. E. Trenkner, D. Smith and N. Segil, _J. Neurosci._ 4:2850 (1984).
18. J. Altman, _J. Comp. Neurol._ 128:431 (1966).
19. S. Fujita, in "Evaluation of Forebrain", R. Hassler & H. Stephen, eds., Plenum Press, New York (1966).
20. J. Edmunson and M. Hatten, _J. Neurosci._ 7:1928 (1987).
21. A. Krystosek and N. W. Seeds, _Exp. Cell Res._ 166:131 (1986).
22. P. G. McGuire and N. W. Seeds, _J. Cell Biol._ 107:374a (1988).
23. G. Moonen, M. Graw-Wagemans and I. Selak, _Nature_ 298:753 (1983).
24. J. P. Quigley, L. Gold, R. Schwimmer and L. Sullivan, _Proc. Nat. Acad. Sci. USA._ 84:2776 (1987).

NEURAL CELL ADHESION MOLECULE AS A REGULATOR OF CELL-CELL INTERACTIONS

Urs Rutishauser

Department of Genetics and Center for Neuroscience
School of Medicine
Case Western Reserve University
Cleveland, Ohio 44106

Progress in the investigation of cell-cell interactions in tissue formation has been particularly rapid in recent years (for review see Rutishauser and Jessell, 1988). A fundamental hypothesis throughout this work has been that these interactions reflect specific molecular complementarity mediated by distinct components at the cell surface. In addition, it is generally assumed that the developmental control of these events occurs through regulation of expression of the molecules directly involved.

But is control of gene expression sufficient to explain the exquisite pattern of cell-cell interactions that occurs during development? Recent work from our laboratory on the neural cell adhesion molecule (NCAM) and its unusual polysialic acid moiety (PSA), suggest that regulation of specific ligands is only one aspect of the developmental control of such interactions. These studies suggest that at least some cell contact-dependent phenomena are a multistep process in which the specific or recognition event requires a prior permissive adhesion event, mediated by a general adhesion system, which brings the two membranes into intimate apposition.

FACILITATION OF INTERACTIONS BY NCAM-MEDIATED ADHESION.

A substantial portion of embryonic tissues are composed of cells that produce, or whose precursors have produced, detectable amounts of NCAM. On this basis alone it is clear that the NCAM-mediated adhesion, with its relatively simple homophilic binding mechanism, is not well suited for identification of individual cells or specification of precise cell-cell interactions. Instead, the molecule may be more accurately described as a versatile "glue" which is used where and when a cell is required to establish adhesive contacts with an NCAM-positive neighbor. As illustrated below, these contacts have important implications for a variety of cell-cell interactions.

The role of NCAM-mediated adhesion in growth of axons. NCAM has been implicated in guidance of axons by the endfeet of radial glial cells. Studies on the distribution of NCAM in the developing visual system of the chick have shown that the molecule is uniformly distributed on axons but preferentially expressed at the endfeet of the radial glia (Silver and Rutishauser, 1984). Moreover, it was demonstrated that the presence of antibodies against NCAM can alter the route of these axons.

There is also evidence that the initial innervation of certain muscles may be regulated by NCAM expression. In the case of the chick hindlimb, NCAM-positive nerves are observed to wait at the periphery of the muscle until the latter, by its own internal program, also produces large amounts of the molecule (Tosney et al., 1986). At that time, there is extensive ramification of the axon into the muscle. This correlation is

Molecular Aspects of Development and Aging of the Nervous System
Edited by J.M. Lauder
Plenum Press, New York, 1990

supported by studies both in vitro and in vivo which indicate that inhibition of NCAM by antibody compromises the extent and pattern of axon-muscle contact (Rutishauser et al., 1983; Landmesser et al., 1988).

Is the growth of axons in these systems directly dependent on NCAM-mediated adhesion? There is evidence in vitro suggesting that the elongation of axons on glial and muscle cells is most sensitive to the action of other adhesion-promoting molecules, namely N-cadherin and an integrin receptor (Tomaselli et al., 1988; Bixby et al., 1987). However, as shown in the following example, there is also evidence that functional cooperativity can occur between NCAM and these adhesion systems.

NCAM binding affects the function of another cell-cell adhesion mechanism. NCAM and the calcium-dependent N-cadherin often exist on the same cell surface (Takeichi, 1987; Hatta et al., 1987; Brackenbury et al., 1981). To investigate a possible functional interdependence of these two adhesion mechanisms, membrane vesicles were prepared such that both CAMs were preserved. Although anti-NCAM does not affect aggregation of cells with only the calcium-dependent mechanism (Brackenbury et al., 1981), under these conditions *both* types of adhesion were blocked by this antibody (Rutishauser et al., 1988). The converse was not true, in that adhesion was only partially blocked by removal of calcium, a treatment that does not affect adhesion mediated by NCAM alone. These data suggest that NCAM function can regulate the action of another class of CAMs when present on the same membrane.

Establishment of junctional communication in the neural plate. Recent studies suggest that in some tissues NCAM-mediated adhesion is a prerequisite for the formation of stable gap junctions (Keane et al., 1988). In primitive neuroepithelium, there is a precise correlation between the ability of two cells to exchange a fluorescent dye and the presence of NCAM on their surfaces. Moreover, while the block of junctional communication by the *src* gene product does not prevent NCAM expression, the addition of anti-NCAM Fab to histotypic cultures of neuroepithelial cells delays the establishment of extensive communication among cells that express NCAM. It is therefore likely that in this tissue NCAM-mediated adhesion is required to hold cells together long enough to allow the assembly of stable junctions.

NCAM binding can influence cell contact-dependent changes in cell phenotype. With both rat and chick sympathetic neurons, cell-cell contact results in increased levels of choline acetyltransferase (ChAT; Adler and Black, 1986; Acheson and Rutishauser, 1988). In studies designed to investigate a potential role for NCAM in this interaction, chick sympathetic neurons were exposed, in the presence or absence of anti-NCAM Fab, to membranes containing NCAM. Under these conditions, membrane contact-mediated increases in ChAT were found to require NCAM binding function. The effects of antibody were specific, in that antibodies against the L1/G4 adhesion molecule did not alter ChAT levels (Acheson and Rutishauser, 1988; Rathjen et al., 1987).

INHIBITION OF CELL INTERACTIONS BY NCAM POLYSIALIC ACID

Whereas the protein portion of NCAM promotes cell-cell binding, the molecule's polysialic acid (PSA) moiety has a negative regulatory effect on cell-cell adhesion. That is, cells or membrane vesicles having NCAM with a relatively high level of sialic acid (NCAM-H) on their surfaces aggregate together more slowly than do those expressing NCAM with low amounts of this carbohydrate (NCAM-L) (Sadoul et al., 1983; Hoffman and Edelman, 1983; Rutishauser et al., 1985). This change in adhesion properties is also reflected in the overall degree of contact between aggregated brain cells (Rutishauser et al., 1988).

Properties and tissue distribution of PSA. The sialic acid associated with NCAM is largely in the form of long linear polymers attached to asparaginyl residues located in a single region of the polypeptide (Crossin et al., 1984). One of the most remarkable features of this carbohydrate is its large apparent excluded volume, which suggests that the carbohydrate portion of NCAM can exert steric effects through a hydrated volume of up to ten times that of the polypeptide chain alone (Rutishauser et al., 1988).

The timing of the changes from the L to H form and back to L has led to the speculation that the different NCAM forms provide additional strength in cell-cell bonding during initial morphogenesis of the nervous system, plasticity in cell interactions during histogenesis, and finally a stabilization of contacts and positions in mature tissues (Sunshine et al, 1987). For example, during early stages the neuroepithelium must withstand mechanical stresses associated with formation of the neural tube, flexures and evaginations. Subsequently, differentiating neurons need more freedom to migrate, extend neurites, and select appropriate targets. Such events involve more selective recognitions whose effectiveness is enhanced by a decreased level of general adhesiveness. The return to the L form in the adult would provide one means of stabilizing the position and connections of fully differentiated neurons.

The effect of PSA on cell contact-dependent changes in cell phenotype. As described above, membrane contact-mediated increases in ChAT were found to require NCAM binding function, as indicated by inhibition with anti-NCAM Fab. In addition, it was observed that this interaction requires the presence of the molecule in its low PSA form. On this basis, it was proposed that NCAM can serve as a permissive regulatory factor in this system in two opposite modes: as NCAM-L, its binding function can promote interaction by holding membranes together, but as NCAM-H it can effect a negative regulation of the ability of cells to transmit the relevant biochemical signal. This conclusion is also supported by the fact that the ability of anti-NCAM Fab to block increased ChAT levels can be reversed by addition of a plant lectin, but again only when the PSA content of the endogenous NCAM is low (Acheson and Rutishauser, 1988).

NCAM PSA content alone can regulate the function of other cell surface ligands. Additional evidence for an influence of NCAM PSA on cell interactions involving molecules other than NCAM has been obtained in studies of the bundling patterns of neurites (Rutishauser et al., 1988). Embryonic chick (E7) spinal cord neurites, whose NCAM is very heavily sialylated, grow as large fascicles on laminin and removal of PSA from these axons actually reduces fasciculation. The simplest explanation of the neurite patterns obtained is that growth cone-substrate adhesion has been enhanced relative to cell-cell adhesion. Consistent with this interpretation is the observation that the effect of endo N occurred even in the absence of NCAM-mediated adhesion, was greater than that produced by anti-NCAM Fab fragments, and could be reversed by antibodies against laminin.

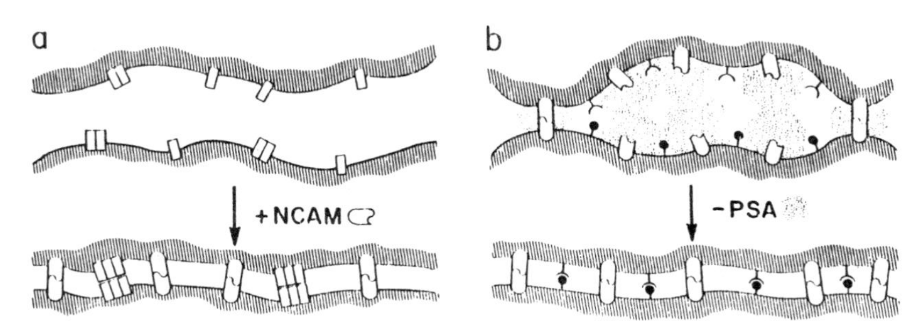

Figure 1. Two mechanisms for the regulation of cell-cell interactions by NCAM. A. Expression of NCAM on cell surfaces and the resultant formation of NCAM-NCAM bonds enhances the probability of junction formation from interacting subunits (rectangles) by increasing the extent or duration of membrane-membrane contact. B. Initiation of cell interactions via specific ligands (ball and socket) by a reduction in NCAM PSA content (stippled areas). The ability of the ligands to engage is enhanced by reducing the excluded volume of carbohydrate between membranes, which impedes close cell-cell contact. PSA is depicted as also compromising interaction between NCAMs, with adhesion occurring between molecules with relatively low amounts of PSA.

PROPOSED MECHANISM FOR NCAM AND PSA REGULATION OF CELL-CELL INTERACTIONS

To account for these phenomena, two mechanisms have been proposed (Figure 1) (Rutishauser et al., 1988). For NCAM with a relatively low PSA content, the molecule's presence would increase the extent or duration of membrane contact and thereby promote other interactions. For molecules with high PSA content, the large volume occupied by the carbohydrate would impede membrane-membrane contact so that the function of some ligands, and probably that of NCAM itself, is hindered.

DISCUSSION

The possibility that the degree or intimacy of membrane apposition can regulate cell contact-dependent events raises a number of interesting questions about the developmental and physiological role of the cell surface. In this capacity, the plasma membrane is not simply a passive surface for the display of receptors whose function is regulated solely through their synthesis. Instead it functions together with a general adhesion molecule, the example here being NCAM, to serve as a permissive "gating" mechanism for the function of contact-dependent interactions.

The function of PSA proposed in this model represents a novel mechanism for control of cell surface events. However, the idea that this chemical structure can serve as a barrier or protective zone around a cell has precedent in other biological systems. For example, it is an abundant component of the surface coat of some bacteria (Vimr et al., 1984), as well as the pelucid zone of fish eggs (Kitajima et al., 1986).

Because the PSA content of NCAM is highly regulated during vertebrate development, the potential exists that this inhibition is itself selective. That is, the ability of PSA to interfere with ligand encounter by steric hindrance would be expected to depend on the density and physical dimensions of different ligands. A possible example of such selectivity would be the outgrowth of axons during development. Many long tract axons express heavily sialylated NCAM during outgrowth. At first glance the presence of an interaction-inhibiting molecule during the formation of specific tracts would seem curious. However, the growing axon must be selective in its contact-dependent behavior, sensing environmental cues that help it choose a path while avoiding inappropriate interactions such as formation of junctions or synapses.

REFERENCES

Acheson, A., Rutishauser, U., 1988, NCAM regulates cell contact-mediated changes in choline acetyltranferase activity in embryonic chick sympathetic neurons, J Cell Biol., 106:479-486.

Adler, J.E., Black, .I.B., 1986, Membrane contact regulates transmitter phenotypic expression, Dev. Brain Res., 30:237-241.

Bixby, J.L., Pratt. J.S., Lilien J., Reichardt, L.F., 1987, Neurite outgrowth on muscle cell surfaces involves extracellular matrix receptors as well as Ca+-dependent and -independent cell adhesion molecules, Proc Natl Acad Sci USA., 84:255-2559.

Brackenbury, R., Rutishauser. U., Edelman, G.M., 1981, Distinct calcium-dependent adhesion systems of chick embryo cells, Proc Natl Acad Sci USA., 78:387-391.

Crossin, K.L., Edelman. G.M., Cunningham, B.A., 1984, Mapping of three carbohydrate attachment sites in embryonic and adult forms of the neural cell adhesion molecule, J Cell Biol., 99: 1848-1855.

Hatta, K., Takagi, S., Fujisawa, H., Takeichi, M., 1987, Spatial and temporal expression of N-cadherin adhesion molecules correlated with morphogenetic processes of chicken embryos, Dev Biol., 120:215-227.

Hoffman, S., Edelman, G.M., 1983, Kinetics of homophilic binding by embryonic and adult forms of the neural cell adhesion molecule, Proc Natl Acad Sci USA., 80:5762-5766.

Keane, R.W., Paramender, P.M., Rose, B., Lowenstein, W.R., Rutishauser, U., 1988, Neural differentiation, NCAM-mediated adhesion and gap junctional communication in neurotransmitters, J Cell Biol., 106:1307-1319.
Kitajima, K., Inoue, Y., Inoue, S., 1986, Polysialoglycoproteins of Salmonidae fish eggs, J Biol Chem., 261(12):5262-5269.
Landmesser, L., Dahm, L., Schultz, K., Rutishauser, U., 1988, Distinct roles for adhesion during innervation of embryonic chick muscle, Dev Biol., 130:645-670.
Rathjen, F.G., Wolff, J.M., Chang, S., Bonhoeffer, F., Raper, J.A., 1987, A novel chick cell surface glycoprotein which is involved in neurofascin-neurite in-interactions, Cell, 51:841-849.
Rutishauser, U., Acheson, A., Hall, A.K., Mann, D.M., Sunshine, J., 1988, The neural cell adhesion molecule (NCAM) as a regulator of cell-cell interactions, Science, 240:53-57.
Rutishauser, U., Grumet, M., Edelman, G.M., 1983, NCAM mediates initial interactions between spinal cord neurons and muscle cell in culture, J Cell Biol., 97:145-152.
Rutishauser, U., Jessell, T.M., 1988, Cell adhesion molecules in vertebrate neural development, Phys Rev., 68(3):819-857.
Rutishauser, U., Watanabe, M., Silver, J., Troy, F.A., Vimr, E.R., 1985, Specific alteration of NCAM-mediated cell adhesion by an endoneuraminidase, J Cell Biol., 101:1842-1849.
Sadoul, R., Hirn, M., Deagostini-Bazin, H., Rougon, G., Goridis, C., 1983, Adult and embryonic mouse neural cell adhesion molecules have different binding properties, Nature, 304:349.
Silver, J., Rutishauser, U., 1984, Guidance of optic axons in vivo by a preformed adhesive pathway on neuroepithelial endfeet, Dev Biol, 106:485-499.
Sunshine, J., Balak, K., Rutishauser, U., Jacobson, M., 1987, Changes in neural cell adhesion molecule (NCAM) structure during vertebrate neural development, Proc Natl Acad Sci USA., 84:5986-5990.
Takeichi, M., 1987, Cadherins: a molecular family essential for selective cell-cell adhesion and animal morphogenesis, TIGS., 3(8):213-217.
Tomaselli, K.J., Neugebauer, K.M., Bixby, J.L., Lilien, J., Reichardt, L.F., 1988, N-cadherin and integrins: Two receptor systems that mediate neuronal process outgrowth on astrocyte surfaces, Neuron 1:33-43.
Tosney, K.W., Watanabe, M., Landmesser, L., Rutishauser, U., 1986, The distribution of NCAM in the chick hindlimb during axon outgrowth and synaptogenesis, Dev Biol, 114:468-481.
Vimr, E.R., McCoy, R.D., Vollger, H.F., Wilkison, N.C., Troy, F.A., 1984, Use of prodaryotic-derived probes to identify poly(sialic acid) in neonatal membranes, Proc Natl Acad Sci USA., 81:1971-1975.

STRUCTURE AND FUNCTION OF THE NEURAL CELL ADHESION MOLECULES NCAM AND L1

Ole Nybroe and Elisabeth Bock

Research Center for Medical Biotechnology

The Protein Laboratory, University of Copenhagen
34, Sigurdsgade, DK-2200 Copenhagen N, Denmark

INTRODUCTION

During development of the nervous system, several morphogenetic processes act together to establish its orderly structure by generating specific nerve connections. The steps involved in neural pattern formation include cell differentiation, cell migration, axonal outgrowth, target recognition and synapse formation. Earlier, these processes were only characterized at the morphological level although it was obvious that adhesive interactions between cell surfaces and the extracellular milieu were of central importance. Several systems mediating cell-cell or cell-substratum adhesion have now been identified on the surface of neural cells and described at the molecular level (Jessell, 1988; Rutishàuser and Jessell, 1988).

Prospective neural cells segregate from non-neural cells during early developmental stages and acquire their identity as components of the nervous system. The initial segregation probably depends on differences in the adhesive properties between cell populations. Experimentally, segregation into two populations can occur when cells transfected with two different cell adhesion molecules are mixed *in vitro* (Nose et al., 1988). This indicates that sorting phenomena early in development could be a relatively simple process supported by only a few adhesion molecules as suggested by Edelman (1984b).

The primary induction of neural cells most likely depends on close contact between the reacting cells. This contact must be maintained for a sufficient period of time, but must also allow the cells to keep their position in the cell sheet, they are a part of. Thus, modulation of cell adhesion may play an important regulating role.

Later during development cells migrate from the division

Molecular Aspects of Development and Aging of the Nervous System
Edited by J.M. Lauder
Plenum Press, New York, 1990

zones to their final locations in the central or peripheral nervous system. A classical example of neural cell migration is the neural crest cells which migrate from their site of origin near the dorsal neural tube to their destination in e.g. peripheral ganglia. During this migration, the two important cell-cell adhesion systems on the migrating cells are turned off (Edelman, 1984a; Hatta et al., 1987). The adhesion system of importance to the migratory cells is cell-substratum adhesion between extracellular matrix components and their receptors on the cell surface (Bronner-Fraser, 1986). Consequently, in this system the adhesive properties of the neural crest cells are to some extent regulated by other cells that, at an earlier stage deposited the extracellular matrix.

Another well characterized migration system is that of the cerebellar granule cell that during development migrate from the external to the internal granule cell layer. Several adhesion systems are involved in this process (Lindner et al., 1986; Chuong et al., 1987; Persohn et al., 1987), reflecting the increasing complexity of adhesion systems during development (Jessell, 1988).

Elongation of axons and the path-finding function of growth cones also seem to be dependent on an intricate interplay between several cell-cell and cell-substratum adhesion systems (Chang et al., 1987; Matsunaga et al., 1988; Neugebauer et al., 1988). Conceivably a changed balance between these systems can affect pathway selection.
It is, however, difficult to envisage how cell migration and axon outgrowth could occur without a driving force to initiate the process and without a mechanism providing directional cues. This could be accomplished by gradients of soluble cues such as growth factors in combination with gradients of surface molecules giving positional information.

CELL ADHESION MOLECULES

The identification of neural cell adhesion molecules has depended on two important techniques: cell adhesion assays and production of monospecific - or monoclonal antibodies. If a single cell suspension is incubated under the appropriate conditions, cells carrying cell adhesion molecules on their surfaces should form aggregates. Such aggregation was observed in cell aggregation assays, and furthermore it became clear that two different types of cell adhesion systems co-existed on the cells: a Ca^{2+}-dependent and a Ca^{2+}-independent. By means of antibodies that blocked aggregation it was then possible to isolate the molecules that mediated adhesion. The Ca^{2+}-dependent adhesion molecule was N-cadherin (Takeichi, 1987) and two major Ca^{2+}-independent molecules were NCAM and L1 (Jessell, 1988; Nybroe et al., 1988).

Jessell (1988) and Rutishauser and Jessell (1988) have suggested a classification of adhesion molecules based on their functional properties. According to this classification, cadherins and NCAM are general adhesion molecules with a rather broad distribution. These molecules are responsible for adhesion at early developmental stages and are later supplemented by more specialized adhesion molecules as L1, that mediates axon outgrowth and fasciculation. Other even more specialized adhesion molecules, e.g. the fascilins and TAG have been described in vertebrate and invertebrate nervous systems.

NCAM

NCAM mediates cell adhesion by a homophilic adhesion mechanism where NCAM on the surface of one cell binds to NCAM on an opposing cell (Rutishauser et al., 1982). In addition to this binding function, NCAM can also bind to heparan sulfate and heparin (Cole and Glaser, 1986). Possibly these two distinct binding mechanisms cooperate at the cell surface (Cole and Glaser, 1986), but heparin does not interfere with NCAM homophilic binding in a solid phase binding assay (Moran and Bock, 1989). NCAM is thus a multifunctional protein that possibly mediates both cell-cell and cell-substratum adhesion.

NCAM is expressed by all major cell types in the nervous system, and simple adhesion assays have demonstrated that NCAM in brain is involved in neuron-neuron, neuron-astrocyte, astrocyte-astrocyte and oligodendrocyte-oligodendrocyte adhesion (Keilhauer et al., 1985; Bhat and Silberberg, 1988). In more complex model systems, NCAM can affect numerous intercellular events which has led to the theory that the function of the molecule is to regulate membrane to membrane contact required to initiate specific interactions between other molecules, see Rutishauser et al. (1988) for review.
Within the nervous system, NCAM appears as several closely related isoforms (Nybroe et al., 1988), and in muscle additional isoforms are expressed (Walsh, 1988). These isoforms are synthesized from at least five different mRNAs that are produced from a single gene by alternative splicing and polyadenylation (Goridis and Wille, 1988; Nybroe et al., 1888; Walsh, 1988). All the brain isoforms are identical in their extracellular NH_2-terminal part, but differ in the COOH-terminal membrane interacting - and cytoplasmic domains (Cunningham et al., 1987; Barbas et al., 1988). Two isoforms: NCAM-A (polypeptide MW 117,000) and NCAM-B (MW 91,000) are transmembrane proteins that differ in the length of their cytoplasmic domains (Gennarini et al., 1984a; Nybroe et al., 1985). The third major isoform NCAM-C (MW 79,000) has no transmembrane domain but is anchored to the membrane by phosphatidylinositol (He et al., 1986; Sadoul et al., 1986). NCAM-C can be released from the membrane and appear as a soluble polypeptide (He et al., 1987) and yet another soluble NCAM component of Mr 170,000 has been described (Bock et al., 1987).

In addition to the above mentioned post-transcriptional modulations, NCAM is subjected to several types of posttranslational modifications. Thus, the fully modulated isoforms migrate in SDS-PAGE with Mr 190,000 (NCAM-A), 140,000 (NCAM-B) and 120,000 (NCAM-C).

NCAM has seven potential sites for addition of aspara-

gin-linked carbohydrate, but probably only five of these are used (Lyles et al., 1984; Cunningham et al., 1987). Three carbohydrate cores are located to a central part of the molecule (Crossin et al., 1984), and one or more of these carries polysialic acid as unbranched alfa2,8 polymers (Finne, 1982). A form of NCAM with high polysialic acid content, which is expressed predominantly by fetal tissue, is 30% sialic acid by weight, whereas a low-sialic acid form expressed at very early embryonic stages and again in post-natal brain contains ca 10% sialic acid. The highly sialylated form of NCAM is less adhesive than the low-sialic acid form probably because the poly-sialic acid creates a highly hydrated domain around the molecule and interfering with adhesion by steric hindrance (Rutishauser et al., 1988).
The carbohydrate epitope recognized by the monoclonal antibodies HNK-1, NC-1 and L2 is present on 15-20% of the NCAM molecules (Kruse et al., 1984). The epitope has been shown to include a terminal 3-sulfate glucuronic acid and has been mapped to a 65 kD NH_2-terminal fragment of NCAM (Cole and Schachner, 1986). Sulfated carbohydrates are probably responsible for the sulfation of NCAM, as this labelling can be completely removed by Endo F treatment (Sorkin et al., 1984).

NCAM-A and NCAM-B are phosphorylated on serine and threonine residues by a process independent of biosynthesis (Gennarini et al., 1984b; Lyles et al., 1984; Sorkin et al., 1984). NCAM-A has more phosphorylation sites than NCAM-B (Linnemann et al., 1985) suggesting that some phosphorylation sites are located to the A-specific part of the cytoplasmic domain. The difference in phosphorylation between NCAM-A and NCAM-B is also emphasized by a difference in phosphoserine-/phosphothreonine ratio (Sorkin et al., 1984). Although it is generally assumed that phosphorylation takes place at the cytoplasmic side of the plasma membrane, a precise mapping of the phosphorylated sites on NCAM has not been performed. Sorkin and co-workers (1984) showed that ^{32}P was not incorporated into Fr1, which contains the major extracellular part of the molecule. However, Ehrlich et al. (1986) have demonstrated that antibodies to NCAM interfere with extracellular phosphorylation of unidentified membrane polypeptides, indicating that NCAM might be a protein kinase or a substrate for ecto-kinases.

To test the possibility that NCAM could function as a protein kinase, we incubated NCAM immunoprecipitates with gamma^{32}P-ATP in reaction buffer, see legend to Figure 1. However no ^{32}P was incorporated into NCAM suggesting that kinase activity was not associated with NCAM immunoprecipitates.

When cultures of fetal rat brain neurons were incubated with gamma^{32}P-ATP, label was incorporated into NCAM-A and NCAM-B, Figure 1, lane A, see legend for experimental procedures. Washing the cells with cold ATP after labelling did not affect the incorporation suggesting that the labelling was specific. The ATP-phosphorylation could be removed by alkaline phosphatase added after labelling, Figure 1, lane B, whereas the enzyme did only marginally affect the labelling of NCAM phosphorylated with inorganic ^{32}P-phosphate, Figure 1, lane C-D. These results suggested that NCAM could be phosphorylated by ecto-kinases. However labelling with Pi was very much stronger than with ATP, and removal of a small percentage of the Pi labelling by phosphatase might not be picked up on the autoradiograms. In order to compare ATP and Pi labelling, peptide maps of ATP-phosphorylated and Pi-

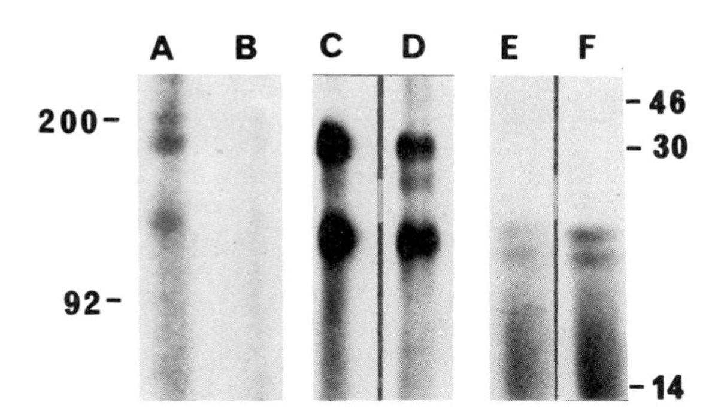

Figure 1.

Lanes A - B: Cultured fetal rat brain neurons grown in 24 well plates were labelled with 80 uCi pr well of gamma^{32}P-ATP for 15 min in a CO_2-incubator. The reaction medium was 145 mM NaCl, 5 mM KCl, 0.8 mM $MgCl_2$, 1.8 mM $CaCl_2$, 20 mM glucose, 10 mM Hepes/NaOH pH 7.4. After labelling the cells were washed 3 times in phosphate free DMEM. Cells were then incubated with 0 units (lane A) or 3 units (lane B) alkaline phosphatase in phosphate free DMEM for 30 min at room temperature. NCAM was isolated and analyzed by SDS-PAGE and autoradiography.
Lanes C - D: Cultured fetal neurons were labelled with 300 uCi pr well of inorganic $^{32}P-PO_4$ for 90 min. Remaining procedure as above. Lane C: 0 units phosphatase, lane D: 3 units phosphatase.
Lanes E - F. Peptide maps of NCAM phosphorylated with ATP (lane E) or Pi (lane F). Enzyme degradation was performed in the stacking gel for 30 min using 150 ug of trypsin.
Position of standard proteins (Mr 200,000; 92,000; 69,000; 46,000; 30,000; 14,000) are indicated in the margin.

phosphorylated NCAM were performed and found to be identical, Figure 1, lane E-F making extracellular labelling less likely.

The function of NCAM phosphorylation is not known. The cytoplasmic domain of NCAM-A is involved in the interaction with cytoskeletal elements (Pollerberg et al., 1987) and phosphorylation might affect this interaction. Also phosphorylation might be involved in the regulation of sidewards mobility in the membrane by charge interactions as demonstrated for chloroplast membrane components (Staehelin et al., 1983).

L1

The neural cell adhesion molecule L1 mediates neuron-neuron and neuron to glia adhesion (Grumet et al., 1984; Keilhauer et al., 1985). The binding mechanism is unknown but it has been suggested that the neuron-neuron binding mechanism is homophilic whereas neuron-glia binding occurs by a heterophilic interaction to an unknown molecule on glial cells (Grumet and Edelman, 1988). Like NCAM, L1 can bind to heparin (M. Olsen, O. Nybroe and E. Bock, unpublished observations) opening the possibility that L1 could have a role for cell-substratum interactions.

L1 is localized to the surface of axons, and the molecule is involved in axon elongation and fasciculation (Stallcup and Bearsley, 1985; Fischer et al., 1986; Chang et al., 1987; Linnemann et al., 1987) and in the initial phases of cerebellar granule cell migration (Lindner et al., 1986; Chuong et al., 1987; Persohn and Schachner, 1987).

L1 is synthesized as a single transmembrane protein with a molecular weight of 138,500 D (Moos et al., 1988). The polypeptide backbone is heavily modulated and migrates with an Mr of ca 210,000 in SDS-PAGE. After synthesis, L1 can break into fragments of Mr 180,000/30,000 or Mr 140,000/80,000 (Faissner et al., 1985; Sadoul et al., 1988). The Mr 180,000 and 140,000 L1 fragments are soluble NH_2-terminal fragments whereas the Mr 30,000 and 80,000 fragments contain the membrane spanning and cytoplasmic domains (Sadoul et al., 1988; Moos et al., 1988). A large proportion of the Mr 180,000 and 140,000 fragments remain attached to the membrane, possibly due to a strong interaction with the truly membrane integrated L1 fragments (Sadoul et al., 1988).

L1 is a member of the immunoglobulin superfamily and carries six Ig homology domains. Furthermore, LI has homology to the type III repeating unit of fibronectin that serves heparin- and cell binding functions in this molecule, and it carries two RDG sequences that might mediate interaction with Integrin-type receptors (Moos et al., 1988).

Complex type N-linked carbohydrates are added to L1, which has not less than 22 potential sites for extracellular glycosylation. Like NCAM, L1 carries the HNK-1/L2 carbohydrate epitope that has been shown to be involved in cell adhesion (Künemund et al., 1988).

The molecule is phosphorylated, predominantly if not exclusively on serine (Salton et al., 1983; Grumet et al., 1984; Faissner et al., 1985; Linnemann et al., 1987). Phosphorylation is associated with the Mr 30,000 and Mr 80,000 fragments both of which carry the cytoplasmic domain and it is generally assumed that L1 is phosphorylated on the cytoplasmic domain.

When cultured fetal rat brain neurons are labelled with gamma^{32}P-ATP, we frequently observe incorporation of label into the Mr 140,000 component, which is entirely extracellular, Figure 2, lane A. This suggests that additional extracellular phosphorylation sites may be present on L1. The kinase(s) responsible for L1 phosphorylation is not characterized. We found that L1-immunoprecipitates could be phosphorylated *in vitro* indicating that a kinase activity is closely associated with L1, Figure 2, lane B. As L1 does not show homology to any known kinase, the observed kinase activity is probably co-isolated with L1 rather than being a function of L1 itself.

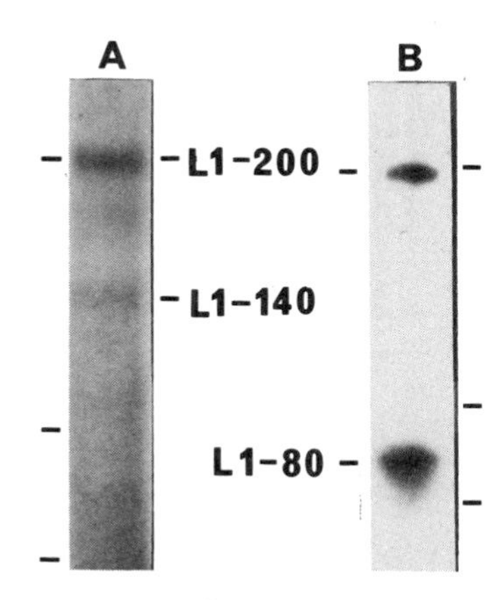

Figure 2

Lane A. Cultured fetal rat brain neurons were labelled with gamma^{32}P-ATP as described and L1 was immunoisolated and analyzed by SDS-PAGE and autoradiography.
Lane B. An immunoprecipitate of rat brain L1 was incubated with 10 uCi gamma^{32}P-ATP for 30 min at 37°C in reaction medium.
Position of standard proteins are indicated by bars in the margin.

The neural adhesion molecules NCAM and L1 undergo several modulations some of which are known to affect their function. This underlines the importance of fine tuning of these adhesion systems during development. Both molecules are phosphorylated but so far the functional role of this modification remains elusive.

ACKNOWLEDGEMENTS

The financial support of The Danish medical Research Council (12-6445) and Frode V. Nyegaard og hustrus Fond is gratefully acknowledged.

REFERENCES

Barbas J.A., Chaix J.C., Steinmetz M., and Goridis C., 1988, Differential splicing and alternative polyadenylation generates distinct NCAM transcripts and proteins in the mouse, EMBO J., 7:625.

Bhat S. and Silberberg D.H., 1988, Developmental expression of neural cell adhesion molecules of oligodendrocytes in vivo and in culture, J. Neurochem., 50:1830.

Bock E., Edvardsen K., Gibson A., Linnemann D., Lyles J.M., and Nybroe O., 1987, Characterization of soluble forms of NCAM, FEBS Lett., 225:33.

Bronner-Fraser M., 1986, An antibody to a receptor for fibronectin and laminin perturbs cranial neural crest development in vitro. Dev. Biol., 117:528.

Chang S., Rathjen F.G., and Raper J.A., Extension of neurites on axons is impaired by antibodies against specific neural cell surface glycoproteins, J. Cell Biol., 104:355.

Chuong C.-M., Crossin K.L., and Edelman G.M., 1987, Sequential expression and differential function of multiple adhesion molecules during the formation of cerebellar cortical layers. J. Cell Biol., 104:331.

Cole G.J. and Glaser L., 1986, A heparin-binding domain from N-CAM is in neural cell-substratum adhesion, J. Cell Biol., 102:403.

Cole C.J. and Schachner M., 1987, Localization of the L2 monoclonal antibody binding site on chicken neural cell adhesion molecule (NCAM) and evidence for its role in NCAM mediated adhesion, Neurosci. Lett., 78:227.

Crossin K.L., Edelman G.M., and Cunningham B.A., 1984, Mapping of three carbohydrate attachment sites in embryonic and adult forms of the neural cell adhesion molecule, J. Cell Biol., 99:1848.

Cunningham B.A., Hemperly J.J., Murray B.A., Prediger E.A., Brackenbury R., and Edelman G.M., 1987, Neural cell adhesion molecule: structure, immunoglobulin-like domains, cell surface modulation, and alternative RNA splicing, Science, 236:799.

Edelman G.M., 1984a, Cell-adhesion molecules. A molecular basis for animal form. Sci. Am., 250:118.

Edelman G.M., 1984b, Cell adhesion and morphogenesis: The regulator hypothesis. Proc. Natl. Acad. Sci., 81:1460.

Ehrlich Y.H., Davis T.B., Bock E., Kornecki E., and Lenox R.H., 1986, Ecto-protein kinase activity on the external surface of neural cells, Nature, 320:67.

Faissner A., Teplow D.B., Kübler D., Keilhauer G., Kinzel V., and Schachner M., 1985, Biosynthesis and membrane topography of the neural cell adhesion molecule L1, EMBO J., 4:3105.

Finne J., 1982, Occurrence of unique polysialosyl carbohydrate units in glycoproteins of developing brain, J. Biol. Chem., 257:11966.

Fischer G., Künemund V., and Schachner M., 1986, Neurite outgrowth patterns in cerebellar microexplant cultures are affected by the antibodies to the cell surface glycoprotein L1, J. Neurosci., 6:605.

Gennarini G., Hirn M., Deagostini-Bazin H., and Goridis C., 1984a, Studies of the transmembrane disposition of the neural cell adhesion molecule N-CAM. The use of liposome inserted radioiodinated NCAM to study its transbilayer orientation, Eur. J. Biochem., 142:65.

Gennarini G., Rougon G., Deagostini-Bazin H., Hirn M., and Goridis C., 1984b, Studies of the transmembrane disposition of the neural cell adhesion molecule N-CAM. A monoclonal antibody recognizing a cytoplasmic domain and evidence for the presence of phosphoserine residues, Eur. J. Biochem., 142:57.

Goridis C. and Wille W., 1988, Three size classes of mouse NCAM proteins arise from a single gene by a combination of alternative splicing and use of different polyadenylation sites, Neurochem. Int., 12:269.

Grumet M., Hoffman S., Chuong C,-M., and Edelman G.M., 1984, Polypeptide components and binding functions of neuron-glia cell adhesion molecules, Proc. Natl. Acad. Sci., 81:7989.

Grumet M. and Edelman G.M., 1988, Neuron-glia cell adhesion molecule interacts with neurons and astroglia via different binding mechanisms, J. Cell Biol., 106:487.

Hatta K., Takagi S., Fujisawa H., and Takeichi M., 1987, Spatial and temporal expression pattern of N-cadherin adhesion molecules correlated with morphogenetic processes of chicken embryos. Dev. Biol., 120:215.

He H.-T., Barbet J., Chaix J.C., and Goridis C., 1986, Phosphatidylinositol is involved in the membrane attachment of NCAM-120, the smallest component of the neural cell adhesion molecule, EMBO J., 5:2489.

He H.-T., Finne J., and Goridis C., 1987, Biosynthesis, membrane association, and release of N-CAM-120, a phosphatidylinositol-linked form of the neural cell adhesion molecule, J. Cell Biol., 105:2489.

Jessell T.M., 1988, Adhesion molecules and the hierarchy of neural development. Neuron, 1:3.

Keilhauer G., Faissner A., and Schachner M., 1985, Differential inhibition of neurone-neurone, neurone-astrocyte and astrocyte-astrocyte adhesion by L1, L2 and N-CAM antibodies, Nature, 316:728.

Kruse J., Mailhammer R., Wernecke H., Faissner A., Sommer I., Goridis C, and Schachner M., 1984, Neural cell adhesion molecules and myelin-associated glycoprotein share a common carbohydrate moiety recognized by monoclonal antibodies L2 and HNK-1, Nature, 311:153.

Künemund V., Jungalwala F.B., Fischer G., Chou D.K.H., Keilhauer G., and Schachner M., 1988, The L2/HNK-1 carbohydrate of neural cell adhesion molecules is involved in cell interactions, J. CEll Biol., 106:213.

Lindner J., Zinser G., Werz W., Goridis C., Bizzini B., and Schachner M., 1986, Experimental modification of postnatal cerebellar granule cell migration in vitro. Brain Res., 377:298.

Linnemann D., Lyles J.M., and Bock E., 1985, A developmental study of the biosynthesis of the neural adhesion molecule, Dev. Neurosci., 7:230.

Linnemann D., Nybroe O., Gibson A., Rohde H., Jørgensen O.S., and Bock E., 1987, Characterization of the biosynthesis, membrane association and function of the cell adhesion molecule L1, Neurochem. Int., 10:113.

Lyles J.M., Linnemann D., and Bock E., 1984, Biosynthesis of the D2-cell adhesion molecule: post-translational modifications, intracellular transport, and developmental changes, J. Cell Biol., 99:2082.

Matsunaga M., Hatta K., Nagafuchi A., and Takeichi M., 1988, guidance of optic nerve fibres by N-cadherin adhesion molecules, Nature, 334:62.

Moos M., Tacke R., Scherer H., Teplow D., Früh K., and Schachner M., 1988, Neural adhesion molecule L1 as a member of the immunoglobulin superfamily with binding domains similar to fibronectin, Nature, 334:701.

Moran N. and Bock E., 1988, Characterization of kinetics of NCAM homophilic binding, FEBS Lett., in press.

Nose A., Nagafuchi A. and Takeichi M., 1988, Expressed recombinant cadherins mediate cell sorting in model systems, Cell, 54:993.

Neugebauer K.M., Tomaselli K.J., Lilien J., and Reichardt L.F., 1988, N-cadherin, NCAM, and Integrins promote retinal neurite outgrowth on astrocytes in vitro, J Cell Biol., 107:1177.

Nybroe O., Albrechtsen M., Dahlin J., Linnemann D., Lyles J.M., Møller C.J., and Bock E., 1985, Biosynthesis of the neural cell adhesion molecule: Characterization of polypeptide C, J. Cell Biol., 101:2310.

Nybroe O., Linnemann D., and Bock E., 1988, NCAM biosynthesis in brain, Neurochem. Int., 12:252.

Persohn E. and Schachner M., 1987, Immunelectromicroscopic localization of the neural cell adhesion molecules L1 and N-CAM during postnatal development of the mouse cerebellum, J. Cell Biol., 105:569.

Pollerberg G.E., Burridge K., Krebs K.E., Goodman S.R., and Schachner M., 1987, The 180 kD component of the neural cell adhesion molecule is involved in cell-cell contacts and cytoskeleton-membrane interactions, Cell Tissue Res., 250:227.

Rutishauser U., Hoffmann S, and Edelman G.M., 1982, Binding properties of a cell adhesion molecule from neural tissue, Proc. Natl. Acad. Sci., 73:577.

Rutishauser U., Acheson A., Hall A.K., Mann D.M., and Sunshine J., 1988, The neural cell adhesion molecule (NCAM) as a regulator of cell-cell interactions, Science, 240:53.

Rutishauser U. and Jessell T.M., 1988. Cell adhesion molecules in vertebrate neural development, Phys. Rev., 68:819.

Sadoul K., Meyer A., Low M.G., and Schachner M., 1986, Release of the 120 kD component of the mouse neural cell adhesion molecule N-CAM from cell surfaces by phosphatidylinositol-specific phospholipase C, Neurosci. Lett., 72:341.

Sadoul K., Sadoul R., Faissner A., and Schachner M., 1988, Biochemical characterization of different molecular forms of the neural cell adhesion molecule L1, J. Neurochem., 50:510.

Salton S.R.J., Shelanski M.L., and Green L.A., 1983, Biochemical properties of the nerve growth factor-inducible large external (NILE) glycoprotein, J. Neurosci., 3:2420.

Sorkin B.C., Hoffmann S., Edelman G.M., and Cunningham B.A., 1984, Sulfation and phosphorylation of the neural cell adhesion molecule, N-CAM, Science, 225:1476.

Staehelin L.A. and Arntzen C.J., 1983, Regulation of chloroplast membrane function: Protein phosphorylation changes the spatial organization of membrane components, J. Cell Biol., 97:1327.

Stallcup W.P. and Beasley L., 1985, Involvement of the nerve growth factor-inducible large external glycoprotein (NILE) in neurite fasciculation in primary culture of rat brain, Proc. Natl. Acad. Sci., 82:1276.

Takeichi M., 1987, Cadherins: molecular family essential for selective cell-cell adhesion and animal morphogenesis, Trends Genet., 3:213.

Walsh F.S., 1988, The N-CAM gene is a complex transcriptional unit, Neurochem. Int., 12:263.

MONOSIALOGANGLIOSIDES AND THEIR ACTION IN MODULATING NEUROPLASTIC BEHAVIORS OF NEURONAL CELLS

Stephen D. Skaper, Guido Vantini, Laura Facci and Alberta Leon

Fidia Research Laboratories
Via Ponte della Fabbrica, 3/A
35031 Abano Terme, Italy

INTRODUCTION

The phenomenon of neuronal plasticity reflects the ability of nerve cells to modify their behaviors under the influence of extrinsic agents. As with any living system, neural tissue represents a dynamic organization, whose elements are continuously changing due to interactions with one another and with their extraneural environment. Neurons are thus subject to influences from the extracellular fluid and matrix, and the other cells with which they are in direct contact. This array of extrinsic influences impinging on the neuron constitutes, in broad terms, what can be called the microenvironment of these cells. Because of their many origins and functions, agents affecting neuronal behaviors represent a crucial and diverse element in determining how nerve cells will respond to cues from the microenvironment. As we shall see later on, these cues can carry either positive or negative signals for the neuron. Our ability to alter the response(s) of neuronal cells to such extrinsic agents will constitute a powerful tool for modulating the neuroplastic behaviors of the former - an important consideration for effecting regeneration and repair processes in the brain. Such is the subject of the present chapter.

NEURONOTROPHIC FACTORS

Before discussing ways to affect the action of extrinsic agents directed to neurons, we will briefly summarize what is known about one very important class of such agents - the neuronotrophic factors. Information concerning neuronotrophic factors derives, in large part, from the discovery and investigation of Nerve Growth Factor (NGF). In general terms, neuronotrophic factors are special proteins produced by the innervation territories of neurons where they are taken up by nerve terminals, and retrogradely transported along the axon to the cell body, to carry out their biological actions. The work of Levi-Montalcini (Levi-Montalcini, 1966, 1987) demonstrated that NGF exerts trophic control of sensory and sympathetic neurons of the peripheral nervous system (PNS). More recently, it has been shown that NGF also provides a trophic function for specific cholinergic neurons in the central nervous system (CNS) (for a review, see Korshing, 1986). This has been well-illustrated in the case of cholinergic basal forebrain neurons, both in vitro (Hefti et al., 1985) and in vivo (Hefti et al., 1984; Mobley et al., 1986). The

Molecular Aspects of Development and Aging of the Nervous System
Edited by J.M. Lauder
Plenum Press, New York, 1990

hippocampus and cortex have been shown to be major sites of NGF synthesis, possessing high levels of NGF messenger RNA (Korshing et al.,1985).

A number of other macromolecules endowed with neuronotrophic activity have been indentified since the discovery of NGF. These include Ciliary Neuronotrophic Factor (CNTF) (Barbin et al., 1984; Manthorpe et al., 1986; Watters and Hendry, 1987), Brain-Derived Neurotrophic Factor (BDNF) (Barde et al., 1982), fibroblast growth factors (Hatten et al., 1988; Unsicker et al., 1987; Walicke, 1988; Walicke et al., 1986; see also Ferrari et al., these proceedings) and epidermal growth factor (Morrison et al., 1987). The reader is referred to recent reviews on this subject for more information (Herschman, 1986; Varon et al., 1988).

WHY STUDY GANGLIOSIDES?

Gangliosides, sialic acid-containing glycosphingolipids, have been recognized for many years as characteristic constituents of the plasma membrane of all mammalian cells (Yamakawa and Nagai, 1978; Ando, 1983) although relatively little is known concerning their function in such membranes. Gangliosides are rich in CNS grey matter, suggesting that these molecules may contribute to the structural-functional features of neuronal membranes. For example, gangliosides undergo changes during neural development (Willinger and Schachner, 1980), in regenerating axons (Sbaschnig-Agler et al., 1984) and in certain neuropathological conditions (Purpura and Baker, 1977), while anti-GM1 ganglioside antibodies inhibit axonal elongation (Sparrow et al., 1984) and neurite regeneration (Spirman et al., 1982). Because of the localization of gangliosides on the external surface of the plasma membrane they have been implicated in a wide variety of phenomena including cell migration, aggregation, neurite outgrowth and and synaptogenesis.

GANGLIOSIDES AND THE POTENTIATION OF NEURONOTROPHIC FACTOR-MEDIATED NEUROPLASTIC RESPONSES OF NEURONS

Agents involved in modulating neuronal behaviors, in particular in response to neuronotrophic influences have been rarely studied, with the possible exception of gangliosides. Among the earliest attempts at understanding the role of the gangliosides in modulating the plastic behaviors of neuronal cells were experiments examining the effects of exogenous ganglioside addition to these cells. Neuronal cells, both primary and clonal, respond to gangliosides with a number of changes characteristic of cell differentiation, including neuritogenesis (Ferrari et al., 1983; Morgan and Seifert, 1979; Roisen et al., 1981; Leon et al., 1984; Katoh-Semba et al., 1984; Doherty et al., 1985; Varon et al., 1986), synaptogenesis (Obata et al., 1977), enhanced expression of neurotransmitter traits (Leon et al., 1988) and an increased synthesis of lipids concurrent with neurite production (Katoh-Semba et al., 1986) and tubulin mRNA (Rybak et al., 1983).

The above types of in vitro studies have yielded two key points with regard to ganglioside action. In cells with a trophic factor requirement for survival and/or neurite outgrowth, gangliosides will only improve the trophic action, but not replace the factor. This neuritogenic response of neurons to ganglioside is a time-related gain, i.e. an earlier onset of neurite regeneration, rather than a permanent increase in the number of neurite-bearing cells (Skaper et al., 1985; Varon et al., 1986). These concepts are clearly shown in Fig. 1, using explant cultures of sympathetic ganglia. NGF is required for neurite development (left panel), an effect which is enhanced by the concurrent presence of GM1 ganglioside (right panel).

These in vitro observations with monosialogangliosides (GM1) have provided impetus for applying the same compounds to rodent in vivo models of neuronal injury. In addition to potentiating neuronal cell responsiveness to exogenous neuronotrophic factors in cell culture, GM1 indeed seems able to improve the outcome, expressed in terms of neurotransmitter traits or even neuronal survival in some cases, following mechanical or chemical types of CNS insult. The literature contains many reports, but too numerous to fully discuss within the present article (cf. Ledeen et al., 1988 and Dal Toso et al., 1988, for recent reviews). Suffice it to say that the ability of GM1 to act in vivo appears to depend upon the extent of the lesion applied, suggesting the need for a minimum level of endogenous trophic support. This idea fits very well with what we already know from in vitro studies, where the facilitating action of GM1 is strictly dependent upon a proper balance between promoting and inhibiting influences acting in concert with the neuronorophic factor (Varon et al., 1986).

Recent findings support the hypothesis that gangliosides can potentiate neuronotrophic factor effects in vivo. Utilizing a PNS model, it was shown that exogenous GM1 facilitates the ability of NGF to antagonize vinblastine-induced sympathectomy in neonatal rats, as measured by evaluating noradrenergic innervation in the heart and spleen; GM1 alone had no effect on vinblastine action (Vantini et al., 1988). Figure 2 diagrammatically presents these results. Note that asialo GM1 (lacking sialic acid) was inactive. In another study, it was reported that both GM1 and NGF prevented the morphological and biochemical changes of lesioned rat basal forebrain neurons, with GM1 and NGF acting synergistically to stimulate choline acetyltransferase activity in cultured septal neurons (Cuello et al., 1988).

GANGLIOSIDES AND EXCITOTOXIC NEURONAL DEATH

Under normal circumstances, the excitatory amino acid glutamate released at the synaptic cleft acts post-synaptically, mediating important plastic behaviors like long term potentiation. Excessive release of glutamate under neuropathological conditions, however, such as cerebral ischemia, anoxia and hypoglycemia is also a good candidate for mediating the subsequent neuronal death (Rothman and Olney, 1986). Glutamate antagonists can be protective when applied to both animal models of ischemia and to in vitro neuronal cultures exposed to either exogenous glutamate or an anoxic environment.

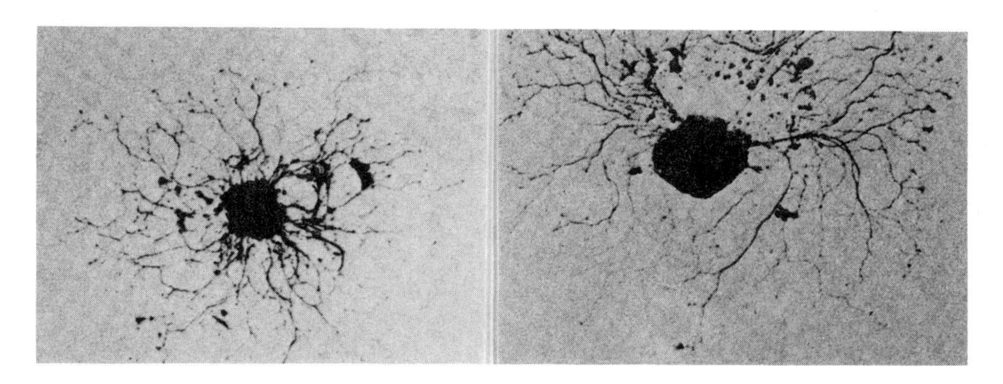

Fig. 1

Embryonic day 11 chicken sympathetic ganglia maintained for 48 hr in the presence of NGF (left) or NGF plus GM1 (right).

Another approach to the problem of excitotoxin-induced CNS injury has relied on gangliosides. The already observed efficacy of gangliosides in improving neural behaviors following brain lesions has led several laboratories to explore the action of these agents in mammalian cerebral ischemia models. Results to date from such studies indicate that monosialogangliosides are indeed able to reduce a number of deficits induced by the ischemic episode (Karpiak et al., 1987; Rubini et al., 1988; Seren et al., 1988; Tanaka et al., 1986). In vitro, pretreatment with gangliosides is reported to protect against the neurotoxicity of exogenous glutamate (Favaron et al., 1988). In our laboratory, we have observed a neuroprotective effect of GM1 for cerebellar granule cells maintained under anoxia, i.e. an environment that simulates a partial ischemic condition in vivo (Facci et al. 1988). Figure 3 shows this effect at the light microscopic level, and that obtained by using exogenous glutamate directly.

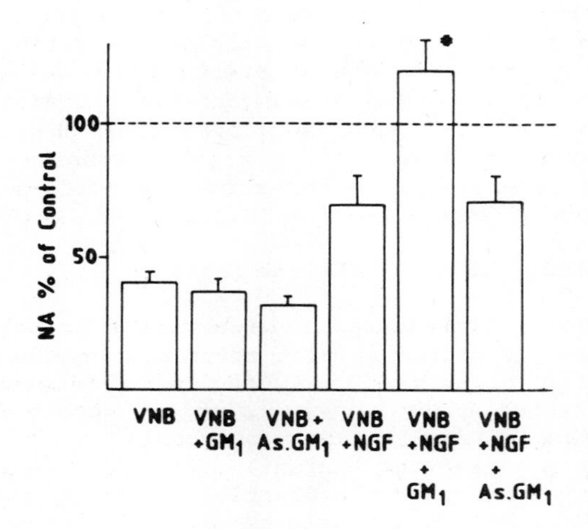

Fig. 2

Effect of vinblastine (VNB) and/or NGF ± GM1 treatments on the noradrenaline (NA) content in heart of 6 day-old rats. Asialo GM1 (As. GM1). * $p < 0.01$ vs. VNB + NGF. Experimental details are essentially as described in Vantini et al. (1988).

CONCLUSION

The capability of a CNS neuron to undergo adaptive changes in response to microenvironmental cues - neuroplasticity - represents a finely tuned balance between promoting and inhibiting influences. Disruption of this mechanism by axotomy or disruption of the blood

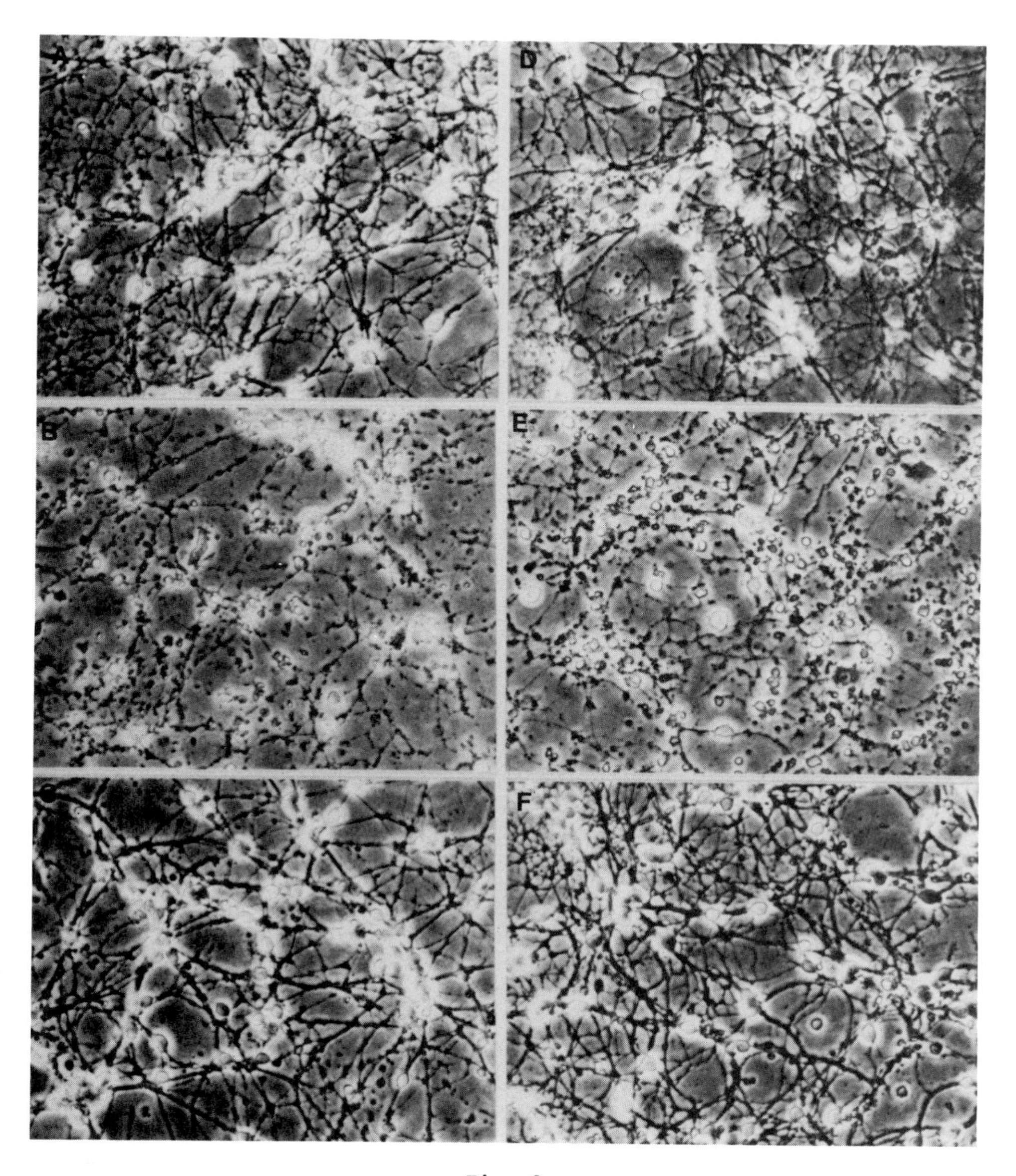

Fig. 3

Ganglioside GM1 prevents anoxia- and glutamate-induced death of cerebellar granule neurons. Cells prepared from 8-day rat pups were maintained in vitro 12 days, followed by GM1 treatment (100 µM) for 2 hr prior to a 5 hr period of O_2 deprivation or a 3 hr pulse of 500 µM L-glutamate. Cultures were photographed 24 hr later. (A,D) Control; (B) anoxia; (C) GM1 + anoxia; (E) glutamate; (F) GM1 + glutamate .

supply are probably the two major causes of neuronal death in the CNS: the first, by deprivation of specific neuronotrophic factors, the second, by hyperstimulation due to excessive amounts of excitatory amino acid transmitters. In some instances, both sequelae may be operative. Functional repair/recovery of the damaged CNS will require, at least in part, interventions that are directed at manipulating the brain's own plastic reactions. The information discussed in this review, we believe, lends encouragement to the potential pharmacological action of gangliosides in improving the recovery of nerve cell distress or impairment due to insufficiency of neuronotrophic influences or excesses of neuronotoxic activities.

REFERENCES

Ando, S., 1983, Gangliosides in the nervous system, Neurochem. Internatl., 5:507.

Barbin, G., Manthorpe, M., and Varon, S., 1984, Purification of the chick eye ciliary Neuronotrophic Factor (CNTF), J. Neurochem., 43:1468.

Barde, Y.-A., Edgar, D., and Thoenen, H., 1982, Purification of a new neurotrophic factor from mammalian brain, EMBO J., 1:549.

Cuello, A.C., Kenigsberg, R., Maysinger, D., and Garofolo, L., 1988, Application of gangliosides and nerve growth factor to prevent cholinergic degeneration in the central nervous system, in: "New Trends in Ganglioside Research: Neurochemical and Neuroregenerative Aspects", R.W. Ledeen, E.L. Hogan, G. Tettamanti, A.J. Yates, and R.K.Yu, eds., Fidia Research Series, Vol. 14, Liviana Press, Padova

Dal Toso, R., Skaper, S.D., Ferrari, G., Vantini, G., Toffano, G., and Leon, A., 1988, Ganglioside involvement in membrane-mediated transfer of trophic information. Relationship to GM1 effects following CNS injury, in: "Pharmacological Approaches to the Treatment of Spinal Cord Injury," D.G. Stein, and B.A. Sabel, eds., Plenum Press, New York.

Doherty, P., Dickson, J.G., Flanigan, T.P., and Walsh, F.J., 1985, Ganglioside GM1 does not initiate, but enhances neurite regeneration of nerve growth factor-dependent sensory neurones, J. Neurochem., 44:1259.

Facci, L., Milani, D., Leon, A., and Skaper, S.D., 1988, Monosialogangliosides antagonize anoxia-induced neuronal injury in vitro, Pharmacol. Res. Comm., 20 (Suppl. 2): 141.

Favaron, M., Manev, H., Alho, H., Bertolino, M., Ferret, B., Guidotti, A., and Costa, E., 1988, Gangliosides prevent glutamate and kainate neurotoxicity in primary neuronal cultures of neonatal rat cerebellum and cortex, Proc. Natl. Acad. Sci. USA, 85:7351.

Ferrari, G., Fabris, M., and Gorio, A., 1983, Gangliosides enhance neurite outgrowth in PC12 cells, Dev. Brain Res., 8:215.

Hatten, M.E., Lynch, M., Rydel, R.R., Sanchez, J., Joseph-Silverstein, J., Moscatelli, D., and Rifkin, D.B., 1988, In vitro neurite extension by granule neurons is dependent upon astroglial-derived fibroblast growth factor, Dev. Biol., 125:280.

Hefti, F., Dravid, A., and Hartikka, J., 1984, Chronic intraventricular injections of nerve growth factor elevate hippocampal choline acetyltransferase activity in adult rats with partial septo-hippocampal lesions, Brain Res., 293:305.

Herschman, H., 1986, Polypeptide growth factors and the CNS, Trends Neurosci., 9:53.

Katoh-Semba, R., Skaper, S.D., and Varon S., 1984, Interaction of GM1 ganglioside with PC12 pheochromocytoma cells: serum and NGF-dependent effects on neurite growth (and proliferation), J.

Neurosci. Res., 12:299.
Karpiak, S.E., Li, Y.S., and Mahadik, S.P., 1987, Gangliosides (GM1 and AGF2) reduce mortality due to ischemia: protection of membrane function, Stroke, 18:184.
Katoh-Semba, R., Skaper, S.D., and Varon, S., 1986, GM1 ganglioside treatment of PC12 cells stimulates ganglioside, glycolipid and lipid, but not glycoprotein synthesis independently from the effects of nerve growth factor, J. Neurochem., 46:574.
Korsching, S., 1986, The role of nerve growth factor in the CNS, Trends Neurosci., 9:570.
Korsching, S., Auburger, G., Heumann, R., Scott, J., and Thoenen, H., 1985, Levels of NGF and its mRNA in the central nervous system of the rat correlate with cholinergic innervation, EMBO J., 4:1389.
Ledeen, R.W., Hogan, E.L., Tettamanti, G., Yates, A.J., and Yu, R.K., "New Trends in Ganglioside Research: Neurochemical and Neuro-regenerative Aspects," Fidia Research Series, Vol. 14, Liviana Press, Padova.
Leon, A., Dal Toso, R., Presti, D., Facci, L., Giorgi, O., and Toffano, G., 1984, Dorsal root ganglia and nerve growth factor: a model for understanding the mechanism of GM1 effects on neuronal repair, J. Neurosci. Res., 12:277.
Leon, A., Dal Toso, R., Presti, D., Benvegnù, D., Facci, L., Kirschner, G., Tettamanti, G., and Toffano, G., 1988, Development and survival of neurons in dissociated fetal mesencephalic serum-free cell cultures: II. Modulatory effects of gangliosides, J. Neurosci., 8:746.
Levi-Montalcini, R., 1966, Nerve growth factor: its mode of action on sensory and sympathetic neurons, Harvey Lect., 60:217.
Levi-Montalcini, R., 1987, The Nerve Growth Factor thirty five years later, Science, 237:1154.
Manthorpe, M., Skaper, S.D., Williams, L.R., and Varon S., 1986, Purification of adult rat sciatic nerve ciliary neuronotrophic factor, Brain Res., 367:282.
Mobley, W.C., Rutowski, J.L., Tennekoon, G.I., Gemski, J., Buchanan, K., and Johnston, M.V., 1986, Nerve growth factor increases choline acetyltransferase activity in developing basal forebrain neurones, Mol. Brain Res., 1:53.
Morgan, J.I., and Seifert, W., 1979, Growth factors and gangliosides: a possible new perspective in neuronal growth control, J. Supramol. Struct., 10:110.
Morrison, R.S., Kornblum, H.I., Leslie, F.M., and Bradshaw, R.A., 1987, Trophic stimulation of cultured neurons from neonatal rat brain by epidermal growth factor, Science, 238:72.
Obata, K., Oide, M., and Handa, S., 1977, Effects of glycolipids on in vitro development of neuromuscular junction, Nature, 266:553.
Purpura, D.P., and Baker, H.J., 1977, Neurite induction in mature cortical neurons in feline GM1-ganglioside storage disease, Nature, 266:553.
Roisen, F.J., Bartfeld, H., Nagele, R., and Yorke, G., 1981, Ganglioside stimulation of axonal sprouting in vitro, Science, 214:577.
Rothman, S., and Olney, J.W., 1986, Glutamate and the pathophysiology of hypoxic-ischemic brain damage, Ann. Neurol., 19:105.
Rubini, R., Fogarolo, F., Biasiolo, F., Fiori, M.G., Seren, S., Lazzaro, A., and Leon, A., 1988, Brain ischemia in rat: pharmacologic evidence for the efficacy of AGF2 (inner ester of monosialoganglioside GM1) as shown by computerized EEG, Pharmacol. Res. Comm., 20 (Suppl. 2): 334.
Rybak, S., Ginsburg, I., and Yavin, E., 1983, Gangliosides stimulate neurite outgrowth and induce tubulin mRNA accumulation in neural cells, Biochem. Biophys. Res. Commun., 116:974.
Sbaschnig-Agler, M., Ledeen, R., and Grafstein, B., 1984, Ganglioside

changes in the regenerating goldfish optic system: comparison with glycoproteins and phospholipids, J. Neurosci. Res., 12:221.

Seren, M.S., Rubini, R., Lazzaro, A., Zanoni, R., Zanetti, A., Fiori, M.G., Toffano, G., and Leon, A., 1988, Monosialoganglioside effects following transitory global cerebral ischemia in rodents, Pharmacol. Res. Comm., 20 (Suppl. 2):355.

Skaper, S.D., Katoh-Semba, R., and Varon, S., 1985, GM1 ganglioside accelerates neurite outgrowth from primary peripheral and central neurons under selected culture conditions, Dev. Brain Res., 23:19.

Sparrow, J.R., Mc Guinness, C., Schwartz, M., and Grafstein, B., 1984, Antibodies to ganglioside inhibit goldfish optic nerve regeneration in vivo, J. Neurosci. Res., 12:233.

Spirman, N., Sela, B.A., and Schwartz, M., 1982, Antiganglioside antibodies inhibit neuritic outgrowth from regenerating goldfish retinal explants, J. Neurochem., 39: 874.

Tanaka, K., Dora, E., Urbanics, R., Greenberg, J.H., Toffano, G., and Reivich, M., 1986, Effects of the ganglioside GM1, on cerebral metabolism, microcirculation, recovery kinetics of ECoG and histology, during the recovery period following focal ischemia in cats, Stroke, 17:1170.

Unsicker, K., Reichert-Preibsch, H., Schmidt, R., Pettmann, B., Labourdette, G., and Sensenbrenner, M., 1987, Astroglial and fibroblast growth factors have neurotrophic functions for cultured peripheral and central nervous system neurons. Proc. Natl. Acad. Sci. USA, 84:5459.

Vantini, G., Fusco, M., Bigon, E., Leon, A., 1988, GM1 ganglioside potentiates the effect of nerve growth factor in preventing vinblastine-induced sympathectomy in newborn rats, Brain Res., 448:252.

Varon, S., Manthorpe, M., Davis, G.E., Williams, L.R., and Skaper, S.D., 1988, Growth factors, in: "Functional Recovery in Neurological Disease", S.G. Waxman, ed., vol. 47, Raven Press, New York.

Varon, S., Skaper, S.D., and Katoh-Semba, R., 1986, Neuritic responses to GM1 ganglioside in several in vitro systems, in: "Neuronal Plasticity and Gangliosides", G. Tettamanti, R. Ledeen, K. Sandhoff, Y. Nagai, and G. Toffano, eds., Fidia Research Series, Vol. 6, Liviana Press, Padova.

Walicke, P.A., 1988, Basic and acidic fibroblast growth factors have trophic effects on neurons from multiple CNS regions, J. Neurosci., 8:2618.

Walicke, P., Cowan, W.M., Ueno, N., Baird, A., and Guillemin, R., 1986, Fibroblast growth factor promotes survival of dissociated hippocampal neurons and enhances neurite extension, Proc. Natl. Acad. Sci. USA, 83:3012.

Watters, D.J., and Hendry, I.A., 1987, Purification of a ciliary neurotrophic factor from bovine heart, J. Neurochem., 49:705.

Willinger, M., and Schachner, M., 1980, GM1 ganglioside as a marker for neuronal differentiation in mouse cerebellum, Dev. Biol., 74:101.

Yamakawa, T., and Nagai, Y., 1978, Glycolipids at the cell surface and their biological functions, Trends Biochem. Sci., 3:128.

SEROTONIN AND MORPHOGENESIS IN THE CULTURED MOUSE EMBRYO

Dana L. Shuey[1], Mark Yavarone[2], Thomas W. Sadler[2], and Jean M. Lauder[1,2]

1 Curriculum in Toxicology, University of North Carolina, School of Medicine, Chapel Hill, NC 27599
2 Department of Cell Biology and Anatomy, University of North Carolina, School of Medicine, Chapel Hill, NC 27599

INTRODUCTION

Morphogenesis is the process whereby embryonic tissues and organs are formed through cell movements, changes in cell shape and cell-cell interactions. Not only are morphogenetic cell movements important in the establishment of tissue form, but they are also believed to play an important role in bringing interactive tissues into contact. For example, it has been demonstrated that the dorsal lip of the blastopore migrating beneath the overlying ectoderm is responsible for induction of the nervous system during gastrulation (Spemann & Mangold, 1924).

SEROTONIN AND MORPHOGENESIS

In recent years, evidence has accumulated to suggest that neurotransmitters may have evolved to their specific functions from more general roles in intercellular communication, and that this evolution may be reflected in development (for review, see Buznikov, 1980). The presence of serotonin (5-HT) and a number of other neurotransmitters has been demonstrated during early stages of development in several invertebrate and vertebrate species (for review, see Lauder, 1988). For example, Buznikov and his colleagues have demonstrated the presence of 5-HT in sea urchin embryos from early cleavage division stages through gastrulation (Buznikov et al., 1964). Various neuropharmacological agents cause defects in embryos at these stages (Buznikov et al., 1970; Buznikov, 1984). A significant delay of cleavage is caused by treatment with 5-HT and specific 5-HT receptor agonists (Renaud et al., 1983). Sensitivity to such

Molecular Aspects of Development and Aging of the Nervous System
Edited by J.M. Lauder
Plenum Press, New York, 1990

compounds can be modulated by calcium ionophores (Buznikov et al., 1984), cyclic nucleotides (Shmukler et al., 1984), and compound lipophilicity (Landau et al., 1981) suggesting the involvement of specific intracellular receptors. It has been hypothesized that in cleavage stage embryos, 5-HT is involved in interblastomere signalling which may contribute to the development of cellular heterogeneity and regulation of cell division, possibly through an effect on contractile elements of the cytoskeleton (Buznikov, 1984; Buznikov & Shmukler, 1981), and formation of the cleavage furrow (Renaud et al., 1983). Treatment of gastrulation stage sea urchin embryos with 5-HT analogs inhibits migration of the primary mesenchyme (Gustafson & Toneby, 1970; 1971) which is associated with inhibition of intracellular contractile element activity (Buznikov, 1984). Similar results have been obtained in early starfish (Buznikov et al., 1984) and other invertebrate embryos (Emanuelsson, 1974). The capacity for specific uptake of biogenic amines has also been demonstrated in unfertilized eggs and early cleavage embryos of the rat (Burden & Lawrence, 1973).

Serotonin and other neurotransmitters have also been associated with morphogenetic events in vertebrate embryos. Treatment of chick embryos with 5-HT and related compounds during early stages of morphogenesis results in disruption of a number of developmental events (Palen et al., 1979), which are associated with impaired ability of embryonic cells to change shape, suggesting an effect on cytoskeletal elements. Accumulation of catecholamines in the neural tube and several non-neural structures has been demonstrated in chick embryos. Uptake of these compounds into the neural tube and notochord is related spatio-temporally with axial flexure, and uptake into the gut mesenchyme appears to be correlated with movement of the intestinal portal (Newgreen et al., 1981). Wallace (1982) observed similar patterns of monoamine uptake in the neural tube and notochord during neurulation in the chick which were correlated with rostro-caudal progression of neural tube closure.

In addition, 5-HT and other neurotransmitters appear to play a role in palate development (for review, see Zimmermen & Wee, 1984). Serotonin has been shown to stimulate mouse palatal mesenchyme migration, as well as palatal shelf elevation and reorientation _in vitro_ (Wee et al., 1979; Zimmerman et al., 1983; Venkatasubramanian & Zimmerman, 1983).

SEROTONIN IN CRANIOFACIAL MORPHOGENESIS AND NEURAL CREST DEVELOPMENT

Recent studies in our laboratory have been undertaken to determine whether 5-HT may play a role in morphogenesis of the mouse embryo (Lauder et al., 1988; Lauder et al., 1987; Lauder & Zimmerman, 1988; Zimmerman & Lauder, 1987). Using the method of whole embryo culture together with immunocytochemistry, we have identified sites of 5-HT uptake in the surface epithelium of the developing craniofacial region in the midgestation **(day 12)** mouse embryo (Lauder et al., 1988). In order to identify possible sites of 5-HT binding in these regions, immunocytochemistry using an antiserum to the 45K form of serotoninbinding protein (SBP; Kirschgessner et al., 1987; Liu et al., 1985; 1987; Tamir, 1983) was performed. SBP-like immunoreactivity (SBP IR) was found in specific regions of craniofacial mesenchyme, adjacent to sites of 5-HT uptake in the epithelial structures of this region. Development of craniofacial skeletal and connective tissue structures involves complex and reciprocal interactions between chondrogenic and osteogenic mesenchyme and overlying surface epithelium (Bee & Thorogood, 1980). The pattern of 5-HT uptake and SBP distribution observed in our studies suggests that 5-HT may be involved in these interactions during craniofacial morphogenesis in the mouse. The observed sites of 5-HT uptake in craniofacial epithelia are transient in nature, having largely disappeared by **day 14** of gestation (Lauder et al., 1988). SBP IR is also no longer observed in the adjacent mesenchyme, but is found in sites of active chondrogenesis in the craniofacial region at this time, further supporting the role of 5-HT in the development of chondrogenic and osteogenic cells.

Skeletal and connective tissue elements of the craniofacial region are derived largely from cranial neural crest cells (Thorogood, 1981; LeDouarin, 1982). These cells arise from the junction between the neural and surface epithelium, undergo an epithelial-mesenchymal transformation and migrate into the primary mesoderm. In the mouse embryo, migration begins prior to neural tube closure, at approximately the 4-6 somite stage (Nichols, 1981). Throughout their extensive migration, cranial neural crest cells are responsive to signals from the overlying surface epithelia, which influence their differentiative potential (Bee & Thorogood, 1980). By the stages of development used in our previous studies (gestational days 12 and 14), the primary migration of cranial neural crest is nearly complete, although secondary migration and tissue rearrangement at the presumptive sites of differentiation is still ongoing (Johnston, 1975). In order to study the possible role of 5-HT in craniofacial development during the major stages of neural crest migration, we have recently conducted studies identifying sites of 5-HT upatke and synthesis as well as SBP IR beginning at the headfold stage (day 9, 4-6 somites; Lindemann Shuey et al., in preparation).

In these studies **day 9 (headfold stage)** mouse embryos were grown in whole embryo culture for 3-4 hours in the presence of 5-HT and related compounds followed by immunocytochemistry using a specific antiserum to 5-HT. Following exposure of embryos to 5-HT in the presence of a monoamine oxidase (MAO) inhibitor, intense 5-HT immunoreactivity (5-HT IR) was observed in the embryonic heart, visceral yolk sac and pharyngeal ectoderm (which is known to interact with and induce cartilage in maxillary and mandibular neural crest derived mesenchyme; Holtfreter, 1968). Little 5-HT IR was observed in embryos treated with 5-HT in the absence of an MAO inhibitor suggesting that these embryos have the capacity to rapidly metabolize 5-HT. Treatment of embryos with 5-HT precursors (L-tryptophan or 5-hydroxytryptophan) gave results similar to 5-HT itself, suggesting that these embryos also have the capacity for

5-HT synthesis. Pretreatment of embryos with 5-HT uptake inhibitors (fluoxetine or amitriptyline) greatly decreased 5-HT IR observed at all sites, suggesting that these are largely sites of uptake. However, 5-HT IR in the heart was variably blocked by uptake inhibitors suggesting that this may be a weak site of 5-HT synthesis at this stage. (Note that by day 10 the heart has a much stronger capacity for 5-HT synthesis, as described below). Immunocytochemistry using an antiserum to the 45K form of SBP revealed that SBP was distributed throughout the mesenchyme, with the exception of the area where neural crest cells were emerging from the neural epithelium. This result suggests that 5-HT and SBP may play a role in the initiation of neural crest cell migration. It is unclear what factors control this process, although the involvement of differential intercellular adhesion (Newgreen, 1985), changes in the extracellular matrix (Pratt et al., 1975), and passive displacement resulting from elevation of the neural tube (Nichols, 1986) have all been implicated.

In addition to the embryonic sites of 5-HT uptake in the headfold stage embryo, intense 5-HT IR was also observed in the ectoplacental cone of these embryos following incubation in medium containing 5-HT (Lauder et al., 1989). Pharmacological treatments with precursors and uptake inhibitors, as described above, indicated that this was due largely to 5-HT uptake. These results suggest that the maternal circulation, which bathes the ectoplacental cone during invasion and implantation may be an important source of 5-HT for the embryo. Placentae from older untreated embryos also exhibited 5-HT IR which seemed to be most intense in tissue adjacent to maternal blood vessels. This suggests that the maternal circulation may provide a longterm source of 5-HT during embryogenesis.

In **day 10 embryos** (approximately 20 somites), uptake of 5-HT was observed in the surface epithelium of the head and visceral arches, which is in close contact with underlying neural crest derived mesenchyme migrating adjacent to it at this stage. These sites were seen after incubation of embryos with 5-HT in the presence of an MAO inhibitor, but not after incubation with 5-HT precursors. As in day 9 embryos, uptake was observed in the pharyngeal ectoderm and heart. At this stage, treatment with precursors and uptake inhibitors strongly indicated that the heart is a site of 5-HT synthesis. Additional sites of 5-HT IR at this stage included the neural tube and thyroid. Neural tube staining was restricted to the hindbrain region and the most dorsal aspects or the neural folds, and appeared to be a remnant of staining observed at day 9 1/2 (12-13 somites) when discrete regions of the neural tube showed very intense 5-HT IR following incubation with 5-HT in the presence of an MAO inhibitor. The significance of this staining is unclear, but because it was not observed at day 9 when the neural folds are elevating and the neural crest is beginning to migrate, we do not beleive it is related to these processes. SBP IR in the day 10 embryo remained widespread throughout the craniofacial mesenchyme with no apparent regions of specific localization.

In the **day 11 embryo** (approximately 30 somites), 5-HT IR in the surface epithelium of the head and visceral arches was much more intense than at day 10. Moreover, these sites were also seen after incubation with 5-HT precursors, which was not the case at day 10. 5-HT IR at these sites was blocked by pretreatment with fluoxetine. Tissue recombination studies using neural crest derived mesenchyme and facial epithelia from various age mouse embryos indicate that these epithelia do not acquire the ability to induce chondrogenesis until day 11 (Hall, 1980). Thus, the changes in 5-HT uptake and synthesis at these sites between days 10 and 11 may be meaningful in terms of the possible involvement of 5-HT in these processes. The surface epithelium in the nasal prominences also stained very intensely at this stage after incubation with 5-HT and precursors, and this staining was restricted to the edges of the epithelium which are invaginating and fusing at this time. 5-HT uptake into the heart and the gut was still observed, and the otocyst, which showed uptake at day 12 but not at day 10, began to exhibit 5-HT IR at day 11. Of particular interest at this stage was the 5-HT IR observed in the developing eye, which is undergoing lens induction. At this stage, under inductive influences from the optic cup, the ectoderm overlying the optic cup thickens and invaginates to form the lens (Twitty, 1955). Following incubation of embryos in medium containing 5-HT and an MAO inhibitor, the invaginating ectoderm exhibited 5-HT IR, but this site could not be demontrated following incubation with 5-HT precursors, and was blocked by pretreatment with fluoxetine, suggesting that it is not a site of synthesis. SBP immunocytochemistry revealed a distribution similar to that observed in the day 12 embryo, i.e. SBP IR was more restricted to the mesenchyme immediately subjacent to sites of 5-HT uptake in the overlying epithelium, compared to day 10 embryos where SBP IR appeared widespread throughout the mesenchyme.

In summary, patterns of 5-HT uptake and SBP distribution during critical stages of craniofacial morphogenesis and cranial neural crest development suggest a role for 5-HT in these events. At **day 9 of gestation**, 5-HT uptake was observed in the pharyngeal ectoderm which is known to interact with and influence the chondrogenic differentiation of neural crest cells of the first visceral arch. SBP at this stage was observed throughout the mesenchyme with the exception of the region where neural crest cells were emerging from the epithelium. At **day 10**, surface epithelia along neural crest cell migratory routes and at presumptive sites of neural crest cell differentiation demonstrated 5-HT uptake which became even more pronounced by **day 11**, the stage at which these epithelia acquire their chondrogenic inductive ability (Hall, 1980). SBP IR, which was initially widespread throughout the mesenchyme, became more localized as sites of 5-HT uptake in overlying epithelia became more intense. Specific 5-HT staining in the invaginating epithelium (surface ectoderm) of the eye and localization in the nasal prominence at the point of invagination and fusion at day 11, suggest that 5-HT may be involved in several types of morphogenetic movements. These studies have demonstrated that embryos at these stages have the capacity for 5-HT uptake and metabolism and that the embryonic heart, the placenta and maternal circulation may provide important sources of 5-HT during embryogenesis.

SEROTONIN IN DEVELOPMENT OF THE HEART

In the course of studies in the craniofacial region, sites of 5-HT IR were discovered in the myocardium of the cultured mouse embryo. In the mouse, the heart tube has formed from fusion of paired primordia by the beginning of gestational day 9 (4-6 somites). During days 9 and 10 of gestation, lateral bending of the heart tube occurs, until by the end of day 10 (27-28 somites), the external contour of the heart foreshadows its division into four chambers.

After culture of day 9 mouse embryos in medium supplemented with 5-HT and staining with 5-HT antiserum, 5-HT IR is apparent throughout the developing ventricle and bulbus arteriosus. A similar pattern of immunoreactivity is seen when the 5-HT precursor L-tryptophan is substituted for 5-HT in the culture medium, indicating that the heart may be a site of 5-HT synthesis at this time. As in the craniofacial region, sites of 5-HT IR in the heart are dynamic. By the beginning of day 10 (21-23 somites), 5-HT IR in the ventricle has almost disappeared, and intense immunostaining is confined to the myocardium of the developing truncoconal region (outflow tract) and atrioventricular canal. These are the regions where endocardial cushions, which form septal and valvular primordia, will develop.

A major event in the formation of the endocardial cushions is the transdifferentiation of endothelial cells into mesenchyme, a process known as "seeding," which has been studied extensively in avian embryos (Markwald, et al., 1976; Fitzharris and Markwald, 1982). Beginning at approximately stage 17 (29-32 somites) some endothelial cells in cushion-forming regions become "activated" to form mesenchyme. These cells hypertrophy, lose their cell-cell contacts, and invade the cardiac jelly interposed between the myocardium and endothelium. The invading cells become the progenitor population for endocardial cushion mesenchyme. (Observations in our laboratory indicate that invasion of cardiac jelly by these cells begins somewhat earlier in the mouse than in the chick, and is underway by the 24-25 somite stage.) These progenitor cells migrate toward the myocardium, apparently undergoing mitosis during migration, until approximately stage 24-25 (c. 49 somites) when the cushions have become evenly populated with mesenchymal cells. Localization of 5-HT IR to the myocardium of the cushion-forming regions shortly before the onset of seeding raises the prospect that 5-HT is involved in the selective ability of the endothelium in these regions to form endocardial cushion tissue. This possibility is supported by the fact that sites of SBP IR have been identified throughout the endocardium, including the progenitor cells which have begun to invade the cardiac jelly at this time. This pattern of 5-HT and SBP IR in the developing cushions persists until the cardiac jelly has become evenly populated with mesenchymal cells (day 12, 45-48 somites). Thus, interaction between 5-HT and SBP may be important in the differentiation, migration, and/or proliferation of cushion tissue cells. Whole embryo and cell culture experiments using 5-HT receptor ligands and inhibitors of 5-HT synthesis or uptake are currently underway to investigate the possible roles of 5-HT in early heart development, including the development of endocardial cushions.

DEFECTS IN DEVELOPMENT CAUSED BY 5-HT AND RELATED COMPOUNDS

In order to test the functional significance of 5-HT during these stages of development, we are currently conducting studies in which mouse embryos are exposed in whole embryo culture to 5-HT, 5-HT uptake inhibitors or specific 5-HT receptor analogs during critical stages of morphogenesis. These embryos are then examined at the light and SEM levels to determine whether a specific pattern of abnormal development emerges. These studies are expected to contribute to our understanding of the functional role of 5-HT in morphogenesis, and because many of these compounds are commonly used psychotherapeutic agents, they may also reveal potential teratogenic effects of clinically relevant drugs administered during pregnancy. Preliminary results indicate that exposure to 5-HT or various 5-HT uptake inhibitors causes developmental defects of the first visceral arch and other craniofacial structures where we have previously observed 5-HT IR following incubation of embryos with 5-HT or precursors. These results suggest that such sites of 5-HT uptake or synthesis may play important roles in craniofacial morphogenesis.

ACKNOWLEDGEMENTS

We wish to thank Robin L. Thomas for excellent technical assistance and Dr. Hadassah Tamir for the gift of SBP antiserum. This work was supported by NIH Grant No. HD22052. DLS was supported by NIH Grant No. 5 T32 ES07126.

REFERENCES

Bee, J. and Thorogood, P. (1980) The role of tissue interactions in the skeletogenic differentiation of avian neural crest cells. Dev. Biol. 78:47-62.

Burden, R.W. and Lawrence, I.E. (1973) Presence of biogenic amines in early rat development. Am. J. Anat. 136:251-257.

Buznikov, G.A. (1980) Biogenic monoamines and acetylcholine in protozoa and metazoan embryos. In J. Salanki and T.M. Turpaev (Eds.), Neurotransmitters, Comparative Aspects, Akademiai Kiado, Budapest, pp. 7-29.

Buznikov, G.A. (1984) The action of neurotransmitters and related substances on early embryogenesis. Pharmac, Ther. 25:23-59.

Buznikov, G.A., Chudakova, I.V. and Zvezdina, N.D. (1964)

The role of neurohumours in early embryogenesis. I. Serotonin content of developing embryos of sea urchin and loach. J. Embryol. Exp. Morph. 12:563-573.

Buznikov, G.A., Kost, A.N., Kucherova, N.F., Mndzhoyan, A.L., Suvorov, N.N. and Berdysheva, L.V. (1970) The role of neurohumours in early embryogenesis. III. Pharmacological analysis of the role of neurohumors in cleavage divisions. J. Embryol. Exp. Morph. 23:549-569.

Buznikov, G.A., Malchenko, L.A., Rakic, L., Kovacevic, N., Markova, L.N., Salimova, N.B., and Volina E.V. (1984) Sensitivity of starfish oocytes and whole, half and enucleated embryos to cytotoxic neuropharmacological drugs. Comp. Biochem. Physiol. 78C:197-201.

Buznikov, G.A., Mileusnic, R., Yurovskaya, M.A. and Rakic, L. (1984) Effect of calcium ionophore A23187 on the sensitivity of early sea urchin embryos to cytotoxic neuropharmacological drugs. Comp. Biochem. Physiol. 79C:425-427.

Buznikov, G.A. and Smukler, Y.B. (1981) The possible role of "prenervous" neurotransmitters in cellular interactions of early embryogenesis: a hypothesis. Neurochem. Res. 6:55-69.

Emanuelsson, H. (1974) Localization of serotonin in cleavage embryos of Ophryotrocha labronica La Greca and Bacci. Roux' Arch. Entw.-mech., 175:253-271.

Fitzharris, T.P. and Markwald, R.R. (1982). Cellular migration through the cardiac jelly matrix: a stereoanalysis by high voltage electron microscopy. Dev. Biol. 92:315-329.

Gustafson, T. and Toneby, M. (1970) On the role of serotonin and acetylcholine in sea urchin morphogenesis. Exp. Cell Res. 62:107-117.

Gustafson T. and Toneby, M. (1971) How genes control morphogenesis. Am. Sci. 59:452-462.

Hall, B.K. (1980) Tissue interactions and the initiation of osteogenesis and chondrogenesis in the neural crest-derived mandibular skeleton of the embryonic mouse as seen in isolated murine tissues and in recombinations of murine and avian tissues. J. Embryol. exp. Morph. 58:251-264.

Holtfreter, J. (1968) On mesenchyme and epithelia in inductive and morphogenetic processes. In R. Fleischmajer and R.E. Billingham (Eds.), Epithelial-Mesenchymal Interactions. Williams and Wilkins, Baltimore, pp. 1-30.

Johnston, M.C. (1975) The neural crest in abnormalities of the face and brain. Birth Defects: Original Article Series 11(7):1-18.

Kirschgessner, A.L., Liu, K.P., Gershon, M.D. and Tamir H. (1987) Co-localization of serotonin with specific serotonin binding proteins in serotonergic neurons of the rat brain and spinal cord. Anat. Rec. 218:73A.

Landau, M.A., Buznikov, G.A., Kabandin, A.S., Teplitz, N.A., and Chernilovskaya, P.E. (1981) The sensitivity of sea urchin embryos to cytotoxic neuropharmacological drugs; the correlations between activity and lipophility of indole and benzole derivatives. Comp. Biochem. Physiol. 69C:359-366.

Lauder, J.M., Lindemann Shuey, D., Thomas, R., Yavarone, M., and Sadler, T.W. (1989) Serotonin in the ectoplacental cone of the cultured mouse embryo. Development (submitted).

Lauder, J.M., Tamir, H., and Sadler T.W. (1988) Serotonin and morphogenesis I. Sites of serotonin uptake and binding protein immunoreactivity in the midgestation mouse embryo. Development 102:709-720.

Lauder, J.M. Tamir, H. and Sadler, T.W. (1987) Sites of serotonin uptake and binding protein immunoreactivity in the cultured mouse embryo: roles in morphogenesis? Soc. Neurosci. Abstr. 13:254.

Lauder, J.M. and Zimmerman, E. (1988) Sites of serotonin uptake in the epithelium of the developing mouse palate, oral cavity and face: possible roles in morphogenesis? J. Craniofac. Genet. Dev. Biol. 8:265-276.

LeDouarin, N.M. (1982) The Neural Crest. Cambridge University Press, Cambridge.

Liu, K.P., Gershon, M.D. and Tamir, H. (1985) Identification, purification, and characterization of two forms of serotonin binding protein from rat brain. J. Neurochem. 44:1289-1301.

Liu, K.P., Hsuing, S.C., Kirschgessner, A.L., Gershon, M.D. and Tamir, H. (1987) Co-localization of serotonin with serotonin binding proteins in the rat CNS. J. Neurochem. (Suppl.) 48:63.

Markwald, R.R., Fitzharris, T.P. and Manasek, F.J. (1976) Structural development of endocardial cushions. Am. J. Anat. 148: 85-93.

Newgreen, D.F. (1985) Control of the timing of commencement of migration of embryonic neural crest cells. Expl. Biol. Med. 10:209-221.

Newgreen, D.F., Allan, I.J., Young, H.M. and Southwell, B.R. (1981) Accumulation of exogenous catecholamines in the neural tube and non-neural tissues of the early fowl embryo. Correlation with morphogenetic movements. W. Roux's Arch. 190:320-330.

Nichols, D.H. (1981) Neural crest formation in the head of the mouse embryo as observed using a new histological technique. J. Embryol. exp. Morph. 64:105-120.

Nichols, D.H. (1986) Formation and distribution of neural crest mesenchyme to the first pharyngeal arch region of the mouse embryo. Am. J. Anat. 176:221-231.

Palen, K., Thorneby, L. and Emanuelsson, H. (1979) Effects of serotonin and serotonin antagonists on chick embryogenesis. W. Roux's Arch. 187:89-103.

Pratt, R.M., Larsen, M.A. and Johnston, M.C. (1975) Migration of cranial neural crest cells in a cell-free hyaluronate-rich matrix. Devl. Biol. 44:298-305.

Renaud, F., Parisi, E., Capasso, A. and DePrisco, P. (1983) On the role of serotonin and 5-methoxytryptamine in the regulation of cell division in sea urchin eggs. Devl. Biol. 98:37-46.

Shmukler, Y.B., Buznikov, G.A., Grigoryev, N.G. and Malchenko, L.A. (1984) Effect of cyclic nucleotides on sensitivity of early sea urchin embryos to cytotoxic neuropharmacological drugs. Bull. Exp. Biol. Med. 97:348-350.

Spemann, H. and Mangold, H. (1924) Uber Induktion von Embryonalanlagen durch Implantation artfremder Organisatoren. Arch Mik. Ant. Entw-Mech. 100:599-638.

Tamir, H. (1983) Serotonin-binding protein: function in synaptic vesicles. Trans. N.Y. Acad. Sci. 41:237-242.

Thorogood, P. (1981) Neural crest cells and skeletogenesis in vertebrate embryos. Histochem. J. 13:631-642.

Twitty, V. (1955) Eye. In B.H. Willier, P.A. Weiss, and V. Hamburger (Eds.), Analysis of Development. W.B. Saunders, Philadelphia and London, pp. 402-414.

Venkatasubramanian, K. and Zimmerman, E.F. (1983) Palate cell motility and substrate interaction. J. Craniofac. Genet. Dev. Biol. 3:143-157.

Wallace, J.A. (1982) Monoamines in the early chick embryo: demonstration of serotonin synthesis and the regional distribution of serotonin-concentrating cells during morphogenesis. Am. J. Anat. 165:261-276.

Wee, E.L., Babiarz, B.S., Zimmerman, S. and Zimmerman E.F. (1979) Palate morphogenesis IV. Effects of serotonin and its antagonists on rotation in embryo culture. J. Embryol. Exp. Morph. 53:75-90.

Zimmerman, E.F., Clark, R.L., Ganguli, S. and Venkatasubramanian, K. (1983) Serotonin regulation of palatal cell motility and metabolism. J. Craniofac. Genet. Dev. Biol. 3:371-385.

Zimmerman, E.F. and Lauder, J.M. (1987) Sites of serotonin uptake in the epithelium of the developing mouse palate, oral cavity and face: possible role in morphogenesis? Teratology 35:39A.

Zimmerman, E.F. and Wee, E.L. (1984) Role of neurotransmitters in palate development. In E.F. Zimmerman

(Ed.), _Current Topics in Developmental Biology: Vol. 19, Palate Development._ Academic Press, New York, pp. 37-63.

NEUROTRANSMITTER ACTIVATION OF SECOND MESSENGER PATHWAYS FOR THE CONTROL OF GROWTH CONE BEHAVIORS

S. B. Kater and L. R. Mills

Department of Anatomy and Neurobiology
Program in Neuronal Growth and Development
Colorado State University, Fort Collins, CO 80523

INTRODUCTION

The neuronal growth cone is now well recognized as one of the primary organelles responsible for the generation, alteration, and regeneration of neuronal form and connectivity (see review by Kater and Letourneau, 1985). The term neuronal growth cone was given by Ramon y Cajal (1890), to define the broad, flattened, lamellapodia and numerous filopodia that characterize the tip of a growing neurite. The growth cone is a highly dynamic structure; the lamellapodium and filopodia continuously move as the growth cone "explores" its environment. Discriminatory behavior is characteristic of motile growth cones; frequently the growth cone turns, branches, or even stops, apparently in response to invisible cues. Once growth cone motility, and hence neurite elongation ceases growth cone morphology also changes; the growth cone rounds up as the lamellapodium and filopodia are withdrawn.

The neuronal growth cone has now been extensively investigated both in situ and in cell culture where it can be more efficiently manipulated. In cell culture the characteristic long filopodia appear to act as transducers searching out the environment. These same filopodia in vivo (e.g, Bentley, and Toroian-Raymond, 1986; Bastiani et al., 1985) seem to play key roles in neuronal pathfinding. While broad lamellapodia and numerous large filopodia have been assumed to be characteristic features of growing neurites many additional observations have been made on elongating neurites with rather more circumscribed growth cone structure. For instance, Lopresti, et al., (1973), have described pioneer growth cones in situ which which possess the characteristic broad, flattened lamellapodia and multiple filopodia. These pioneer neurons are followed by an additional set of neurons whose growth cones display quite simple morphologies. Perhaps, pioneering growth cones which must select appropriate paths, and navigate to appropriate targets, deploy the broadened lamellapodium and long filopodia as primary agents of pathfinding. Interestingly, we have observed that an individual neuron can display over different periods of its lifetime different shaped growth cones (Wong, et al., 1981). In fact, it is now clear that a variety of stimuli both endogenous i.e., to individual neurons, and exogenous i.e., environmental cues, can play significant roles in the alteration of growth cone form and function.

Global stimuli can affect all of the growth cones of a given neuron. Alternatively, very local cues can act on individual growth cones (Haydon et al., 1985; Gunderson and Barrett,

Molecular Aspects of Development and Aging of the Nervous System
Edited by J.M. Lauder
Plenum Press, New York, 1990

1980; Letourneau, 1978). Experimental evidence provides strong support for the idea that growth cones can behave quite autonomously. In culture growth cones can survive for several days, perform typical growth cone behaviors, and respond to environmental signals even when severed from the parent neuron (Shaw and Bray; 1973; Guthrie et al; 1989). Stimuli that would inhibit outgrowth of growth cones connected to the cell body, can also inhibit isolated growth cones (McCobb and Kater, 1988). Observations such as these lend considerable weight to the idea that these organelles may indeed act autonomously in their primary functions of pathfinding in vivo.

INSTRUCTIONAL COMPONENTS OF GROWTH CONE BEHAVIOR

For some time it has been clear that a diverse variety of stimuli, both intrinsic and extrinsic, can alter growth cone form and function. These stimuli have largely appeared as unrelated but, as will be discussed below ,we now recognize a common link between them at the level of intracellular second messengers. The longest known of the extrinsic signals has been the simple surfaces or substrata upon which neurons grow. Neurons growing on, for instance, polylysine as opposed to collagen, display quite different morphologies (Bray, 1973; Mattson et al., 1988).

Another major set of stimuli that influence growth cone behavior are trophic factors. The best established of these is Nerve Growth Factor (NGF) which is known to have multiple effects ranging from direct effects on growth cone behavior and morphology, to the induction of particular genes (Thoenen and Barde, 1980). Interestingly, while NGF has been assumed to act as a diffusible molecule, recent evident indicates the likelihood that NGF might, in fact, also act when substrate bound (Gunderson, 1985). This raises interesting possibilities for additional roles for diffusible molecules in growth cone navigation.

Our recent work has introduced a whole new class of molecules that regulate growth cone behaviors and thus act as as guidance cues for neurite outgrowth. Neurotransmitters are now known to regulate a wide range of growth cone behaviors in a wide range of nervous systems. For instance, within the snail, Helisoma, serotonin can inhibit outgrowth and characteristic motile growth cone morphology of selected neurons. Another subset of neurons is totally unaffected by the presence of serotonin (Haydon, et al., 1984,1987). An additional overlapping but not identical subset of neurons, can be inhibited in their outgrowth by dopamine (McCobb et al.,1988). Similar results have now been obtained on hippocampal pyramidal neurons where we have found that glutamate can have inhibitory effects on growth cone elongation. Here, we have the additional advantage that we can compare growth cones of axons with those of dendrites. In this system as well, the action of the neurotransmitter is highly selective for specific growth cones. However, it is not that the growth cones of different neurons show different responsiveness to glutamate but rather that different classes of growth cone on the same neurons show differences. Namely, the dendritic growth cones of hippocampal pyramidal neurons are inhibited in their outgrowth and can actually be pruned back by the presence of glutamate whereas axonal growth cones continue their motile processes (Mattson, et al., 1988). Taken together with the now numerous observations in other species (e.g., Lankford, et al., 1988; Lipton, et al.; 1988), the work on Helisoma has demonstrated that an entirely new class of molecules can act as guides to neuronal growth cones. This, then, extends the idea that these basic agents of information transfer in mature nervous systems also play formative roles in neuronal development (Lauder, 1987).

Another recent observation demonstrates that simple mechanical stimuli can significantly alter growth cone morphology (Mills, et al., 1988). For instance, the application of a brief mechanical stimulus to the growth cone (e.g., a gentle puff of medium, or a fine touch with

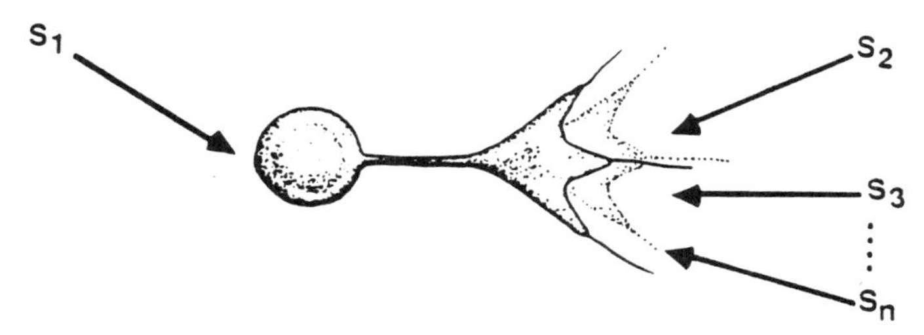

Figure 1. Neuronal growth cones potentially integrate multiple environmental as well as intracellular signals. Such cues include substrate-bound bound molecules, neurotransmitters, trophic factors, mechanical stimuli, and even the electrical activity generated within the neuron. These signals can act either locally at the level of the growth cone (e.g. as do neurotransmitters) or have more global effects on the entire neuron as would, for instance, electrical activity which propagates throughout the cell. In vitro, it has been possible to show that each of these separate stimuli can effect the behavior of growth cones. In vivo, it is clear that many of these stimuli could be present simultaneously and that the neuron would then integrate the effects of all stimuli and transform this net signal into a change in growth cone behavior.

a microscopic probe), as well as permanent physical barriers (e.g., a surface scratch in the tissue culture dish), can reliably evoke discrete turning and/or branching, behaviors. Thus, a brief mechanical stimulus can radically change the course of the growth cones elongation path, while a more sustained physical stimulus can induce bifurcation, as well as

dramatically alter the direction of outgrowth. Such purely physical navigational aids may well not act alone. More likely they are integrated with other signals e.g. chemical, to determine the ultimate sequence of neuromorphogenetic events.

Intrinsic signals can also have major effects upon growth cone behavior. For some time "activity" has been discussed by neurobiologists as a potentially prominent sculptor of neuronal architecture and connectivity. A direct assessment of this idea has come from our work which examined the role of activity upon the growth cones of individual neurons in cell culture. The rational for this approach was that the behavior of the growth cone was ultimately responsible for the genesis of particular neural architectures and that the generation of action potentials was a basic model of neuronal "activity". Using a very straightforward paradigm of a whole cell patch electrode upon the surface of the cell body, it was possible to generate action potentials experimentally while optically monitoring the behavior of several growth cones of that neuron. These experiments demonstrated that motile growth cones can be abruptly inhibited in their elongation and pathfinding behaviors by the generation of action potentials. This inhibition was reversible: following cessation of stimulation, motility and growth cone morphology return to normal and the growth cone resumes active elongation (Cohan and Kater, 1986). These results indicate potentially important roles for neuronal circuitry in the generation of neuronal morphology. Quite different neuronal geometries would be expected in neurons participating in actively firing circuits as apposed to those belonging to quiescent ensembles of neurons.

NEURONS INTEGRATE THE EFFECTS OF MULTIPLE SIGNALS

While it is clear that individual signals can have profound effects on the behavior of the neuronal growth cone, it is equally clear that many of these signals co-exist both during development and in the mature organism (Figure 1). Clearly, growth cones must be able to

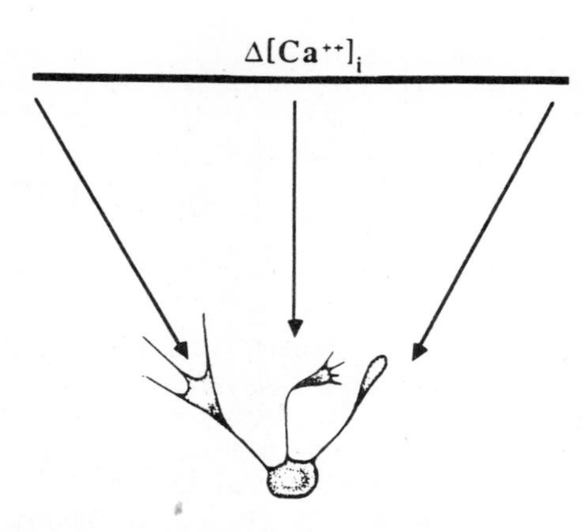

Figure 2. Growth cone behavior changes with local changes in intracellular free calcium concentration. An array of different growth cone morphologies and behaviors is associated with different levels of intracellular calcium. At the usual calcium set point associated with outgrowth, characteristic broadened lamellipodia and long filopodia are observed. With a slight reduction in intracellular calcium, growth cones become considerably reduced but motility increases. Either increasing intracellular calcium significantly or reducing it to lower levels both result in a nonmotile form of the growth cone. Preliminary evidence (Mills, et al., 1988) indicates that even directionality may be regulated by intracellular calcium concentrations.

integrate signals provided by unique substrata, specific trophic factors , mechanical stimuli, and the presence or absence of certain neurotransmitters, all at the same time. Perhaps the best model for this kind of integration of multiple behavioral cues comes from how mature neurons integrate multiple convergent inputs.. Our best example of the effects of multiple stimuli on growth cone behavior reveals how the presence of a second neurotransmitter can profoundly alter the effect of a first. In Helisoma, a primary effect of serotonin is the inhibition of growth cone motility in a specific set of neurons. The presence of acetylcholine however, can negate the serotonin-induced inhibition of growth cone motility (McCobb, et al., 1988). This is not an isolated phenomenon, but rather integration of the effects of multiple transmitters seems to be a general principle. In rat hippocampal pyramidal cells in cell culture, motility of dendritic growth cones is retarded by glutamate. These growth-inhibiting effects of glutamate can be negated, by the inhibitory transmitter, GABA (Mattson and Kater,1988). That is to say, it appears that in both of these examples, the net affect on growth cone behavior is a composite of the actions of more than one environmental stimulus.

GROWTH CONE BEHAVIOR CHANGES WITH LOCAL CHANGES IN INTRACELLULAR CALCIUM

By employing the fluorescent indicator, fura-2, it has been possible to measure intracellular calcium levels in neurons from a variety of sources. The emerging picture is that intracellular calcium concentrations profoundly effect the behavior and morphology of neuronal growth cones (Kater, et al. 1988). A neuronal growth cone on a given neuron may exhibit: (1) characteristic long-filopodia with a broadened lamellapodium, (2) a decreased number of filopodia and reduced lamellapodium, or (3) the rounded, phase-bright growth cone with no filopodia that is characteristic of the nonmotile growth cones of Helisoma neurons (Wong et al., 1981; Haydon et al., 1984). Each of these forms is a function of specific intracellular calcium concentrations (Figure 2).

For each of the stimuli that regulate growth cone behavior, some transduction mechanism must exist to link the stimuli with the intracellular machinery controlling motility. For many of these stimuli, that second messenger is intracellular calcium (Figure 3). We know, for instance, that the generation of action potentials can greatly increase intracellular calcium in growth cones. Similarly, we know that serotonin greatly increases the concentration of calcium,whereas the presence of acetylcholine can negate the rise in intracellular calcium (McCobb, et al., 1988). The equation below shows that the mechanism by which neurons integrate multiple stimuli is by summing the net effects of various stimuli on changes intracellular calcium with their own intracellular basal calcium level.

$$[Ca^{++}]_i = [Ca^{++}]_{Basal} + \Delta[Ca^{++}]_{S1} + \Delta[Ca^{++}]_{S2} + \Delta[Ca^{++}]_{Sn}$$

The primary point is that the net intracellular concentration and its spatial distribution is going to be a composite. This composite could vary for an individual neuron when it received different combinations of stimuli and also be different for different neurons receiving the same stimuli. For example, if different neurons have different basal calcium levels, then a given change in calcium concentration could be expected to produce very different effects on those two neurons. In each case the summation of various stimuli, some of which raise intracellular calcium and some of which lower intracellular calcium would produce a net effect through a final common determinant--- free intracellular calcium.

This model assumes that the effects of many, if not all, regulatory signals summate at the level of intracellular calcium. There is, however a variety of evidence that indicates that other second messengers e.g., cyclic nucleotides, and protein kinase C (Forscher et al., 1987; Mattson et al., 1988) can regulate growth cone behavior perhaps independently of effects on intracellular calcium. Whether such mechanisms are ancillary, or subordinate, to the mechanisms proposed here is yet to be determined. Nevertheless it is clear that intracellular calcium does play a key role in the regulation of neuronal growth cone behavior and ultimately in the genesis of neuronal architecture

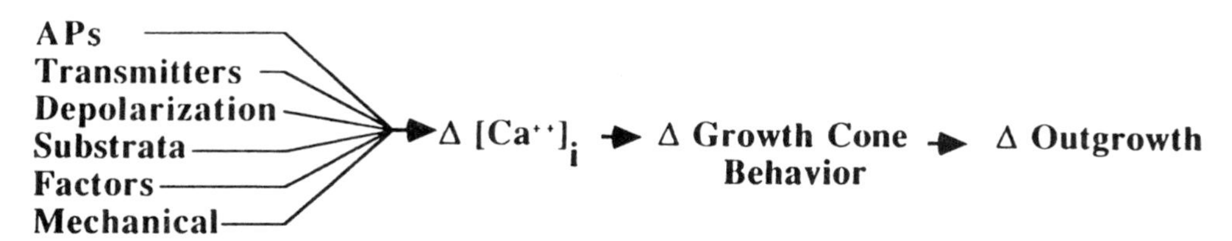

Figure 3. It is now known that a variety of stimuli ranging from action potentials to mechanical deformation result in discrete and predictable changes in intracellular calcium. The neuronal growth cone and its form and function are highly sensitive to intracellular concentrations and, thus, these stimuli acting through changes in intracellular calcium act to change growth cone behavior. This in turn changes the outgrowth in neurons and alters the ultimate architecture and connectivity patterns in which these neurons participate.

THE DETERMINANTS OF INTRACELLULAR CALCIUM CONCENTRATION

The regulation of intracellular calcium concentration is extremely complex involving multiple regulatory processes. While a variety of agents can predictably alter intracellular calcium concentration, and thus growth cone behavior, there is no reason to assume that all of these agents acts on the same component process. Rather, it seems reasonable that

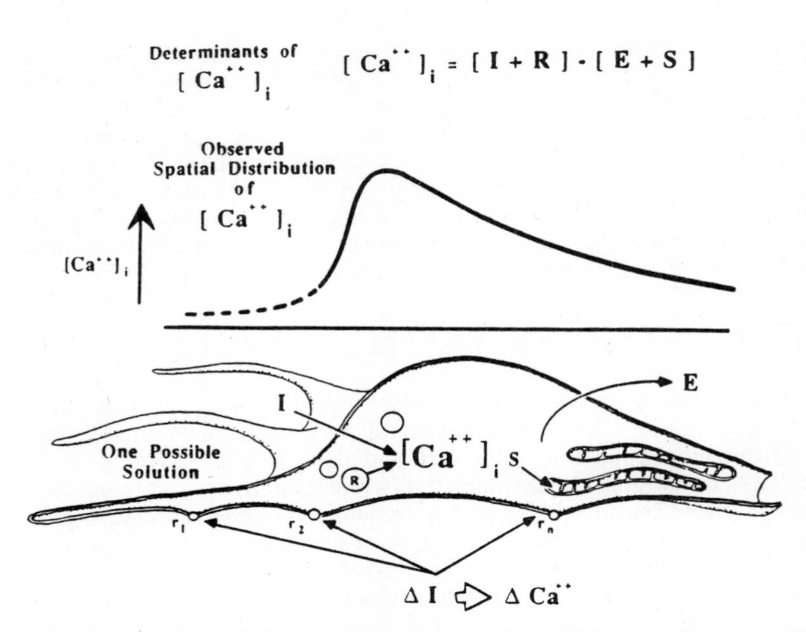

Figure 4. Factors determining calcium concentration within growth cones. The top equation shows the factors determining calcium concentrations within the growth cone. Calcium levels are increased by Influx and Release from intracellular stores. Levels are decreased by Efflux (pumping), and Sequestration (both cytoplasmic buffering and intracellular organelles).

The middle panel shows the observed calcium rest levels in a growth cone as determined by previous fura-2 mapping.

The bottom panel gives one possible distribution of mechanisms which could regulate intracellular calcium concentration and might explain the observed gradient of distribution of intracellular calcium within the neuronal growth cones. If influx were localized primarily to the leading edge of the main body of the growth cone and/or efflux and sequestration were localized primarily to the growth cone neurite junction, calcium concentrations would be higher at the leading edge than in the neurite. An important implication of the existence of this gradient is the possibility that anything that would perturb the gradient might have profound effects on the behavior of the growth cone. For example, if neurotransmitter receptors were localized on the leading edge of the growth cone, activation of these receptors would directly increase the influx of calcium at that point, and increase the gradient. Activation of other receptors at the trailing edge of the growth cone (e.g. receptors interacting with extracellular matrix molecules) could reduce the gradient by increasing influx at that point. Taken together, while speculative, these ideas offer the opportunity to test directly how particular environmental and intrinsic cues affect the behavior of the neuronal growth cone by specifically affecting components of calcium regulating mechanisms.

different environmental cues act upon different aspects of the overall calcium regulatory system.

Our present view (see Figure 4) is that free intracellular calcium concentration is a function of calcium influx and release of calcium from intracellular stores, minus the efflux of calcium through pumping systems, and sequestration by the buffering capacity of the cytoplasm and organelles such as mitochondria. Accordingly, different environmental cues might well regulate different aspects of this balance. For instance, some neurotransmitters may well act to alter calcium influx through membrane channels, while other neurotransmitters, operating through other second messengers might alter calcium release from intracellular stores. These are but two examples of an array of possibilities that might extend to the modulation of all of the component of calcium homeostasis.

In this review we have only referred to the overall concentration of calcium within a growth cone. At a finer grain, however, we can see that even within a single growth cone calcium is not likely to be homogeneously distributed. For example, we presently have 'calcium maps' of individual growth cones which imply the existence of gradients of calcium concentration (Guthrie et al., 1988). While this observation remains to be understood in terms of its underlying causes, it is clear that it could play a very important role in growth cone behavior. It is well known that there are very different underlying components of growth cone elongation. We envision that processes like actin assembly, tubulin polymerization, and vesicle fusion may each be regulated by distinct and different calcium concentrations. Specific distributions of calcium (the 'calcium map') appear to overly the distribution of cytoskeletal and vesicular apparatus employed for motility. This reinforces the idea that precise local concentrations of calcium may indeed regulate the underlying calcium-dependent processes for motility.

A complex regulatory role for the calcium map is important because we now know that environmental signals know to change growth cone behavior also change the calcium map. We have observed such changes with a variety of agents ranging from action potentials through neurotransmitters , depolarization of membrane potential (Guthrie et al., 1988; Kater et al., 1988; Kater and Guthrie, 1988), changes in substrata, growth factors (unpublished observations), and even mechanical stimuli (Mills et al., 1988) . A change in <u>any</u> of the regulatory components of calcium homeostasis would be expected to cause a change in these maps. It now remains to partition and dissect this system and define how each of these specific agent affects a specific component of the calcium homeostatic system in order to bring about the changes in calcium which ultimately result in changes in neuronal growth cone behavior.

SUMMARY

The generation and regeneration of neuronal form and connectivity both undoubtedly rely upon the integration of intrinsic and extrinsic information of many kinds. Our work has demonstrated that the concentration and spatial distribution of intracellular calcium is a key locus of integration of such information. Through a delicate balance of mechanisms that raise free calcium and mechanisms that lower free calcium, a steady state level is achieved that appears to have significant regulatory control over neuronal growth cone behavior. Cues, both internal and external, alter intracellular calcium levels, and consequently alter growth cone behavior. It is through the alteration of the various components of calcium homeostasis that we envision the complexities of neuronal architecture and connectivity may be fine-tuned throughout the life histories of neuronal ensembles.

REFERENCES

Bastiani, M.J., Doe, C.Q., Helfand, S.L., and Goodman, C.S., 1985, Neuronal specificity and growth cone guidance in grasshopper and <u>Drosophila</u> embryos. <u>TINS</u>:8 257.

Bentley, D., and Toroian-Raymond. A. ,1986, Disoriented pathfinding by pioneer neurons growth cones deprived of filopodia by cytochalasin treatment. <u>Nature (Lond.)</u> 323:712.

Bray, D. 1973, Branching patterns of individual sympathetic neurons in culture. J.Cell. Biol. 56:702.

Cohan, C. S. and Kater, S.B., 1986, Suppression of neurite elongation and growth cone motility by electrical activity. Science 232:1638.

Forscher,C., Kaczmarek, L. K., Buchanan, J., and Smith, S. J.,1987, Cyclic AMP induces changes in the distribution and transport of organelles within growth cones of *Aplysia* bag neurons. *J.* Neurosci., 7:3600.

Gunderson, R. W., and Barrett, J. N.,1980, Characterization of the turning response of dorsal root neurites towards nerve growth factor. J. Cell Biol. 87:546.

Gunderson, R.W., 1985, Sensory neurite growth cone guidance by substrate adsorbed nerve growth factor. J. Neurosci. Res. 13: 199.

Guthrie, P.B., Mattson, M.P., Mills, L.R., and Kater, S.B., 1988,"Calcium homeostasis in molluscan and mammalian neurons: Neuron-selective set-point of calcium rest concentration." Soc.Neurosci.Abstr.

Guthrie, P. B., Lee, R.B., and Kater,S.B., 1989, A Comparison of Neuronal Growth Cone and Cell Body Membrane: Electrophysiological and Ultrastructural Properties. in press, J. Neurosci.

Hadley R. D., D. A. Bodnar and S. B. Kater, 1985, Formation of electrical synapses between isolated, cultured Helisoma neurons requires mutual neurite elongation. J. Neurosci. 5:3145.

Haydon, P. G., Cohan, C. S., McCobb, D. P. Miller, H. R. and Kater, S. B.,1985, Neuron-Specific growth cone properties as seen in identified neurons of Helisoma. J. Neurosci. Res. 13:135.

Haydon, P. G., McCobb, D. P., Kater, S. B.,1987, The regulation of neurite outgrowth, growth cone motility, and electrical synaptogenesis by serotonin. J. Neurobiol. 18:197.

Kater, S.B., and Guthrie, P.B., 1988, The neuronal growth cone calcium regulation of a presecretory structure. In "Sectretion and its control" eds. C.M. Armstrong and G.S. Oxford Wiley Interscience New York

Kater, S. B. and Letourneau, P., editors, 1985: "The Biology of the Neuronal Growth Cone". Alan R. Liss, New York.

Kater, S. B., Mattson,M., . Cohan, C., Connor,J., 1988, Calcium Regulation of the Neuronal Growth Cone, Trends in Neurosci. 11:315.

Lankford, K. L., DeMello, F. G., and Klein, W. L. 1988 , D1 dopamine receptors inhibit growth cone motility in cultured retinal neurons; evidence that neurotransmitters act as morphogenetic growth regulators in the developing nervous system. Proc.Natl.Acad.Sci.USA 85:45671.

Lauder, J. M.,1987,Neurotransmitters as morphogenetic signals and trophic factors. in: "Model systems of development and aging of the nervous system" edited by A. Vernadakis, A. Privat, J. M. Lauder, P. S. Timiras and E. Giacobini, Martinus Nijhoff Publishing, Boston, p. 219.

Lipton, S. A., Frosch, M. P., Phillips, M. D., Tauck, D. L., and Aizenman, E.,1988,

Nicotinic agonists enhance process outgrowth by retinal ganglion cells in culture. Science 239:1293.

Lopresti, V., Macagno, E. R., and Levinthal, C. 1973, Structure and development of neuronal connections in isogenic organisms: cellular interactions in the development of the optic lamina of *Daphnia*. Proc.Natl.Acad.Sci.USA 70:433.

Mattson, M. P., Dou, P., and Kater, S.B.,. 1988, Outgrowth-regulating actions of glutamate in isolated hippocampal pyramidal neurons. J. Neurosci. 8: 2087.

Mattson, M. P., and Kater, S.B., 1988, Excitatory and inhibitory neurotransmitters in the generation and degeneration of hippocampal neuroarchitecture. Brain Research 478:337.

Mattson, M. P., Taylor-Hunter, A. and Kater, S.B., 1988, Neurite outgrowth in individual neurons of a neuronal population is differentially regulated by calcium and cyclic AMP. J. Neurosci. 8:1704.

Mattson, M.P., Guthrie,P.B., and Kater, S.B., 1989, Intracellular messengers in the generation and degeneration of hippocampal neuroarchitecture. J. Neurosci. Res. 21:447.

McCobb, D. P., Cohan,C. S, Connor,J.A., and Kater,S.B., 1988, Interactive effects of serotonin and acetylcholine on neurite elongation. Neuron 1:375.

McCobb, D. P. and Kater, S.B., 1988, Dopamine and serotonin inhibition of neurite elongation of different identified neurons. J. Neurosci. Res. 19:19.

McCobb, D. P., and Kater, S.B., 1988, Membrane voltage and neurotransmitter regulation of neuronal growth cone motility. Developmental Biology. 130:599.

Ramon y Cajal, S., 1890,: A quelle epoque apparaissent les expansions des cellules nerveuses de la moelle epiniere du poulet. Anat.Anz. 5:609.

Shaw, G., and Bray, D., 1977, Movement and extension of isolated growth cones. Exp.Cell Res. 104:55.

Thoenen, H., and Barde, Y.A., 1980, Physiology of nerve growth factor. Physiol. Rev. 60:1284.

Wong, R. G., Hadley, R, Kater, S.B., and G. Hauser, G., 1981, Neurite outgrowth in molluscan organ and cell cultures: the role of conditioning factor(s). J. Neurosci. 1:1008.

Acknowledgements

The authors wish to thank the constructive comments made on this manuscript by Drs. P. Guthrie and M.Murrain and also the technical assistance of D. Giddings and B. Bertram. This work was supported by NIH grants NS24683, NS15350, and a gift from the Monsanto Corporation.

DIBUTYRYL CYCLIC AMP TREATMENT OF ASTROCYTES IN PRIMARY CULTURES AS A SUBSTITUTE FOR NORMAL MORPHOGENIC AND 'FUNCTIOGENIC' TRANSMITTER SIGNALS

Leif Hertz
Department of Pharmacology
University of Saskatchewan
Saskatoon, Saskatchewan, S7N OWO Canada

INTRODUCTION

It has now become very well established that compounds which in the mature organism function as neurotransmitters may play a major developmental role as "morphogens" (Turing, 1952) during development (Lauder, 1988). In some cases this occurs in regions where the compound in question is not utilized as a transmitter in the mature animal, but in other cases a neurotransmitter seems to be a morphogen for specifically those cells which are destined to become its target cells. Such a system has also been known for a long time in the autonomic nervous system where, for example, innervation directly affects receptor clustering. There is good reason to believe that the same agents also have a differentiating effect on functional (biochemical and biophysical) parameters, i.e., also act as "functiogens".

Cultured astrocytes express beta adrenergic (Gilman and Nirenberg. 1971; McCarthy et al., 1988) as well as alpha adrenergic (Ebersolt et al., 1981; Bockaert and Ebersolt, 1988) receptors, but it is only recently that it has been established that mature astrocytes in vivo also possess such receptors (Aoki et al., 1987); the receptor density on neurons seems to be considerably lower (McCarthy et al., 1988). The adrenergic innervation of astrocytes is of relevance for astrocytic functions by modulating glucose metabolism (Browning et al., 1972, Cummins et al., 1983, Kala et al., 1989), protein phosphorylation (McCarthy et al., 1988; Browning, 1988) and probably also channel-mediated K^+ transport (Bender et al., 1988). In vivo, astrocytes seem to be functionally innervated by noradrenergic fibers which spread from locus coeruleus over the entire cerebrum and cerebellum; most of these fibers do not terminate in genuine synapses but in varicosities (Beaudet and Descarries, 1984; see, however, also Foote et al., 1983) from which the transmitter reaches its target cells by diffusion. Besides astrocytes the main target cells seem to be microvessels (Hartman et al., 1979; Harik et al., 1980), and the whole system is very similar to the sympathetic autonomic nervous system in the periphery which also innervates both vasculature and a parenchymal cell type.

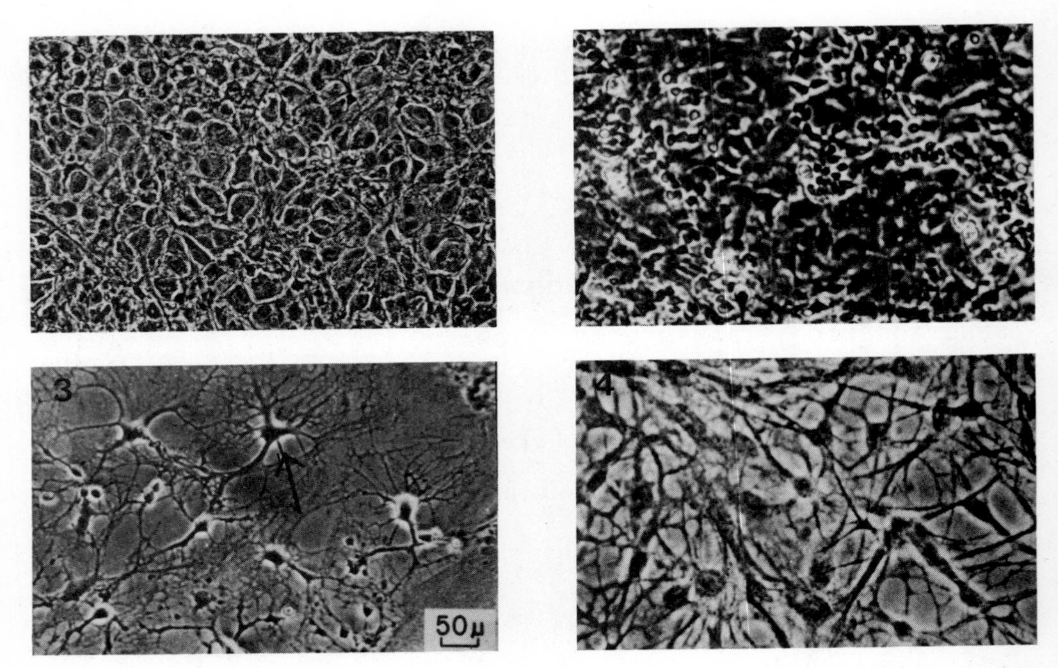

Fig.1. Primary culture of mouse astrocytes grown for approximately 3 weeks in tissue culture medium with serum. Note confluent homogeneous layer of cobblestone-like cells. From Hertz et al., 1982. Fig. 2. Corresponding culture of rat astrocytes seeded more densely and treated with brain extract. Note second population of small round cells on top of astrocytic monolayer. From Sensenbrenner et al., 1982. Fig. 3. Corresponding culture of rat astrocytes treated with 100 μM norepinephrine and 0.5 mM aminophylline. Note extension of processes. From Narumi et al., 1978. Fig. 4. Corresponding culture of rat astrocytes treated with 10 μM forskolin. Note extension of processes. From Pollenz and McCarthy, 1986. Figs. 5-6. Corresponding rat culture treated at day 3, i.e., before confluency with 1 mM monobutyryl cyclic AMP. Note stellate cells of different morphologies. From Shapiro, 1973. Fig. 7. Corresponding culture of rat cells treated at day 9, i.e., after confluency with 1.0 mM dibutyryl cyclic AMP (dBcAMP). Note non-stellate morphology and homeogeneity. From Lim et al., 1973. Fig. 8. Corresponding culture of rat astrocytes treated with 1.0 mM dBcAMP. From Moonen et al., 1975. Figs. 9-10. Cultures of 3-week-old mouse astrocytes which from the age of 14 days was treated with 0.25 mM dBcAMP in a slightly modified Dulbecco's medium in the presence (Fig. 9) or absence (Fig. 10) of serum. Note stellate appearance in Fig. 10, but not in Fig. 9. From Hertz, 1978 (Fig. 9) and Hertz et al., 1978b (Fig. 10). Fig. 11 Corresponding but less densely seeded, culture (colony culture) of mouse astrocytes treated for at least 1 week with 0.25 mM dBcAMP in the presence of serum. Note stellate morphology. From Fedoroff et al., 1981. Fig. 12. Corresponding culture of rat astrocytes treated with 1.0 mM dBcAMP, but otherwise in a similar medium as shown in Fig. 3. Note identical cell morphology in the two Figs. From Kimelberg et al., 1982. Fig. 13. Secondary cultures of rat astrocytes treated with 1.0 mM dBcAMP. From Facci et al., 1987. Fig. 14. Primary culture of rat astrocytes treated with 0.5 mM dBcAMP. From MacVicar, 1987.

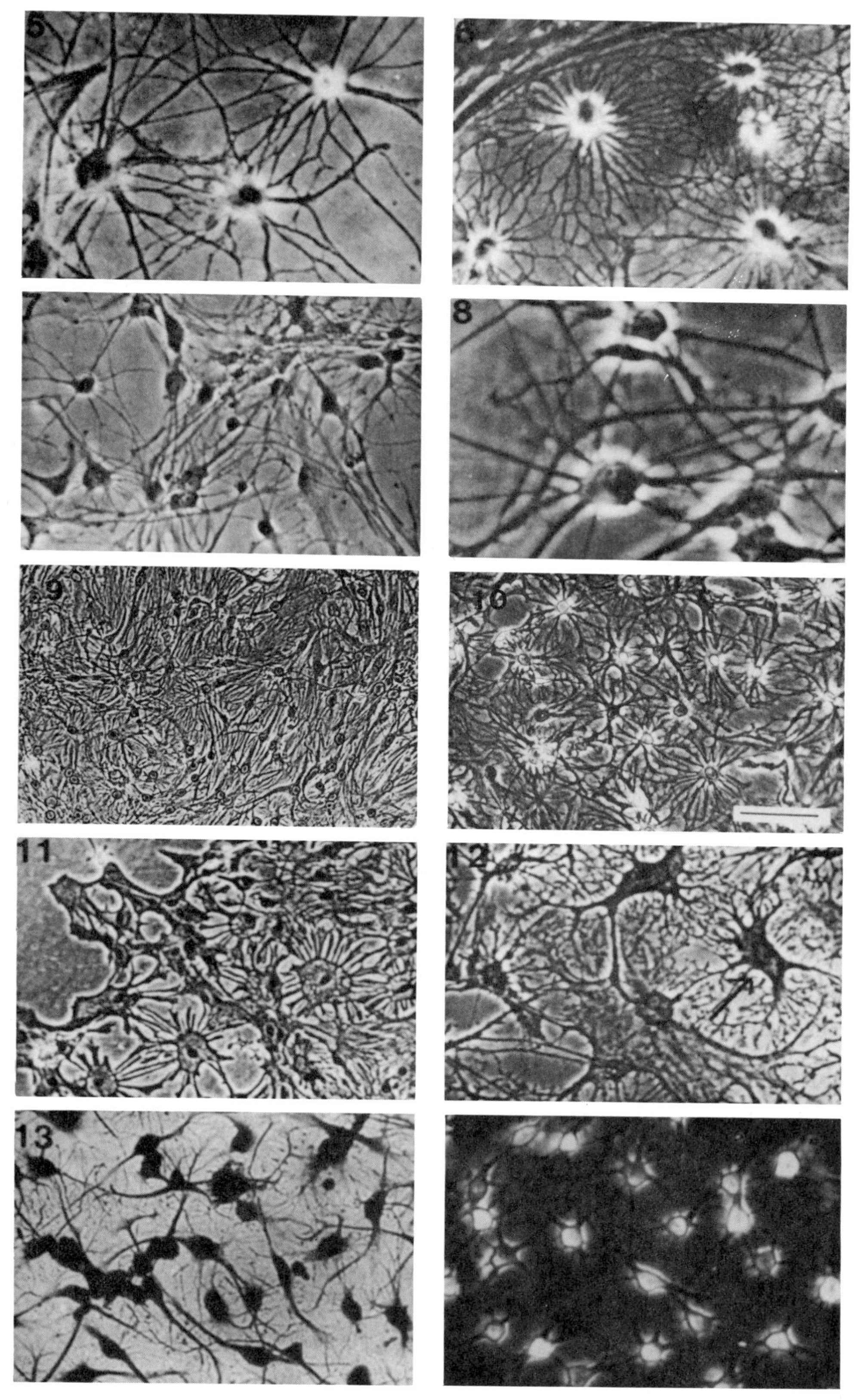
5
6
7
8
9
10
11
12
13

The "birthday" of most locus coeruleus cells in the rat is between 10 and 13 days of gestation (Lauder and Bloom, 1974). Beginning neocortical noradrenergic innervation has been observed as early as embryonic day 16 but the development of the neocortical fibers continues through the first 2 postnatal weeks (Foote et al., 1983). Astrocytes proliferate and differentiate at a late, mainly postnatal stage in both mouse and rat (Privat, 1975). Thus, the morphogenic signals they receive must operate postnatally. Primary cultures of mouse and rat astrocytes, which routinely are prepared from newborn animals and contain no neurons, therefore, do not receive noradrenergic signals which may be essential for their normal differentiation. This may well mean that these cells morphologically and functionally remain undifferentiated, unless noradrenergic signals normally received in vivo are replaced by modulation of the external medium. Since exposure to beta adrenergic agonists, leads to an increased intracellular concentration of cyclic AMP (cAMP) it might be possible to mimic this effect, e.g., by including dibutyryl cyclic AMP (dBcAMP) in the culturing medium and thus increase the intracellular level of cAMP (e.g., Facci et al., 1987). Some effects of dBcAMP on astrocytes will therefore be discussed below, although it is not possible to treat the whole field comprehensively within the framework of a short review.

dBcAMP AS A MORPHOGEN

Primary cultures of astrocytes prepared from the perinatal mouse or rat cerebral hemispheres consist mainly of flat, cobblestone-like epithelial cells (Figs. 1-2). There may also be a second population of small round, loosely attached cells which in the presence of fetal calf serum spontaneously transform into small, process bearing cells which stain intensely for glial fibrillary acid protein (GFAP). There seems to be general consensus that the flat epithelial cells are "type-1-like" astrocytes. The cells which spontaneously develop into process bearing cells are progenitor cells which, depending upon the culturing conditions, may develop into either "type-2-like" astrocytes or oligodendrocytes. The number of cells belonging to the second population is negligible at a low plating density (Fig. 1), but can be quite large (Fig. 2) at higher seeding densities and especially in the presence of brain extract (Labourdette et al., 1980; Sensenbrenner et al., 1982).

Under conditions favouring the maintenance and growth of "type-1-like" astrocytes they become confluent after 1-2 weeks and constitute as much as 95% of the total cell population (Hertz et al., 1985; Juurlink and Hertz, 1985). They are the cell type known to display large amounts of adrenergic receptors, especially $beta_1$ receptors. It can therefore be expected that beta adrenergic agonists might be a morphogen for these cells. That this is, indeed, the case can be seen from Fig. 3 which shows that the cobblestone like cells in the presence of the beta adrenergic agonist isoproterenol and a diesterase inhibitor (theophylline) extend a multitude of processes (of different morphology than "type-2-like" cells) and become very similar to protoplasmic astrocytes in vivo. However, there appears to be some variability in this phenomenon and high concentrations of norepinephrine may be required. Forskolin which stimulates the adenylate cyclase directly has a similar effect (Fig. 4).

Before it was known that exposure to isoproterenol or

forskolin leads to a transformation of "type-1-like" astrocytes to process bearing cells, it had been described that addition to the medium of monobutyryl- or dibutyryl cyclic AMP (dBcAMP) leads to extension of processes. This effect was first demonstrated by Shapiro (Figs. 5-6) and by Lim et al. (Fig. 7), both in 1973. Shapiro added the cAMP analog monobutyryl cyclic AMP to immature (3-day-old) cultures prepared from the brains of new-born rats, and observed that some of the cells transformed to process bearing cells of different morphology. Lim exposed older, confluent cultures of rat astrocytes to dBcAMP and obtained a more homogenous population of cells like those shown in Fig. 7. A similar procedure was used by Moonen et al. (1975), who also observed cells of somewhat different morphologies, including that shown in Fig. 8. They performed a thorough study of the effect of dBcAMP under different conditions and observed the peculiar phenomenon that the rapidity with which the cells transformed to a large extent depended upon the tissue culture medium used. Thus, cells grown in Eagle's basal medium (BME) transformed within minutes, whereas those maintained in Dulbecco's medium underwent a very slow (days) extension of processes. The latter probably corresponds more closely to an in vivo differentiation (Moonen et al., 1975; Fedoroff et al., 1981). The morphology of the process bearing cells also depends upon other factors, such as the presence or absence of serum in the medium. Hertz and co-workers reported the formation of the cells shown in Fig. 9 after addition of dBcAMP in the presence of serum, but a somewhat different morphology (Fig. 10) in the absence of serum. The morphology of the cells in the presence of serum is very similar to that observed by Goldman and Chiu (1984a) in almost identical cultures. Fedoroff has especially studied cells in non-confluent colony cultures (Fig. 11). Whether or not the cultures are confluent appears to affect cell morphology since the cells most often illustrated by Fedoroff and co-workers under conditions which otherwise are virtually identical to those used by Hertz (Fig. 9), have the morphology illustrated in Fig. 11. Fig. 12 shows the morphology demonstrated by Kimelberg et al. (1982) in the presence of 1.0 mM dBcAMP and it should be noted that this morphology is identical to that (Fig. 3) obtained by the same group in the presence of isoproterenol (see cells indicated by arrows in Figs. 3 and 12). Fig. 13 illustrates the morphology in secondary rat cultures observed by Varon and co-workers and Fig. 14 that reported by MacVicar (1987) after exposure to dBcAMP. When Figs. 5-14 are considered together, it becomes obvious that process extension in rodent type-1 astrocytes after exposure to dBcAMP has been unanimously reported by all authors but that there are differences, probably related to species, age, type of medium, type of serum, cell density (which is known to affect the morphology and the cytoskeleton of non-treated cultures (Goldman and Chiu, 1984b)), as well as several other factors, including the length of the time of exposure to dBcAMP (Goldman and Chiu, 1984a). Under most conditions, the cells do not become typically star shaped (stellate) even in cultures of relatively low density or in areas where the cells are well separated from each other. The only exceptions are in colony cultures (Figs. 1 and 11) and in the absence of serum (Fig. 10). It is likely that also other dBcAMP induced characteristics, e.g., those related to cell function, might differ under different culturing conditions. This might help to explain why Hertz and associates believe that addition of dBcAMP to cultures in the presence of serum has an effect which is relatively similar to that seen in normal in vivo development, whereas Fedoroff and co-workers interpret the effects they observe as a differentiation towards reactive-like astrocytes. Our conclusion does not mean that astrocytes in primary cultures display no abnormal features (Juurlink and Hertz, 1985). Thus, their content of GFAP (measured by rocket immunoelectrophoresis) is much higher that that of the brain in vivo (Bock et al., 1975; Hertz et al., 1978a) and the ratio between vimentin and GFAP is also higher than should be expected in mature astrocytes (Goldman and Chiu, 1984a). However, this applies to primary cultures of "type-1-like" astrocytes regardless of whether or not they have been treated with dBcAMP, and the increase in GFAP after addition of dBcAMP is rather slow (see below).

The presence of calcium is essential for the morphological response to dBcAMP and during exposure to inorganic and some organic calcium blockers (not including the dihydropyridine antagonists of the voltage dependent L-channel) the morphological alteration seen in Fig. 14 after exposure to dBcAMP is completely abolished and the cells retain their cobblestone like appearance (MacVicar, 1987). Very interestingly, R.K.H. Liem (personal communication) has recently demonstrated that a similar morphogenic effect of dBcAMP in an astrocytoma cell line is abolished after removal of GFAP by gene deletion. This seems to be the first demonstration of a physiological role of this astrocyte specific protein, and it would be extremely important to know if deletion of the GFAP gene also affects other parameters of the astrocytic response to elevation of cAMP.

Also some procedures that do not affect intracellular contents of cAMP have been found to cause formation of processes in "type-1-like" astrocytes in primary cultures. Thus, Facci et al. (1987) observed that astrocytes in primary cultures respond to lysophosphatidylserine with extension of processes. However, the morphology of these cells is distinctly different from the morphology observed by the same authors in the presence of dBcAMP or forskolin (for details, see Facci et al., 1987). Similarly, Kato et al. (1988) have observed "stellation" after exposure to alpha- and beta-D-N-acetylneuraminyl cholesterols although the cAMP level remained unaltered. Again, there seems to be differences between this response and that evoked by dBcAMP. Finally, Mobley et al. (1987) have observed transformation of cobblestone like astrocytes in primary cultures to process-bearing cells after treatment with phorbolester, but the morphology of these cells does not resemble any of those shown in Figs. 3-14.

dBcAMP AS A FUNCTIOGEN

Receptors and Second Messengers

The most striking effect of dBcAMP on receptor expression is a pronounced decline in beta-adrenergic activity, measured as isoproterenol induced formation of cyclic AMP (Table 1). This is not an unspecific response to aging since untreated cells show only a minor reduction of the response after 4-5 months in culture. The apparent small reduction of cAMP formation in the absence of dBcAMP probably reflects experimental variation. An increase of protein phosphorylation by calcium-calmodulin dependent protein kinase (see below) suggests that also the phosphoinositol system might be altered after culturing with dBcAMP. In addition, there is a marked decline in high affinity binding of serotonin to intact cultures (Whitaker-Azmitia, 1988).

Table 1. Production of cAMP (picomol/mg protein during a 10 min. period) by primary cultures of astrocytes grown in the absence or presence of 0.25 mM dBcAMP and measured under control conditions and in the presence of 1 μM isoproterenol.

	Control		Isoproterenol	
Age of Culture	-dBcAMP	+dBcAMP	-dBcAMP	+dBcAMP
3-6 weeks	104.5±11.1*	56.0±9.3	909±133	125.6±40.9
4-5 months	46.9**	57.8±1.7	641.4	85.7±3.1

* Mean ± SEM; ** Averages of 2 experiments.

Table 2. Contents of proteins and activities of enzymes in primary cultures of mouse astrocytes grown in the absence of dBcAMP for 3 weeks or in its presence during the last week.

	-dBcAMP	+dBcAMP
GFAP, arbitrary units/mg protein	1.00	1.83*[1]
GFAP, arbitrary units/cell	1.00	1.55*[1]
	1.00	1.04**[2]
Vimentin, arbitrary units/cell	1.00	1.26**[2]
Protein, picog/cell	384.6	326.1[1]
GS, nmol/min per mg protein	23.4	28.1[1]
GLDH, nmol/min per mg protein	3.0	3.4[1]
AAT, nmol/min per mg protein	198	151[1]
GABA-T, nmol/min per mg protein	0.93	1.64*[3]
LDH nmol/min per mg protein	1580	2330*[4]
CA, Wilbur-Anderson units	2.8	6.9*[5]
MAO A, nmol/hr per mg protein	4.35	5.70*[6]
MAO B, nmol/hr per mg protein	0.38	1.03*[6]

1) Recalculated from Hertz et al., 1978a; 2) From Goldman and Chiu, 1984a; 3) From Schousboe et al., 1977; 4) From Nissen and Schousboe, 1979; 5) From Schousboe et al., 1980; 6) From Yu and Hertz, 1982.

*Statistically significant difference ($P<0.05$);
** Larger amounts after culturing for 2 weeks

Protein Synthesis and Phosphorylation

Immunocytochemical staining for GFAP, which in untreated cells (Fig. 15) often is relatively weak and diffuse (to some extent depending upon cell density (Goldman and Chiu, 1984b)), shows a distinct network of fibers extending into the processes and surrounding the nucleus (Fig. 16) in dBcAMP treated cells. Even then, the staining is less intense than in the small type-2 astrocytes, one of which is also illustrated in the Fig. At the same time the immunochemical staining for vimentin also becomes more distinct (Fedoroff et al., 1987). Nevertheless, the total

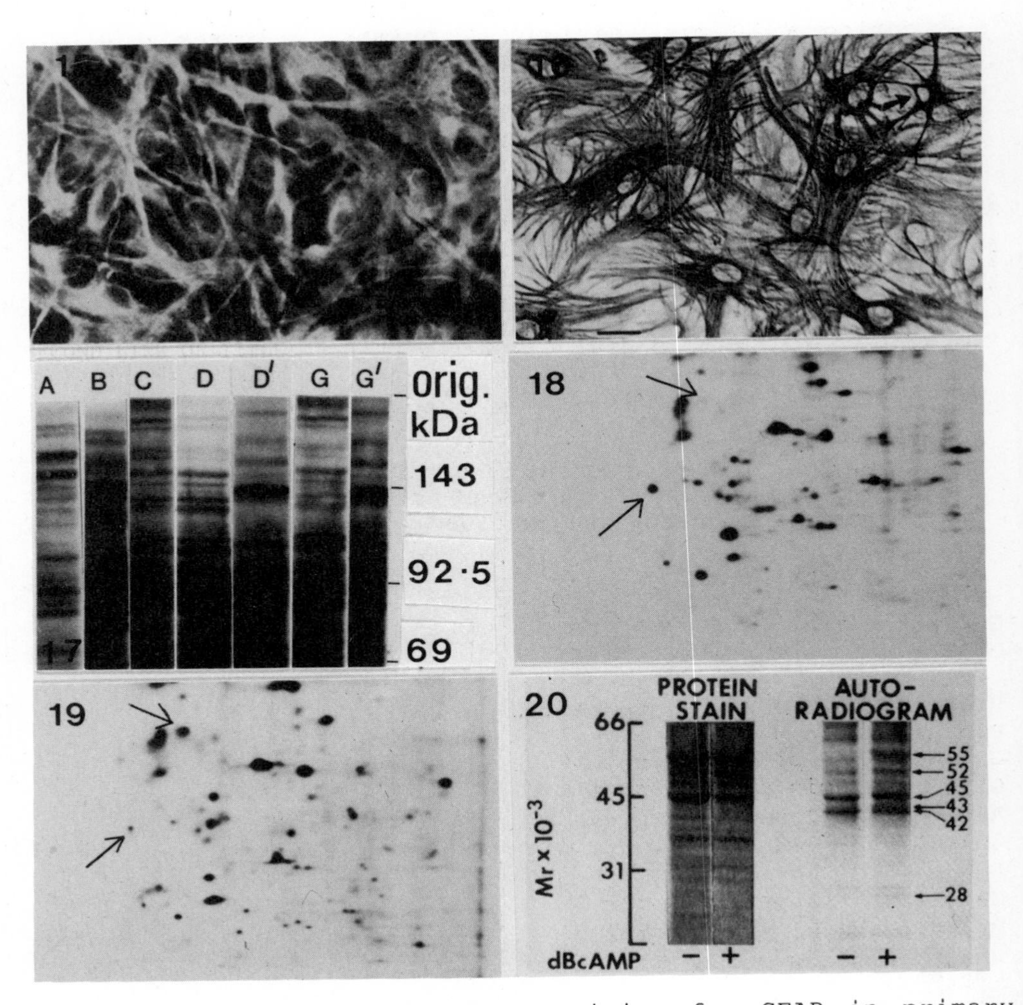

Fig. 15. Immunofluorescent staining for GFAP in primary culture of rat astrocytes grown in the absence of dBcAMP. From Chiu and Goldman, 1984. Fig. 16. Immunological staining for GFAP in primary culture of mouse astrocytes treated with dBcAMP. Note also small, intensely staining type-2 astrocyte (arrow). From Juurlink and Hertz, 1985. Fig. 17. Autoradiograms showing incorporation of [^{14}C]leucine into proteins from astrocyte cultures of varying ages (A:9; B:13; C:21; G:63 days) and grown in the absence or presence (C' and G') of 0.25 mM dBcAMP. From White and Hertz, 1981. Figs. 18-19. Autoradiograms showing incorporation of [^{14}C]leucine into proteins from astrocytic cultures grown in the absence (18) or presence (19) of dBcAMP separated by SDS-PAGE followed by isoelectric focusing on a pH gradient from 7.0 to 3.5. Note dissimilarities indicated by arrows. Unpublished experiments by F. White and L. Hertz.. Fig. 20 Gel electrophoreses of proteins from primary cultures of rat astrocytes grown in the presence or absence of dBcAMP (left) and autoradiograms of incorporation of labelled phosphate into proteins (right). Note similarities in protein pattern and dissimilarities in phosphorylation, indicated by arrows. From Neary et al., 1987.

amounts of GFAP and vimentin increase slowly (Goldman and Chiu, 1984a) and less than two-fold in cells treated for one week (Table 2). The increase in GFAP occurs both in the presence and absence of serum (Hertz et al., 1978a). The total amount of protein per cell is, in contrast, slightly decreased by dBcAMP.

Incorporation of labelled amino acids into protein tends to increase slightly after exposure to dBcAMP (White and Hertz, 1981; Murthy and Hertz, 1987). The pattern of individual proteins into which leucine incorporation occurs undergoes some marked alterations, i.e., deletion of some proteins and addition of others. In our hands (Fig. 17) there are profound spontaneous changes during the first 2-3 weeks in culture, but the response to dBcAMP is different from these and identical at 3 weeks and after prolonged culturing (White and Hertz, 1981); however, a different conclusion was reached by Bridoux et al. (1986). In the presence of forskolin these authors observed changes which were reminiscent of those exerted by dBcAMP, but not identical. White and Hertz (1982 and unpublished) have also studied the incorporation into low molecular proteins in some detail by extraction of proteins from cultures exposed to labelled leucine and separation of the proteins first by isoelectric focusing on a pH gradient from 7.0 to 3.5 and then by SDS-PAGE gel electrophoresis on a 7 1/2% polyacrylamide gel and visualization of radioactive protein by autoradiography (Fig. 18). Although most proteins were found to be synthesized to a comparable extent in neurons (not shown) and astrocytes, primary cultures of astrocytes synthesized at least 8 major proteins which are not synthesized by neurons (White and Hertz, 1982). The synthesis of these proteins was altered, mainly quantitatively, by treatment with dBcAMP (Fig. 19). This method allowed distinction between approximately 50 proteins. However, detailed studies by de Vellis and coworkers, using computer analysis, have demonstrated synthesis of up to 1000 different proteins in primary cultures of astrocytes (Bohn et al., 1988). Similar techniques will have to be employed to study differences in protein synthesis between neurons and astrocytes in primary cultures in more detail as well as the consequences of exposure of astrocytic cultures to dBcAMP.

It is not only synthesis of certain proteins but also their phosphorylation (Neary et al., 1987, 1988) which can be altered by culturing with dBcAMP (Fig. 19). This probably reflects the observation by McCarthy and co-workers (e.g., McCarthy et al., 1988) and Browning (1988) that phosphorylation of such proteins as GFA and vimentin is enhanced by exposure to beta adrenergic agonists. Treatment with dBcAMP also results in a four-fold increase in calcium-calmodulin dependent phosphorylation of a 59 kDa protein (Neary et al., 1988).

<u>Enzyme Activities</u>

From Table 2 it can be seen that, at least in mouse cultures, the activities of the three glutamate metabolizing enzymes, glutamine synthetase (GS), glutamate dehydrogenase (GLDH) and aspartate aminotransferase (AAT) are virtually unaffected by dBcAMP (Hertz et al., 1978a). The GABA transaminase (GABA-T) activity increases (Schousboe et al., 1977), and there are modest increases in the cytosolic and mitochondrial NADP linked isocitrate dehydrogenase isozymes (Juurlink and Hertz, 1985). There is also an increase in lactate dehydrogenase (LDH) activity and a shift in lactate dehydrogenase isozymes towards a more mature pattern (Nissen and Schousboe, 1979). Na+,K+-ATPase activity goes up (Moonen and Franck, 1977; Kimelberg et al., 1978; Hertz, 1982). Protein kinase C activity is unaltered (Neary et al., 1986).

It has been disputed whether carbonic anhydrase (CA), which is a glial specific enzyme (Giacobini, 1962) is localized only in oligodendrocytes or in both astrocytes and oligodendrocytes. Studies by Church et al.(1980) and Sapirstein and Hertz (see Schousboe et al., 1980) have demonstrated that the carbonic enhydrase activity in primary cultures of astrocytes is too large to be explainable by a possible minor contamination with oligodendrocytes. The activity of this enzyme increases several fold after culturing with dBcAMP (Table 2). A comparable situation (Table 2) exists in the case of monoamine oxidase B (MAO B), whereas there is little change in MAO A activity (Yu and Hertz, 1982).

Metabolism

The rate of oxygen consumption in primary cultures of astrocytes increases from slightly above 200 nmol/min per mg protein to ≈ 300 nmol/min after culturing with dBcAMP (Hertz and Hertz, 1979). Formation of $^{14}CO_2$ from labeled glucose occurs only slowly, probably due to isotope dilution in a large number of metabolites, and is not measurably affected. Astrocytes also use other metabolic substrates, including glutamate. Glutamate oxidation increases by about 30% after exposure to dBcAMP (G. Kala and L. Hertz, unpublished). This might explain a decreased glutamate content in cultured astrocytes in spite of an unaltered concentration in the incubation medium (Table 3). In contrast, the rate of glutamine formation from glutamate appears to be reduced (Potter et al., 1982). The contents of alanine are increased both in the cells and in the media (Table 3). The mechanism behind this effect might be an enhanced glycolysis, due to the increase in LDH activity, followed by transamination to alanine. The branched chain amino acid isoleucine is also decarboxylated at an increased rate (Murthy and Hertz, 1987), but its metabolism is not nearly as fast as that of glutamate. The branched chain amino acids may be of special metabolic importance because they are converted to tricarboxylic acid cycle intermediates. Since isoleucine is an essential amino acid its only source is the incubation medium, and the isoleucine content of the conditioned medium, i.e., the medium in which the cultures have been incubated (under our conditions for 3-4 days) is considerably less than in the fresh medium (Table 3). This difference, representing isoleucine utilization, is significantly increased by dBcAMP. Leucine is also utilized at an increased rate but less actively than isoleucine. There are no significant effects on valine. Arginine is decreased in the cells but not in the medium, maybe reflecting an altered uptake. Both cell and medium contents of other esential amino acids are unaltered (results not shown).

Hydrolysis of glycogen in astrocytes is enhanced by adrenergic agonists (Browning et al., 1972; Kala et al., 1989). In spite of the decrease in cAMP production in response to adrenergic agonists, the potency of isoproterenol and norepinephrine is not reduced in dBcAMP treated cells (V.S.R. Kala and L. Hertz, unpublished results).

Table 3. Contents of amino acids in cells (nmol/mg protein) and conditioned media (μM). The concentrations in fresh media are shown in parenthesis.

Amino acid	Cells nmol/mg protein		Medium μM	
	-dBcAMP	+dBcAMP	-dBcAMP	+dBcAMP
glutamate (0)	81.1±7.3	55.6±4.0	857±82.8	866±95.2
alanine (0)	8.9±0.9	13.7±1.5	193±20.9	277±24.2
leucine (800)	11.1±1.3	7.1±0.3	542±35.8	393± 7.2
isoleucine (800)	9.1±1.0	5.2±0.3	401±42.2	275± 7.4
valine (800)	14.1±1.3	12.0±0.8	705±36.0	635±24.9
arginine (1200)	14.4±1.4	8.9±0.7	689±17.4	775±0.42

* Results are means ± SEM. L. Hertz and M. Farmer, unpublished.

Membrane transport

Substrates, metabolites and inorganic ions cross cell membranes by either diffusion (often through channels) or active, carrier-mediated uptake. Inorganic ions are often transported actively in one direction and passively in the other, and amino acids, including the two transmitter amino acids glutamate and GABA, are accumulated into neural cells by active, sodium-dependent uptake.

The carrier-mediated transport of glutamate and especially of GABA into astrocytes is affected by treatment with dBcAMP. Both K_M and V_{max} values for glutamate are increased after one week of treatment with dBcAMP (Hertz et al., 1978a). The net effect, calculated from the Michaelis-Menten equation is that the uptake at an extracellular glutamate concentration of 50 μM is increased by about 30%. Cotransport with sodium is unaltered, i.e., one to one. The V_{max} for GABA uptake is greatly increased but, probably even more importantly, the cotransport mechanism with sodium changes from cotransport of one molecule GABA with one sodium ion in untreated cultures to cotransport with two sodium ions in dBcAMP treated cultures (Larsson et al., 1986). This allows the establishement of a much steeper gradient between intracellular and extracellular GABA concentrations, i.e., a much more efficient removal of transmitter GABA from the extracellular clefts.

Accumulation of the potassium ion (K^+) by carrier mediated and channel mediated uptake is extremely rapid into astrocytes, i.e., 2 μmol/mg protein per min or, with an assumed water space of 9 μl/mg protein ≈ 200 μmol per ml (200 mM) per min (Hertz, 1986). It is not affected by dBcAMP.

The concentration of free calcium ion, Ca^{2+}, is low inside cells, partly due to sequestration and partly due to active extrusion. Entry of Ca^{2+} along a concentration gradient seems to play a major role in cell signaling. One of the voltage-sensitive Ca^{2+} channels, the L-channel, is found in astrocytes treated with dBcAMP but not in untreated cells (MacVicar, 1984, 1988; Chun et al., 1986, MacVicar and Tse, 1988). Nimodipine and nitrendipine are organic Ca^{2+}-channel inhibitors which, at low concentrations (nM), are specific for the L-channel. It is in agreement with the absence of this channel in untreated cultures that a high-affinity (K_D 2.5 nM) binding of nitrendipine to astrocytes is found only after exposure to dBcAMP (Hertz, 1989), whereas both treated and untreated astrocytes have a low-affinity binding (K_d values 209 and 218 nM, respectively). Moreover, a K^+-stimulated uptake of Ca^{2+}, which is potently inhibited by nimodipine (Hertz et al., 1989) is present only after dBcAMP treatment.

CONCLUDING REMARKS

From the present review it can be concluded that dBcAMP has profound morphogenic and functiogenic effects on astrocytes in primary cultures. It seems also quite clear that the effects can be altered by even minor changes in medium and that cell density and length of exposure are important factors. It would probably be naive to believe that dBcAMP treatment completely substitutes for all normal transmitter effects on astrocytic development. In contrast to the continued and uniform exposure to dBcAMP in culture, transmitter signals in vivo will be limited in time, different at different locations, and able to create concentration gradients. Intracellular messengers other than dBcAMP may also be involved. Some features of astrocytes in primary cultures are clearly abnormal, first and foremost the pronounced expression of GFA and the lack of vimentin deletion during the differentiation. However, the GFA content is far too high even in untreated cultures and its increase in treated cultures, although impressive immunochemically, is only two-fold. Also, although treatment with dBcAMP does increase vimentin staining and content, the ratio between GFA and with vimentin remains unchanged (Goldman and Chiu, 1984a).

For the evaluation of the effects evoked by dBcAMP it is important to compare characteristics of dBcAMP treated cells and of untreated cells with those of mature and immature normal astrocytes. There seems to be little doubt that dBcAMP treated astrocytes morphologically are much more similar to protoplasmic astrocytes in situ (Figs. 1-16). With respect to a functiogenic role of dBcAMP there is good evidence that astrocytes obtained by microdissection or gradient centrifugation also express L-channels for Ca^{2+} (Newman, 1985) and saturable binding sites for nitrendipine (L. Hertz, M. Farooq and W.T. Norton, unpublished experiments). Preliminary experiments have suggested that the latter probably include high affinity binding sites. Up until now, attempts to demonstrate voltage-sensitive channels in astrocytes in brain slices have, however, not been successful (Walz and MacVicar, 1988). This could be due to deterioration of Ca^{2+} channels in the slices, since Ca^{2+} channels depend upon intact energy metabolism (Kostyuk, 1984). Studies of cerebral astrocytes prepared by gradient centrifugation support the concept of cotransport of GABA with at least two sodium ions (Sellstrom et al., 1980), but astrocytes from the cauda equina of the frog transport GABA together with only one sodium ion (Hertz and Schousboe, 1980). However, these astrocytes may be as deprived of neuronal signals as are astrocytes in primary cultures. Intuitively, a potassium induced Ca^{2+} uptake and a more efficient GABA (and glutamate) uptake would appear consistent with the assumed normal functions of mature astrocytes, interacting with neurons (Hertz, 1989).

If dBcAMP is substituting for at least some of the normally occurring transmitter signals it follows that such signals exert a morphogenic and functiogenic influence in vivo. One way in which it may be possible to obtain further information about such neuronal-astrocytic interactions would be co-culturing of developing cortical astrocytes with locus coeruleus neurons. Recently Hansson and Ronnback (1988) have elegantly demonstrated that striatal astrocytes differentiate both morphologically and functionally when co-cultured with neurons from substantia nigra, a natural projection area. Thus, a role of monoamines as morphogens and functiogens for astrocytic development has been established in a closely related system.

REFERENCES

Aoki, C., Joh, T.H., and Pickel, V.M., 1987, Ultrastructural localization of immunoreactivity for beta-adrenergic receptors in cortex and neostriatum of rat brain, Brain Res., 437: 264-282.

Beaudet, A., and Descarries, L., 1984, Fine structure of monoamine axon terminals in cerebral cortex, in: "Monoamine Innervation of Cerebral Cortex", L. Descarries, T.R. Reader, and H.H. Jasper, eds., Alan R. Liss, New York, pp. 77-93.

Bender, A.S., Hertz, L., Woodbury, D.M. and White, H.S., 1988, Potent modulation of potassium (K+) uptake into astrocytes by protein kinase C (PKC), Abstracts, Soc. Neurosc. 14: 1057.

Bock, E., Jorgensen, O.S., Dittman, L. and Eng, L.F., 1975, Determination of brain-specific antigens in short-term cultivated rat astroglia cells and in rat synaptosomes, J. Neurochem., 25: 867-870.

Bockaert, J., and Ebersolt, C., 1988, α-Adrenergic receptors on glial cells, in: "Glial Cell Receptors," H.K. Kimelberg, ed., Raven Press Ltd., New York, pp. 35-51.

Bohn, M.C., Walencewicz, A., Lynch, M., and deVellis, J., 1988, Identification of glucocorticoid regulated proteins in purified rat cerebral astrocytes by quantitative 2D-gel electrophoresis, Abst. Soc. Neurosci., 14: 1057.

Bridoux, A.M., Fages, C., Couchie, D., Nunez, J., and Tardy, M., 1986, Protein synthesis in astrocytes: "Spontaneous" and cyclic AMP-induced differentiation, Dev. Neurosci., 8: 31-43.

Browning, E.T., 1988, Hormone and second messenger regulated protein phosphorylation by cultured rat astrocytes: Cytoskeletal intermediate filament phosphorylation, in: "Glial Cell Receptors," H.K. Kimelberg, ed., Raven Press, New York, pp. 23-24.

Browning, E.T., Schwartz, J.P., and Breckenridge, B.McL., 1972, Norepinephrine sensitive properties of C-6 astrocytoma cells, Mol. Pharmacol., 10: 162-174.

Chiu, F.-C., and Goldman, J.E., 1984, Synthesis and turnover of cytoskeletal proteins in cultured astrocytes, J. Neurochem., 42, 166-174.

Chun, L.L.Y., Barres, B.A., and Corey, D.P., 1986, Induction of calcium channel in astrocytes by cAMP, Abst. Soc. Neurosci., 12: 1246.

Church, G.A., Kimelberg, H.K., and Sapirstein, V.S., 1980, Stimulation of carbonic anhydrase activity and phosphorylation in primary astroglial cultures by norepinephrine, J. Neurochem., 34: 873-879.

Cummins, C.J., Lust, W.D., and Passonneau, J.V., 1983, Regulation of glycogen metabolism in primary and transformed astrocytes in vitro, J. Neurochem., 40: 128-136.

Ebersolt, C., Perez, M., and Bockaert, J., 1981, α_1 and α_2 adrenergic receptors in mouse brain astrocytes from primary cultures, J. Neurosci. Res., 6: 643-652

Facci, L., Skaper, S.D., Levin, D.L., and Varon, S., 1987, Dissociation of the stellate morphology from intracellular cyclic AMP levels in cultured rat brain astroglial cells: Effects of ganglioside G_{M1} and lysophosphatidylserin., J. Neurochem., 566-573.

Fedoroff, S., White, R., Subramanyan, L., and Kalnins, V.I., 1981, Properties of putative astrocytes in colony cultures of mouse neopallium, in: "Glial and Neuronal Cell Biology," V. Fedoroff, ed., Liss, New York, pp. 1-19.

Fedoroff, S., Ahmed, I., Opas, M. and Kalnins, V.I., 1987, Organization of Microfilaments in astrocytes that form in the presence of dibutyryl cyclic AMP in cultures, and which are similar to reactive astrocytes in vivo, Neuroscience, 22: 255-266.

Foote, S.L., Bloom, F.E. and Aston-Jones, G., 1983, Physiol. Rev. 63: 844-914.

Giacobini, E., 1962, A cytochemical study of the localization of carbonic anhydrase in the nervous system, J. Neurochem., 9: 169-177.

Gilman, A.G., and Nirenberg, J., 1971, Effect of catecholamines on the adenosine 3':5'-cyclic monophosphate concentrations of cloned satellite cells of neurons, Proc. Natl. Acad. Sci. USA, 68: 2165-2168.

Goldman, J.E., and Chiu, F.C., 1984a, Dibutyryl cyclic AMP causes intermediate filament accumulation and actin reorganization in astrocytes, Brain Res., 306: 85-95.

Goldman, J.E., and Chiu, F.C., 1984b, Growth kinetics, cell shape, and the cytoskeleton of primary astrocyte cultures, J. Neurochem., 42: 175-184.

Hansson, E., and Ronnback, L., 1988, Neurons from substantia nigra increase the efficacy and potency of second messenger arising from striatal astroglia dopamine receptor, Glial, 393-397.

Harik, S.I., Sharma, V.K., Wetherbee, J.R., Warren, R.H., and Banerjee, S.P., 1980, Adrenergic receptors of cerebral microvessels, Eur. J. Pharmacol., 61: 207-208.

Hartman, B.K., Swanson, L.W., Raichle, M.E., Preskorn, S.H., and Clark, H.B., 1979, Central adrenergic regulation of cerebral microvascular permeability and blood flow; anatomic and physiologic evidence, Adv. Exp. Med. Biol., 131: 113-126.

Hertz, E., and Hertz, L., 1979, Polarographic measurement of oxygen uptake by astrocytes in primary cultures using the tissue culture flask as the respirometer chamber, In Vitro, 15: 429-436.

Hertz, L., 1978, Kinetics of adenosine uptake into astrocytes, J. Neurochem., 31: 55-62.

Hertz, L., 1982, Astrocytes, in: "Handbook of Neurochemistry", 2nd Ed., Vol. 1, A. Lajtha, ed., Plenum Pres, New York, pp.319-355.

Hertz, L., 1986, Potassium as a signal in metabolic interactions between neurons and astrocytes., in: "Dynamic Properties of Glial Cells, Cellular and Molecular Aspects," T. Grisar, G. Franck, L. Hertz, W.T. Norton, M. Sensenbrenner and D.M. Woodbury, eds., Pergamon Press, pp. 215-224.

Hertz, L., 1989, Functional interactions between neurons and glial cells, in "Regulatory Mechanisms of Neuron to Vessel Communication in Brain ", S. Govoni, F. Battaini and M.S. Mangoni, eds., Springer, Heidelberg, in press.

Hertz, L., and Schousboe, A., 1980, Interactions between neurons and astrocytes in the turnover of GABA and glutamate, Brain Res. Bull 5 Suppl.2: 389-395.

Hertz, L., Bock, E., and Schousboe, A., 1978a, GFA content, glutamate uptake and activity of glutamate metabolizng enzymes in differentiating mouse astrocytes in primary cultures, Dev. Neurosci., 1: 226-238.

Hertz, L., Schousboe, A., Boechler, N., Mukerji, S. and Fedoroff, S., 1978b, Kinetic characteristics of the glutamate uptake into normal astrocytes in cultures. Neurochem. Res. 3: 1-14.

Hertz, L., Juurlink, B.H.J., Fosmark, H., and Schousboe, A., 1982, Astrocytes in primary cultures, in: "Neuroscience Approached Through Cell Culture," S.E. Pfeiffer, ed., CRC Press, Boca Raton, FL., pp. 175-186.

Hertz, L., Juurlink, B.H.J., and Szuchet, S., 1985, Cell cultures, in: "Handbook of Neurochemistry," 2nd ed., A. Lajtha, ed., Plenum Press, New York, Vol. 8, pp. 603-661.

Hertz, L., Bender, A.S., Woodbury, D.M., and White, H.S., 1989, Potassium-stimulated calcium uptake in astrocytes and its potent inhibition by nimodipine, J. Neurosci, Res., 22: in press.

Juurlink, B.H.J. and Hertz, L., 1985, Plasticity of astrocytes in primary cultures: An experimental tool and a reason for methodological caution. Dev. Neurosci. 7: 263-277.

Kala, V.S.R., Richardson, J.S., and Hertz, L., 1989, Adrenergic stimulation of glycogenolysis in astrocytes, Trans. Am. Soc. Neurochem., 20: in press.

Kato, T., Ito, J.I., Tanaka, R., Suzuki, Y,m Hirabayashi, Y., Matsumoto, M., Ogura, H., and Kato, K., 1987, Sialosyl cholesterol induces morphological and biochemical differentiations of glioblasts without intracellular cyclic AMP level rise. Brain Res. 438: 277-285.

Kimelberg, H.K., Narumi, S., and Bourke, R.S., 1978, Enzymatic and morphological properties of primary rat brain astrocyte cultures, and enzyme development in vivo, Brain Res., 153: 55-77.

Kimelberg, H.K., Stieg, P.E., and Mazurkiewics, J.E., 1982, Immunocytochemical and biochemical analysis of carbonic anhydrase in primary astrocyte cultures from rat brain, J. Neurochem., 39: 734-742.

Kostyuk, P.G., 1984, Metabolic control of ionic channels in the neuronal membrane, Neurosci., 13: 983-989.

Labourdette, G., Roussel, G., and Nussbaum, J.L., 1980, Oligodendroglia content of glial cell primary cultures, from newborn rat brain hemispheres, depends on the initial plating density, Neurosci. Lett., 18: 203-209.

Larsson, O.M., Hertz, L., and Schousboe, A., 1986, Uptake of GABA and nipecotic acid in astrocytes and neurons in primary cultures: changes in the sodium coupling ratio during differentiation, J. Neurosci. Res., 16: 699-708.

Lauder, J.M., 1988, Neurotransmitters as morphogens, Progr. Brain Res: 365-387.

Lauder, J.M. and Bloom, F.E., 1974, Ontogeny of monoamine neurons in the locus coeruleus, raphe nuclei and substantia nigra of the rat I. Cell differentiation, J. Comp. Neurol. 155: 469-482.

Lim, R., Mitsunobu, K., and Li, W.K.P., 1973, Maturation-stimulating effect of brain extract and dibutyryl cyclic AMP on dissociated embryonic brain cells in cultures, Exp. Cell Res., 79: 243-246.

MacVicar, B.A., 1984, Voltage-dependent calcium channels in glial cells, Science, 226: 1345-1347.

MacVicar, B.A., 1987, Morphological differentiation of cultured astrocytes is blocked by cadmium or cobalt, Brain Res., 420: 175-177.

MacVicar, B.A., and Tse, F.W.Y., 1988, Norepinephrine and cyclic adenosine 3':5'-cyclic monophosphate enhance a nifedipine-sensitive calcium current in cultured rat astrocytes, Glia, 1: 359-365.

McCarthy, K.D., Salm, A., and Lerea, L.S., 1988, Astroglial receptors and their regulation of intermediate filament protein phosphorylation, in: "Glial Cell Receptors," H.K. Kimelberg, ed., Raven Press, New York, pp. 1-22.

Mobley, P.L., Scott, S.L. and Cruz, E.G., 1986, Protein kinase C in astrocytes: a determinant of cell morphology, Brain Res., 398: 366-369.

Moonen, G., and Franck, G., 1977, Potassium effect on Na^+, K^+-ATPase activity of cultured newborn astroblasts during differentiation, Neurosci. Lett., 4: 263-267.

Moonen, G., Cam, Y., Sensenbrenner, M., and Mandel, P., 1975, Variability of the effects of serum-free medium, dibutyryl-cyclic AMP or theophylline on the morphology of cultured new-born astrocytes, Cell Tissue Res., 163: 365-372.

Murthy, Ch.R.K., and Hertz, L., 1987, Acute effects of ammonia on branched chain amino acid oxidation and incorporation into proteins in astrocytes and in neurons in primary cultures, J. Neurochem., 49: 735-741.

Narumi, S., Kimelberg, H.K., and Bourke, R.S., 1978, Effects of norepinephrine on the morphology and some enzyme activities of primary monolayer cultures from rat brain, J. Neurochem., 31: 1479-1490.

Neary, J.T., Norenberg, L.O.B, Norenberg, M.D., 1986, Calcium-activated phopholipid-dependent protein kinase and protein substrates in primary cultures of astrocytes, Brain Res., 385: 420-424.

Neary, J.T., Gutierrez, M.P., Norenberg, L.O.B., and Norenberg, M.D., 1987, Protein phosphorylation in primary astrocyte cultures treated with and without dibutyryl cyclic AMP, Brain Res., 410: 164-168.

Neary, J.T., Norenberg, L.O.B., and Norenberg, M.D., 1988, Protein phosphorylation in astrocytes: A possible role in epilepsy, in: "Biochemical Pathology of Astrocytes," Norenberg, M.D., Hertz, L., and Schousboe, A., eds., Alan R. Liss, New York, pp. 519-533.

Newman, E.A., 1985, Voltage-dependent calcium and potassium channels in retinal glial cells, Nature, 317: 809-811.

Nissen, C., and Schousboe, A., 1979, Activity and isoenzyme patterns of lactate dehydrogenase in astroblasts cultured from brains of newborn mice, J. Neurochem., 32: 1787-1792.

Pollenz, R.S., and McCarthy, K.D., 1986, Analysis of cyclic AMP-dependent changes in intermediate filament protein phosphorylation and cell morphology in cultured astroglia, J. Neurochem., 47: 9-17.

Potter, R.L., Yu, A.C., Schousboe, A., and Hertz, L., 1982, Metabolic fate of (U-^{14}C)-labelled glutamate in primary cultures of mouse astrocytes as a function of development, Dev. Neurosci., 5: 278-284.

Privat, A., 1975, Postnatal gliogenesis in the mammalian brain, Internat. Rev. Cytol. 40: 281-323.

Schousboe, A., Hertz, L., and Svenneby, G., 1977, Uptake and metabolism of GABA in astrocytes cultured from dissociated mouse brain hemispheres, Neurochem., Res., 2: 217-229.

Schousboe, A., Nissen, C., Bock, E., Sapirstein, V., Juurlink, B.H.J., and Hertz, L., 1980, Biochemical development of rodent astrocytes in primary cultures, in: "Tissue Culture in Neurobiology," E. Giacobini, A. Vernadakis, and A. Shahar, eds., Raven Press, New York, pp. 397-409.

Sensenbrenner, M., Barakat, I., J.P. Delaunoy, Labourdette, G. and Pettman, B., 1982, in "Neuroscience Approached Through Cell Culture", S.E. Pfeiffer, ed., vol 1: pp. 87-105.

Sellstrom, A., Henn, F., Estborn, L., Hansson, E., and Hamberger, A., 1980, On the GABA transport in fractions of astrocytes and nerve-terminals, Brain Res. Bull, 5: 95-99.

Shapiro, D.L., 1973. Morphological and biochemical alterations in foetal rat brain cells cultured in the presence of monobutyryl cyclic AMP, Nature, 241: 203-204.

Turing, A.M., 1952, The chemical basis of morphogenesis, Trans. R. Soc. Lond. Ser. B, 237: 37-72.

Walz, W., and MacVicar, B.A., 1988, Electrophysiological properties of glial cells: comparison of brain slices with primary cultures, Brain Res., 443: 321-324.

Whitaker-Azmitia, P.M., 1988, Astroglial serotonin receptors, in: "Glial Cell Receptors," H.K. Kimelberg, ed., Raven Press, New York, pp. 107-120.

White, F.P., and Hertz, L., 1981, Protein synthesis by astrocytes in primary cultures, Neurochem. Res., 6: 353-364.

White, F.P., and Hertz, L., 1983, Comparison of proteins synthesized by astrocytes and neurons in primary cultures, Abstracts, Soc. Neurosci., 8: 244.

Yu, P.H., and Hertz, L., 1982, Differential expression of type A and type B monoamine oxidase of mouse astrocytes in primary cultures, J. Neurochem., 39: 1493-1495.

THE ONTOGENY OF ADRENERGIC FIBERS IN RAT SPINAL CORD

Holli Bernstein-Goral[1] and Martha Churchill Bohn[2]

[1]Department of Anatomy and Cell Biology
Georgetown University,School of Medicine
3900 Reservoir Road, N.W.,Washington, D.C. 20007

[2]Department of Neurobiology and Anatomy
University of Rochester Medical Center
Rochester, NY 14642

INTRODUCTION

Preganglionic neurons play a pivitol role in the integration of descending central signals involved in sympathetic and parasympathetic regulation. However, the number of neurotransmitter systems which converge in the preganglionic cell column has made study of the complex synaptology of the preganglionic neuron and the development of its innervation exceedingly difficult. Although many neurotransmitters have been identified in the region of preganglionic neurons, it was not until recently that the termination of phenotypically defined fibers was demonstrated on preganglionic neurons which project to identified sympathetic targets. This chapter reviews studies which have focused on the development and synaptology of one of these projections which arises in adrenergic neurons located in the medulla oblongata and which terminates in the thoracic region of spinal cord. Study of adrenergic neurons has been facilitated by the specific expression of the epinephrine-synthesizing enzyme, phenylethanolamine N-methyltransferase (PNMT), which distinquishes adrenergic neurons from other types of catecholamine-synthesizing cells. Study of this particular projection is interesting due to the implication of adrenergic neurons in a wide array of autonomic and endocrine functions including cardiovascular regulation (Saavedra et al., 1979; Ross et al., 1984b; Goodchild et al., 1984; Morrison et al., 1988), reproduction (Coen and Coombs, 1983; Sheaves et al., 1985; and the stress response (Saavedra and Torda, 1980; Mezey et al., 1984; Liposits et al., 1986 a,b).

ONTOGENY OF THE BULBOSPINAL ADRENERGIC PROJECTION

In the brain, the majority of adrenergic neurons are located in the medulla oblongata where they are organized into three cell groups designated C1-C3 (Hökfelt et al., 1974; Howe et al., 1980; Kalia et al., 1985a,b). A small population of PNMT-immunoreactive (PNMT-IR) neurons has also been identified in the paraventricular nucleus of the hypothalamus, as well as in the retina (Ross et al., 1984a; Ruggiero et al., 1985; Foster et al., 1985a; Kaiser et al, 1987), although classification of these neurons as "adrenergic" is controversial since it appears that other catecholamine synthetic enzymes are not expressed in all of these neurons (Mefford, 1988).

Molecular Aspects of Development and Aging of the Nervous System
Edited by J.M. Lauder
Plenum Press, New York, 1990

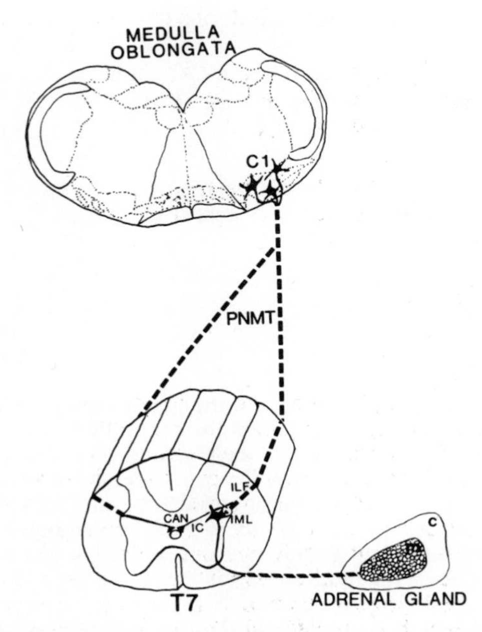

Figure 1. Schematic diagram of the adrenergic bulbospinal projection emanating from PNMT-immunoreactive neurons in the C1 cell group of the rostral ventrolateral medulla. PNMT-immunoreactive fibers project to the sympathetic preganglionic nuclei of thoracic (T7) and upper lumbar spinal segments, overlapping the distribution of sympathetic and adrenal preganglionic neurons (IML and CAN). Adrenal preganglionic neurons send axons which innervate the chromaffin cells of the adrenal medulla (m). Abbreviations: IML, intermediolateral nucleus; CAN, central autonomic nucleus; ILF, intermediolateral funiculus; IC, intercalated nucleus; c, adrenal cortex; m, adrenal medulla.

Adrenergic neurons in the brainstem have widespread projections throughout the brain and spinal cord which include terminal fields in the hypothalamic paraventricular, preoptic, and dorsomedial nuclei (Hökfelt et al., 1974, 1988; Swanson et al., 1981; Sawchenko and Swanson, 1982; Tucker et al., 1987; Hornby and Piekut, 1987), the locus coeruleus (Astier et al., 1987; Pieribone et al., 1987, 1988; Guyenet and Young, 1987), the parabrachial complex and raphe nuclei (Ruggiero et al., 1985) and several limbic structures including the amygdala and frontal cortex (Beato et al., 1987). PNMT-IR fibers also terminate locally in the medulla oblongata in the nucleus tractus solitarius and dorsal motor nucleus of the vagus (Hökfelt et al., 1974). Of particular interest is a descending adrenergic projection to thoracic and upper lumbar segments of spinal cord which specifically overlaps the distribution of sympathetic preganglionic neurons (figure 1; Ross et al., 1981a,b; 1984c; Bernstein-Goral and Bohn, 1988, 1989).

The expression of PNMT in the brain occurs much earlier than in the adrenal medulla (Foster et al., 1985b; Bohn et al., 1981, 1986; Teitelman et al., 1979; Verhofstad et al., 1979). The first PNMT-IR neuroblasts appear in the presumptive C1 cell group located on the floor of the rhombencephalon on gestational day 14 in the rat embryo (E14). Neuroblasts migrating within the wall of the rhombencephalon also express PNMT at this stage. PNMT-IR neurons do not appear in the dorsal adrenergic cell groups, C2 and C3 until late in gestation (Bohn et al., 1986).

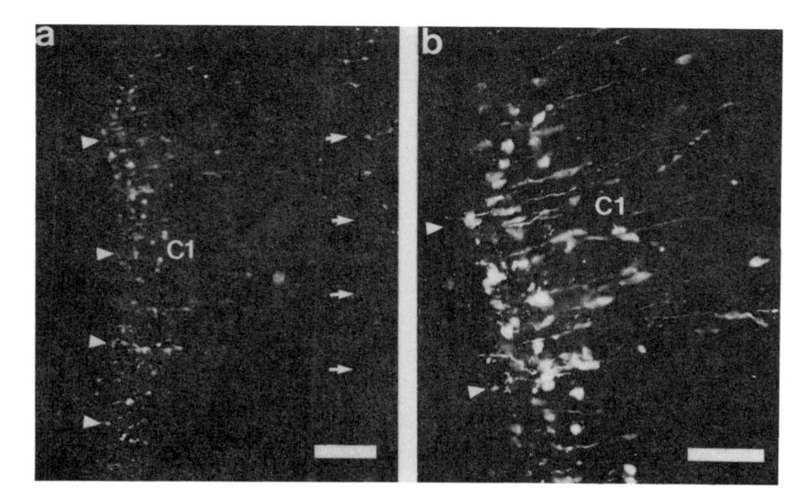

Figure 2. PNMT-immunofluorescent neurons in E15 embryo in the sagittal plane of the ventral medulla oblongata (arrow heads).
a) Presumptive C1 adrenergic cell group (arrow heads) in the ventrolateral medulla with beaded PNMT-IR fibers located in the dorsal medulla near the ventricle (arrows on right).
b) High magnification of elongated bipolor PNMT-IR neuroblasts in ventral C1 cell group close to the ventral pial surface (arrowheads). Bars = a) 50um; b) 25um.

Within one day following expression of PNMT in the brain stem (E15), ascending and descending adrenergic fiber bundles which emanate from bipolar neuroblasts in the ventral lateral medulla are observed (figure 2). At this age, PNMT-IR fibers are present in the lateral funiculus of cervical through upper thoracic spinal segments (figure 3). On E16, PNMT-IR fibers turn perpendicular to the myelinated axons in the lateral funiculus and enter the intermediate gray matter in the lateral horn (mantle region) (figure 3). The early appearance of PNMT-IR fibers in fetal thoracic spinal cord is followed by a rapid increase in the number of PNMT-IR fibers due either to in-growth of new adrenergic fibers or increased levels of

PNMT in existing fibers. Epinephrine has been detected as early as E14 in both embryonic brain and spinal cord (Foster et al.1985c, 1987; Bernstein-Goral and Bohn, 1988).

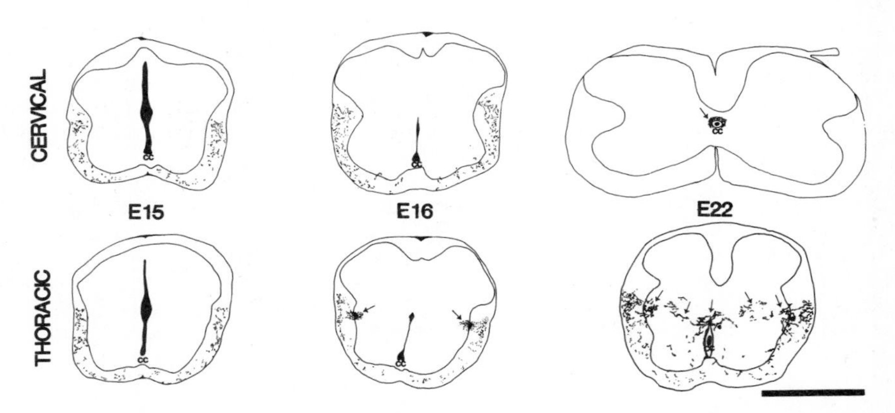

Figure 3. Camera lucida illustrations of the embryonic distribution of PNMT-IR fibers in transverse sections of cervical and thoracic spinal cord in E 15, E 16 and E 22 embryo. Arrows indicate PNMT-IR fibers in embryonic gray matter. cc: central canal. Bar = 0.5 mm. (reprinted by permission of A.R. Liss, Inc., Bernstein-Goral and Bohn, 1988)

PNMT-IR fibers display a lateral to medial increase in fiber density with age (figure 3). They distribute initially in the intermediate gray matter of thoracic spinal cord concentrated along the longitudinal cell column in the principle sympathetic preganglionic nucleus, the intermediolateral nucleus (IML). Just prior to birth, PNMT-IR fibers aggregate to form prominent transverse fiber bundles which span the intermediate gray matter from the lateral horn to the region dorsomedial to the central canal (figure 3). A secondary longtudinal PNMT-IR fiber bundle is formed at older postnatal ages. This intercepts the transverse fiber bundles at 'nests' of preganglionic neurons in IML (figure 5). Observed together, these two PNMT-IR axonal trajectories form the periodic 'ladder-like' pattern for PNMT-IR fibers in the longitudinal plane of thoracic spinal cord in the adult rat (figure 5; Ross et al., 1984c; Bernstein-Goral and Bohn, 1988).

The ontogenic increase in the number of PNMT-IR fibers is paralleled by increasing levels of epinephrine. Both PNMT-IR fiber density and epinephrine levels reach a maximum between postnatal day 10-15 and then gradually decline to adult levels (figures 4 and 5, Bernstein-Goral and Bohn, 1988). This peak of epinephrine is not derived from the adrenal gland, since epinephrine levels in spinal cord are unaffected by acute bilateral adrenalectomy (Bernstein-Goral and Bohn, 1988). Other studies have reported that PNMT activity in the medulla and adrenergic receptors in thoracic spinal cord also peak during the second postnatal week followed by a prolonged decline during the postnatal period to adult levels (Nakamura and Nakamura, 1978; Saavedra, 1979; Ross et. al., 1982; Lau et al., 1985). The adrenergic fiber loss in late postnatal development may be a result of naturally occurring cell death or transient expression of the adrenergic phenotype in a population of bulbospinal neurons in brain stem. It is also possible that retraction or redistribution of axons or collaterals in spinal cord contributes to adrenergic fiber reorganization. Considering the apparent hyperinnervation during the neonatal period and subsequent decline in adrenergic fiber density and epinephrine into adulthood, the bulbospinal adrenergic pathway may exert developmental cues during formation of spinal sympathetic nuclei.

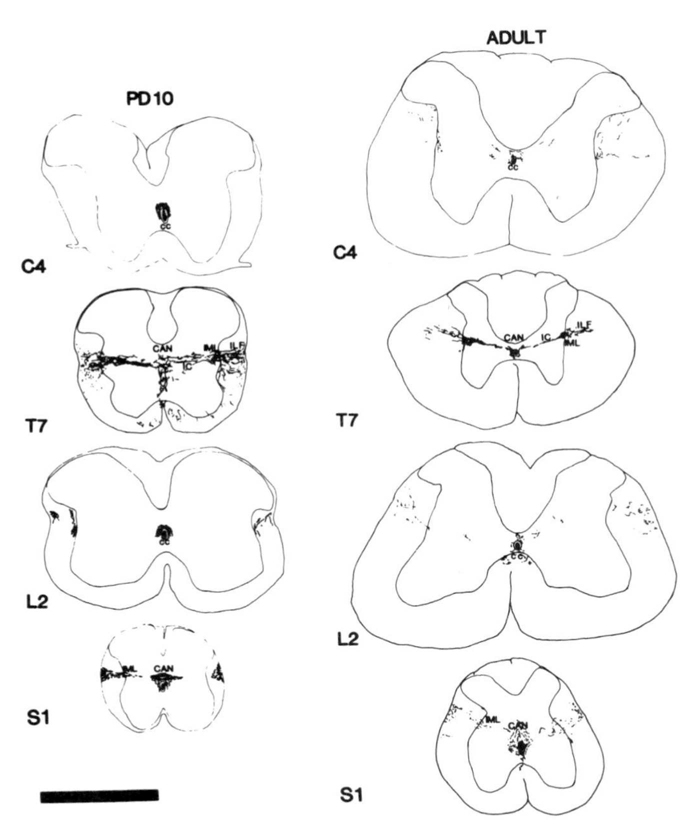

Figure 4. Camera lucida illustrations of postnatal day 10 and adult distribution of PNMT-IR fibers in transverse sections of cervical (C4), thoracic (T7), lumbar (L4) and sacral (S1) spinal cord. IML, n. intermediolateralis pars principalis; ILF, n. intermediolateralis pars funicularis; IC, nucleus intercalatus; CAN, central autonomic nucleus; cc, central canal. Bar = 1 mm. (reprinted by permission of A.R. Liss, Inc., Bernstein-Goral and Bohn, 1988)

RELATIONSHIP OF BULBOSPINAL PNMT-IR FIBERS AND SPINAL CORD TARGETS

PNMT-IR fibers in the developing and adult spinal cord are intimately, although not exclusively, associated with sympathetic preganglionic nuclei (figure 4; Hökfelt et al., 1974; Ross et al., 1981a,b; 1984c; Bernstein-Goral and Bohn, 1988). These fibers also project to parasympathetic preganglionic neurons in sacral spinal cord segments (figure 4; Bernstein-Goral and Bohn, 1988). At both thoracic and sacral levels, the distribution of PNMT-IR fibers closely follows the nested organization of preganglionic cell bodies and their complex dendritic arborization (figure 4). Adrenergic fibers are present in the region adjacent to the ependymal cells of central canal throughout spinal cord, but PNMT-IR fibers are not observed in dorsal or ventral horn at any age (Bernstein-Goral and Bohn et al., 1988). The specificity of the adrenergic bulbospinal projection provides anatomical support for the postulated involvement of adrenergic neurons in cardiovascular and adrenal medullary regulation (Ross et al., 1984c; Morrison et al., 1988).

The relationship between PNMT-IR fibers and a target specific population of sympathetic preganglionic neurons, those that project to the chromaffin cells of the adrenal medulla, has been examined by combining immunocytochemistry with retrograde labeling. Adrenal preganglionic

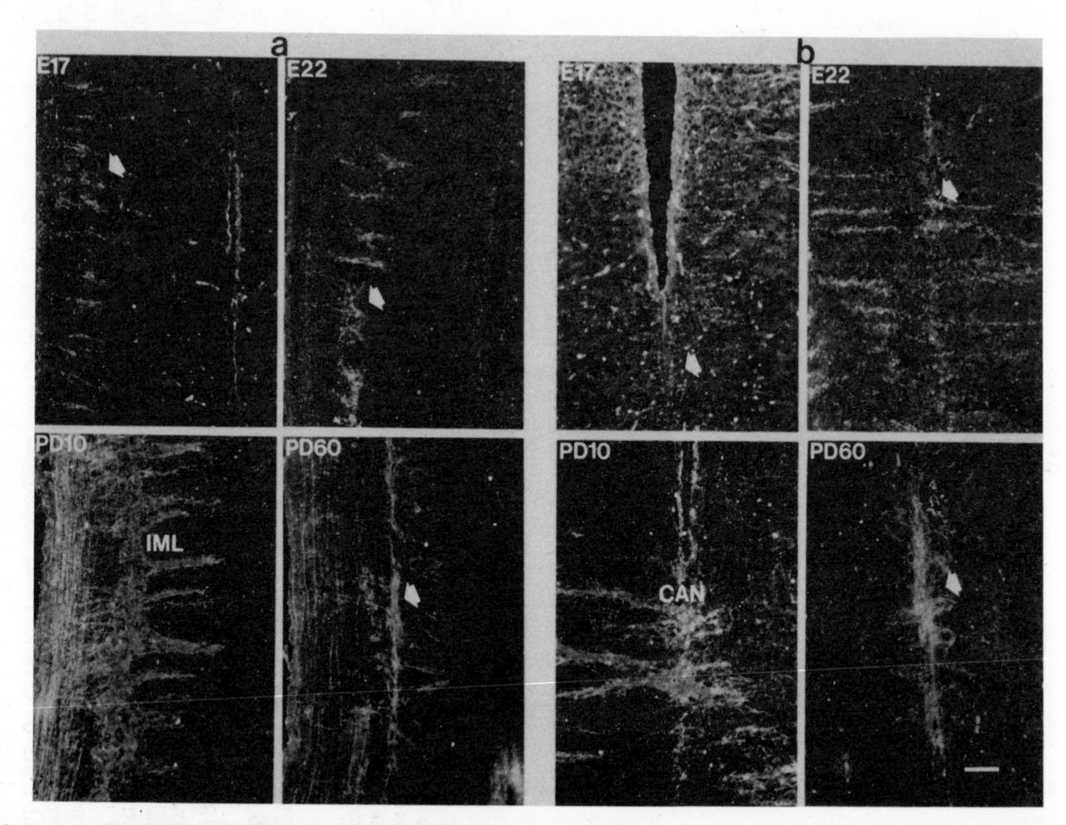

Figure 5. Darkfield photomicrograph of the PNMT-IR in horizontal spinal cord sections through the (a) intermediolateral nucleus (IML) and (b) central autonomic nucleus (CAN) at embryonic E17 and E22), postnatal (PD10) and adult (PD60) ages. arrowhead at IML or CAN; cc, central canal. Bar = 100 um. (reprinted by permission of A.R. Liss, Inc., Bernstein-Goral and Bohn, 1988)

neurons located in mid-thoracic spinal segments (Schramm et al., 1975; Strack et al., 1988; Bernstein-Goral, 1988) were retrogradely labeled with either the fluorescent tracer, True Blue or horseradish peroxidase (HRP). PNMT-IR fibers were observed to be intimately associated with adrenal preganglionic neurons in IML throughout development (figure 6; Bernstein-Goral and Bohn, 1988, 1989). Interestingly, the plexus of PNMT-IR varicose fibers which encase adrenal preganglionic neurons were found to be unexpectedly dense in the neonatal rat. In the neonate, large PNMT-IR varicosities were observed in apposition to adrenal preganglionic cell bodies (figure 6a,b), whereas in the adult rat, PNMT-IR fibers were observed to course through and around nests of preganglionic neurons in IML, but did not appear to follow the contours of cell bodies as closely (figure 6c,d). In the adult rat, large PNMT-IR varicosities were more associated with tapering dendrites as they coursed either rostrocaudally in IML or mediolaterally toward the central autonomic nucleus dorsal to the central canal (Bernstein-Goral and Bohn, 1989). These observations suggest that the precise synaptic relationship between PNMT-IR fibers and adrenal preganglionic neurons may change during maturation of the spinal cord.

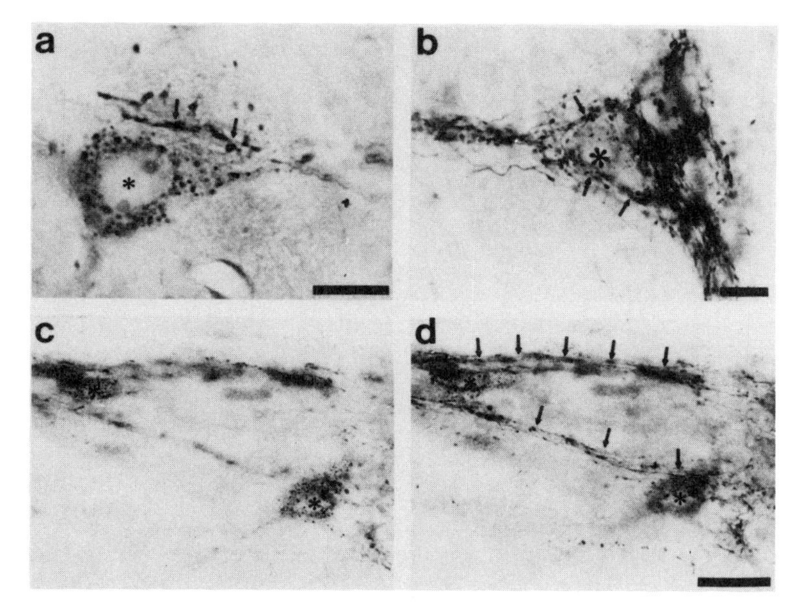

Figure 6. PNMT-IR varicose axons and associated HRP labeled adrenal preganglionic neurons in IML at (a) postnatal day 7, (b) postnatal day 30 and (c-d) postnatal day 90. (a) 1 um semi-thin plastic section with PNMT-IR varicosities (at arrows) closely apposed to an HRP filled cell body (asterisks). (b) A 30 um thick section with HRP-labeled cell body (asterisks) surrounded by PNMT-IR fibers (arrows). (c) Two retrogradely labeled adrenal preganglionic cell bodies in a 30 um thick section in adult IML. (d) Same field as (c) in a different plane of focus, arrows follow PNMT-IR fibers bundles coursing through the neuropil to intercept retrogradely labeled cell bodies. Bars = a) 20 um; b-d) 25 um. (reprinted by permission of Pergamon Press, Bernstein-Goral and Bohn, 1989)

ADRENAL PREGANGLIONIC NEURONS ARE SYNAPTIC TARGETS OF THE ADRENERGIC BULBOSPINAL PROJECTION

Light microscopic observations suggest that PNMT-IR fibers are in a good position to directly influence adrenal preganglionic neurons (figure 6; Bernstein-Goral and Bohn, 1988). To determine whether the close appositions between adrenergic terminals and adrenal preganglionic neurons represent synaptic contacts, this relationship was examined at the ultrastructural level. Adrenal preganglionic perikarya and their proximal dendrites were retrogradely labeled with HRP and tissue sections were doubly stained for PNMT-IR. The maturation of adrenergic boutons was examined in the neonatal (7-9 postnatal days), postnatal (24-30 postnatal days) and adult (60-90 postnatal days) rat IML (Bernstein-Goral and Bohn, 1989). As early as the first postnatal week, PNMT-IR boutons filled with small spherical electron-lucent vesicles were observed to form classical symmetrical synaptic specializations in IML (Bernstein-Goral and Bohn, 1989). Direct axo-somatic and axo-dendritic synaptic contacts on retrogradely labeled adrenal preganglionic somata and dendrites were observed in young rats (figure 7 and 8; Bernstein-Goral and Bohn, 1989; Bernstein-Goral, 1988). Axo-somatic synaptic associations between PNMT-IR boutons and choline acetyltransferase-immunoreactive neurons in IML were also observed in the young rat (Kohno et al., 1988). In the adult rat, PNMT-IR synapses on preganglionic soma or proximal dendrites are rarely observed (Milner et al., 1988; Bernstein-Goral and Bohn, 1989). The major termination of PNMT-IR boutons in the neuropil of IML of adult rat appears to be on small diameter dendrites. Both symmetrical and asymmetrical PNMT-IR synapses have been observed in the adult rat (Milner et al., 1988; Bernstein-Goral and Bohn, 1989).

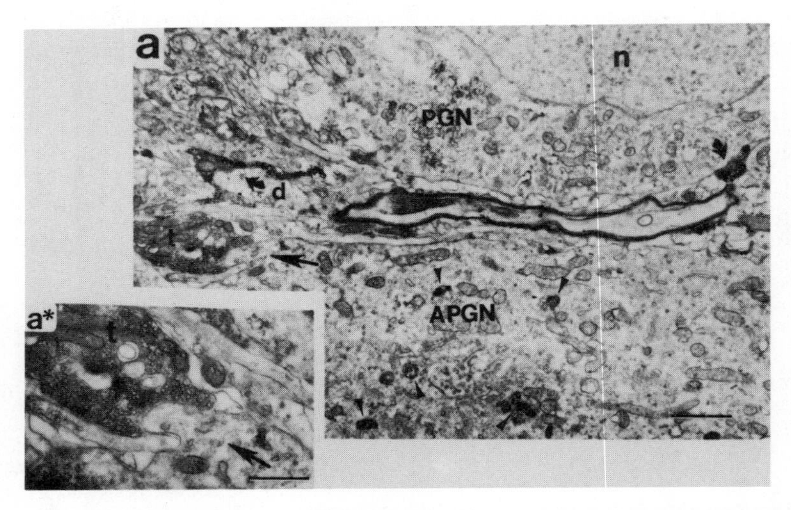

Figure 7. Low magnification electron micrograph of PNMT-IR terminals contacting retrogradely labeled adrenal preganglionic neurons in IML at postnatal day 30. a) PNMT-IR bouton (t) forming a direct synaptic contact (arrow) onto an adrenal preganglionic neuronal soma (APGN) labeled with HRP (arrowheads). Other PNMT-IR boutons contact an adjacent unidentified preganglionic soma and a dendrite in neuropil (curved arrows). a*) Insert on left shows a higher magnification of bouton (t) in synaptic association with a somatic protrusion. Bars = a) 2 um; a*) 1 um. (reprinted by permission of Pergamon Press, Bernstein-Goral and Bohn, 1989)

These ultrastructural observations demonstrate that the morphology of adrenergic synaptic associations with adrenal preganglionic neurons changes during postnatal development. Since PNMT-IR axo-somatic synapses are present in young rats and are virtually absent in the mature rat, we have proposed that, initially, the exuberant adrenergic innervation of immature preganglionic neurons is on perikarya and proximal dendritic processes. With the subsequent decrease in PNMT-IR fiber density, adrenergic axo-somatic input is lost and redistributed to distal dendrites. This may result from synapses following the growth of dendrites with which they are associated. Interestingly, this adrenergic re-organization occurs in IML at a time when the longitudinal component of the preganglionic dendritic arbor is forming (Schramm et al., 1976). Since many of the axo-somatic contacts formed by PNMT-IR boutons are on large somatic protrusions (figure 7), it is possible that these protrusions represent precursors to budding dendrites which will form the longitudinal dendritic bundle. Similar synaptic re-arrangements in which synapes form initially (and transiently) on the soma and proximal dendrites and later on distal dendrites and spines as the dendritic arbor is extended during development have been described for other developing neurons, for example, climbing fiber input to cerebellar Purkinje cells, motor neurons in ventral horn, postganglionic sympathetic neurons (Larramendi, 1969; Vaughn et al., 1974, 1977; Smolen and Raisman, 1980; Forehand, 1985; Smolen and Beaton-Wimmer, 1986). It is not known whether the early adrenergic innervation plays an active role in promoting and directing the extension of longitudinal dendritic arbor, but this is an intriguing possibility.

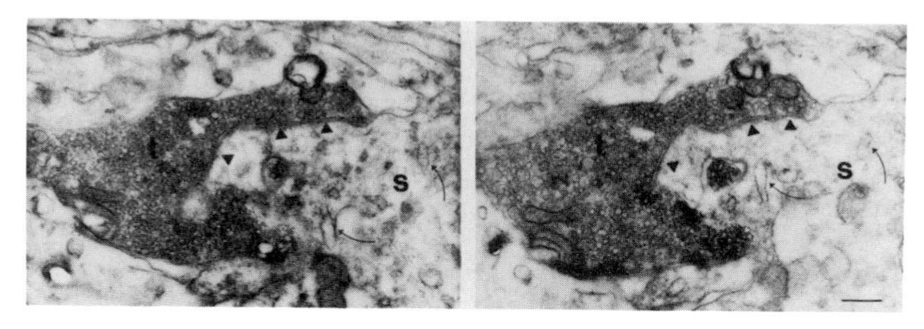

Figure 8. Two serial photomicrographs of PNMT-IR bouton (t) presented in figure 7a*, forming a synaptic contacts onto a somatic protrusion of a HRP-labeled adrenal preganglionic neuron in IML. PNMT-IR boutons form symmetrical synaptic specializations (arrow heads) with a large somatic protrusion (s) on adrenal preganglionic perikarya. Thin curved arrows indicate endoplasmic reticulum in soma. Bar = 0.5 um. (reprinted by permission of Pergamon Press, Bernstein-Goral and Bohn, 1989)

THE ADRENERGIC BULBOSPINAL PATHWAY AND ADRENAL MEDULLARY FUNCTION

The central and peripheral autonomic circuitry which mediates neurogenic control of the adrenal medulla is morphologically and functionally immature in the neonatal rat, despite the fact that adrenal chromaffin cells are capable of exocytotic release of adrenal catecholamine in response to hypoxia (Seidler and Slotkin, 1985, 1986a,b). This early reflex response is mediated by a transient non-neurogenic mechanism which disappears by the end of the first postnatal week (Seidler and Slotkin, 1985, 1986a,b). During this period, preganglionic axons form synaptic contacts with adrenal chromaffin cells (Makhail and Mahran, 1965; Daikoku et al., 1977), and preganglionic axonal transmission commences, enabling centrally mediated reflex catecholamine and peptide secretion from the adrenal medulla (Seidler and Slotkin, 1986; Kirby and McCarty, 1987; LaGamma and Adler, 1988).

Several important changes occur in the central autonomic pathways which may contribute to the onset of neurogenic adrenal regulation including the observations that: 1) bulbospinal adrenergic fibers hyperinnervate the IML, 2) PNMT-IR terminals establish synaptic contacts with adrenal preganglionic neurons, and 3) major morphological changes occur in the neuropil in IML and the shape of preganglionic neurons. Since adrenergic re-modeling occurs during a critical period with regard to the onset of neurogenic sympatho-adrenal regulation, it is reasonable to speculate that the changes in adrenergic synaptic associations and concomintant changes in epinephrine levels, provide essential developmental cues which influence the differentiation of adrenal preganglionic neurons and the maturation of adrenal medullary function. Previous studies have demonstrated that the normal development of sympathetic preganglionic neurons, as well as peripheral sympathetic ganglia and the adrenal medulla, require intact descending pathways in spinal cord, including early forming catecholaminergic bulbospinal pathways (Black et al., 1976; Hamill et al., 1977, 1983; Lawrence et al., 1981; Gauthier and Reader, 1982; Gagner et al., 1983; Lau et al., 1985, 1987). Preganglionic axonal outgrowth, synaptogenesis and differentiation of chromaffin cells in the adrenal medulla depends upon the integrity of descending afferent input to adrenal preganglionic neurons in spinal cord (Ross et al., 1980, 1981, 1983; Lau et al., 1987). A precise sequence of anterograde and retrograde interactions is required for proper development of the sympathetic nervous system (Black et al., 1976; Johnson et al., 1977; Schafer et al., 1983). Based on the observations of the studies reviewed here, central adrenergic projections are in a position to participate in one link of this chain.

ACKNOWLEDGMENTS

The authors thank Mrs. Dorothy Herrera, Mrs. Nancy Dimmick and Ms. Kim Coene for help in preparing this manuscript. This work was supported by a National Institute of Health Grant NS20832, a Research Career Development Award NS00910 to M.C.B. and aided by a grant from the Familial Dysautonomia Foundation.

REFERENCES

Astier, B., Kitahama, K., Denoroy, L., Jouvet, M., Renoud B. (1987). Immunohistochemical evidence for the adrenergic medullary longitudinal bundle as a major ascending pathway to the locus coeruleus. Neuroscience Lett. 74(2):132-1.

Beato, K.K., Burke, W.J., Joh, T.H., and Haring, J.H. (1987). Cortical epinephrine projections demonstrated by retrograde tracing combined with tyrosine hydroxylase and phenylethanolamine N-methyltransferase immunocytochemistry. Soc. Neurosci. Abst. 13:366.1

Bernstein-Goral, H.L. (1988). The ontogeny of adrenergic fibers in rat spinal cord. Ph.D. Dissertation. State University of New York at Stony Brook.

Bernstein-Goral, H., and Bohn, M.C. (1988). Ontogeny of adrenergic fibers in rat spinal cord in relationship to adrenal preganglionic neurons. J. Neurosci. Res. 21:333-351.

Bernstein-Goral, H., and Bohn, M.C. (1989). Phenylethanolamine N-methyltransferase-(PNMT) immunoreactive terminals synapse on adrenal preganglionic neurons in the rat spinal cord. Neurosc. (in press).

Black, I.B., Bloom, E.M., Hamill, R.W. (1976). Central regulation of sympathetic neuron development. Proc. Natl. Acad. Sci. USA 73(10):3575-3578.

Bohn, M.C. Goldstein, M., Black, I.B. (1981). Role of glucocorticoids in expression of the adrenergic phenotype in rat embryonic adrenal gland. Dev. Biol. 82:1-10.

Bohn, M.C., Goldstein, M., Black, I.B. (1986). Expression and development of phenylethanolamine N-methyltransferase (PNMT) in rat brain stem: studies with glucocorticoids. Dev. Biol. 114:180-193.

Coen, C.W., Coombs, M.C. (1983). Effects of manipulating catecholamines on the incidence of the preovulatory surge of luteninizing hormone and ovulation in the rat: Evidence for a necessary involvement of hypothalamic adrenaline in the normal or midnight surge. Neurosci. 10:187-206.

Daikoku, S., Kinutani, M., Sako, M., (1977). Development of adrenal medullary cells in rats with reference to synaptogenesis. Cell Tiss. Res. 179;77-86.

Forehand, C.J. (1985). Density of somatic innervation on mammalian autonomic ganglion cells is inversely related to dendritic complexity and preganglionic convergence. J. Neurosci. 5(12):3403-3408.

Foster, G.A., Schultzberg, M., Goldstein, M., Hökfelt T. (1985a). Differential ontogeny of three putative catecholamine cell types in the postnatal rat retina. Develop. Brain Res. 22:187-196

Foster, G.A., Schultzberg, M., Goldstein, M., Hökfelt, T. (1985b). Ontogeny of phenylethanolamine N-methyltransferase- and tyrosine hydoxylase-like immunoreactivity in presumptive adrenaline neurons of the foetal rat central nervous system. J. Comp. Neurol. 236:348-381.

Foster G.A., Schultzberg, M., Dahl, D., Goldstein, M., Verhofstad, A.A.J. (1985c). Ephemeral existence of a single catecholamine synthetic enzyme in the olfactory placode and the spinal cord of the embryonic rat. Int. J. Devel. Neuroscience. 3(6):597-608.

Foster, G.A., Sundstrom, E., Helmer-Matyjek, E., Goldstein, M., Hökfelt T. (1987). Abundance in the embryonic brainstem of adrenaline during the absence of detectable tyrosine hydroxylase activity. J. Neurochem. 48:202-207.

Gagner, J., Gauthier, S., Sourkes, T.L. (1983). Participation of spinal monoaminergic and cholinergic systems in the regulation of adrenal tyrosine hydoxylase. Neuropharm. 22:45-53.

Gauthier, P., and Reader, T.A. (1982). Adrenomedullary secretory response to midbrain stimulation in rat: effects of depletion of brain catecholamines or serotonin. Can. J. Physiol. Pharmacol. 60;1464-1474.

Goodchild, A.K., Moon, E.A., Dampney, R.A.L., Howe, P.R.C. (1984). Evidence that adrenaline neurons in the rostral ventolateral medulla have a vasopressor function. Neuroscience Lett. 45:267-272.

Guyenet, P., and Young, B. (1987). Projections of nucleus paragigantocellularis lateralis to locus coeruleus and other structures in rat. Brain Res. 406:171-184.

Hamill, R.W., Bloom, E.M., Black, I.B. (1977). The effect of spinal cord transection on the development of cholinergic and adrenergic sympathetic neurons. Brain Res. 134:269-278.

Hamill, R.W., Cochard, P., Black, I.B. (1983). Long-term effects of spinal transection on the development and function of sympathetic ganglia. Brain Res. 266:21-27.

Hökfelt, T., Foster, G.A., Johansson, O., Schultzberg, M., Holets, V., Ju, G., Skagerberg, G., Palkovits, M. (1988). Central phenylethanolamine N-methyltransferase-immunoreactive neurons: Distribution, projections, fine structure, ontogeny, coexisting peptides. In; Epinephrine in the Central Nervous System. (J. Stolk, D. Uprichard, K. Fuxe, eds.) Oxford U. Press. pp. 10-45.

Hökfelt, T., Fuxe, K., Goldstein, M., Johansson, O. (1974). Immunohistochemical evidence for the existence of adrenaline neurons in the rat brain. Brain Res. 66:235-251.

Hornby, P.J., and Piekut, D.T. (1987). Catecholamine distribution and relationship to magnocelluar neurons in the paraventricular nucleus of the rat. Cell Tiss. Res. 248(2):239-246.

Howe, P.R.C., Costa, M., Furness, J.B., Chalmers, J.P. (1980). Simultaneus demonstration of phenylethanolamine N-methyltransferase immunofluorescent and catecholamine fluorescent nerve cell bodies in the rat medulla oblongata. Neurosci. 5:2229-2238.

Johnson, Jr., E.M., Caserta, M., Ross, L.L. (1977). Effects of destruction of the postganglionic sympathetic neurons in neonatal rats on development of choline acetyltransferase and survival of preganglionic cholinergic neurons. Brain Res. 136:455-464.

Kaiser, K.P., Karten, H.J., Katz, B., and Bohn, M.C. (1987) Catecholaminergic horizontal and amacrine cells in the feret retina. J. Neurosci. 7:3996-4007.

Kalia, M., Woodward, D.J., Smith, W.K., Fuxe, K. (1985a). Rat medulla oblongata. IV. Topographical distribution of catecholaminergic neurons with quantitative three-dimensional computer reconstruction. J. Comp. Neurol. 233:350-364.

Kalia, M., Fuxe, K., Goldstein, M. (1985b). Rat medulla oblongata. III. Adrenergic (C1 & C2) neurons, nerve fibers and presumptive terminal processes. J.Comp. Neurol. 233:333-349.

Kirby, R.F., and McCarty, R. (1987). Ontogeny of functional sympathetic innervation to the heart and adrenal medulla in the preweanling rat. J. Autonom. Nerv. Syst. 19:67-75.

Kohno, J., Shinoda, K., Kawai, Y., Ohuchi, T., Ono, K., Shinotani, Y. (1988). Interaction between adrenergic fibers and intermediate cholinergic neurons in the rat spinal cord: A new double-immunostaining method for correlated light and electron microscopic observations. Neuroscience 25(1):113-212.

LaGamma, E.F., and Adler, J.H. (1988). Development of transynaptic regulation of adrenal enkephalin. Dev. Brain Res. 39:177-182.

Larramendi, L.M.H. (1969). Analysis of synaptogenesis in the cerebellum of the mouse. In: Neurobiology of cerebellar evolution and development, Amer. Med. Ass. Educ. & Res. Fdn., Chicago. pp. 803-843.

Lau, C., Pylypiw, A., Ross, L. (1985). Development of serotonergic and adrenergic receptors in the rat spinal cord: effects of neonatal chemical lesions and hyperthyroidism. Dev. Brain Res. 19:57-66.

Lau, C., Ross, L.L., Whitmore, W.L., Slotkin, T.A. (1987). Regulation of adrenal chromaffin cell development by the central monoaminergic system: Differential control of norepinephine and epinephrine levels and secretory responses. Neurosci. 22(3):1067-1075.

Lawrence, J.M., Hamill, R.W., Cochard, P., Raisman, G., Black, I.B. (1981). Effects of spinal cord transection on synapse numbers and biochemical maturation in rat lumbar sympathetic ganglia. Brain Res. 212:83-88.

Liposits, Zs., Phelix, C., Paull, W.K. (1986). Electron microscopic analysis of tyrosine hydroxylase, dopamine-B-hydroxylase and phenylethanolamine N-methyltransferase immunoreactive innervation of the hypothalamic paraventricular nucleus in the rat. Histochemistry 84:105-120.

Liposits, Zs., Phelix, C., Paull, W.K. (1986). Adrenergic innervation of corticotropin releasing factor (CRF)-synthesizing neurons in the hypothalamic paraventricular nucleus of the rat. Histochemistry 84:201-205.

Makhail, Y., and Mahran, Z. (1965). Innervation of the cortical and medullary portions of the adrenal gland of the rat during postnatal life. Anat. Rec. 152:431-438.

Mefford, I.N. (1988). Are there epinephrine neurons in rat brain? Brain Res. Reviews 12:383-395.

Mezey, E., Kiss, J.L., Skirboll, L.R., Goldstein, M., Axelrod, J. (1984). Increase of corticotropin-releasing factor staining in rat paraventricular nucleus neurons by depletion of hypothalamic adrenaline. Nature 310:140-141.

Milner, T.A., Morrison, S.F., Abate, C., Reis, D.J. (1988). Phenylethanolamine N-methyltransferase-containing terminals synapse directly on sympathetic preganglionic neurons in the rat. Brain Res. 448:205-222.

Morrison, S.F., Milner, T.A., Reis, D.J. (1988). Reticulospinal vasomotor neurons of the rat rostral ventrolateral medulla: Relationship to sympathetic nerve activity and the C1 adrenergic cell group. J. Neurosci. 8:1286-1301.

Nakamura, K., and Nakamura, K. (1978). Role of brainstem and spinal noradrenergic and adrenergic neurons in the development and maintenance of hypertension in spontaneously hypertensive rats. Naunyn-Schmiedeberg's Arch. Pharmacol. 305:127-133.

Pieribone, V., Aston-Jones, G., Bohn, M.C. (1988). Most adrenergic afferents to locus coeruleus originate in the C1 ventrolateral medullary cell group: a fluorescent double labeling study. Neurosci. Lett. 85:297-303.

Pieribone, V.A., Aston-Jones, G., Bohn, M., Bernstein-Goral, H. (1987). Double labeling using flurogold reveals neurotransmitter identity of afferents to locus coeruleus. Soc. Neurosci. Abstr. 13:1458.

Ross, C.A., Armstrong, D.A., Ruggiero, D.A., Pickel, V.M., Joh, T.H. Reis, D.J. (1981a). Adrenaline neurons in the rostral ventrolateral medulla innervate thoracic spinal cord: a combined immunocytochemical and retrograde transport demonstration. Neuroscience Lett. 25:257-262.

Ross, C.A., Ruggiero, D.A., Reis, D.J. (1981b). Projections to the spinal cord from neurons close to the ventral surface of the hindbrain in the rat. Neuroscience Lett. 25:145-148.

Ross, C.A., Ruggiero, D.A., Meeley, M.P., Park, D.H., Joh, T.H., Reis, D.J. (1984a). A new group of neurons in hypothalamus containing phenylethanolamine N-methyltransferase (PNMT) but not tyrosine hydroxylase. Brain Res. 306:349-353.

Ross, C.A., Ruggiero, D.A., Park, D.H., Joh, T.H., Sved, A.F., Fernandez-Pardal, J., Saavedra, J.M., Reis, D.J. (1984b). Tonic vasomotor control by the rostral ventrolateral medulla: effects of electrical or chemical stimulation of the areas containing C1 adrenaline neurons on arterial pressure, heart rate and plasma catecholamines and vasopressin. J. Neuroscience 4(2):474-494.

Ross, C.A., Ruggiero, D.A., Joh, T.H., Park, D.H., Reis, D.J. (1984c). Rostral ventrolateral medulla: selective projections to the thoracic autonomic cell column from the region containing C1 adrenaline neurons. J. Comp. Neurol. 228:168-185.

Ross, L.L., Smolen, A.J., Cherry, J. (1980). Spinal cord transection interferes with normal development of sympathetic preganglionic neurons. Soc. Neurosci. Abs. 6:385.

Ross, L.L., Smolen, A.J., Cherry, J. (1981). Supraspinal pathways regulate the development of the normal pattern and density of innervation of the adrenal medulla. Anat. Rec. 199:127A.

Ross, L.L., Pylypiw, A., Chmelewski. W. (1982). Development of catecholamines and adrenergic receptors in the rat spinal cord. Soc. Neurosci. Abstr. 8:175.

Ross, L.L., Smolen, A.J., McCarthy, L. (1983). Supraspinal pathways regulate the mitotic activity of adrenal medulla cells. Anat. Rec. 205:167-168.

Ruggiero, D.A., Ross, C.A., Anwar, M., Park, D.H., Joh, T.H. Reis D.J. (1985). Distribution of neurons containing phenylethanolamine N-methyltransferase in medulla and hypothalamus of the rat. J. Comp. Neurol. 239:127-154.

Saavedra, J.M., Kvetnansky, R., Kopin, I.J. (1979). Adrenaline, noradrenaline, and dopamine levels in specific brainstem areas of acutely immobilized rats. Brain Res. 160:271-280.

Saavedra, J.M., Torda, T. (1980). Increased brainstem and decreased hypothalamic adrenaline-forming enzyme after acute and repeated immobilization stress in the rat. Neuroendocrinol. 31:142-146.

Sawchenko, P.E., and Swanson, L.W. (1982). Immunohistochemical identification of neurons in the paraventricular nucleus of the hypothalamus that project to the medulla or to the spinal cord in the rat. J. Comp. Neurol. 205:260-272.

Schafer, T., Schwab, M.E., Thoenen, H. (1983). Increased formation of preganglionic synapses and axons due to a retrograde trans-synaptic action of nerve growth factor in the rat sympathetic nervous system. J. Neurosci. 3(7):1501-1510.

Schramm, L.P., Adair, J.R., Striblin, J.M., Gray, L.P. (1975). Preganglionic innervation of the adrenal gland of the rat: a study using horseradish peroxidase. Exp. Neurol. 49:540-553.

Schramm, L.P. Stribling, J.M., Adair, J.R. (1976). Developmental reorientation of sympathetic preganglionic neurons in the rat. Brain Res. 106:166-171.

Seidler, F.J., and Slotkin, T.A. (1985). Adrenomedullary function in the neonatal rat: responses to acute hypoxia. J. Physiol. (London) 358:1-16.

Seidler, F.J., and Slotkin, T.A. (1986a). Non-neurogenic adrenal catecholamine release in the neonatal rat: exocytosis or diffusion? Dev. Brain Res. 28:274-277.

Seidler, F.J., and Slotkin, T.A. (1986b). Ontogeny of adrenomedullary responses to hypoxia and hypoglycemia: Roll of splanchnic innervation. Brain Res. Bull. 16:11-14.

Sheaves, R., Laynes, R., and MacKinnon, C.B. (1985). Evidence that central epinephrine neurons participate in the control and regulation of neuroendocrine events during the estrous cycle. Endocrinol. 116:542-546.

Smolen, A., and Raisman, G. (1980). Synapse formation in the rat superior cervical ganglion during normal development and after neonatal deafferentiation. Brain Res. 181:315-323.

Smolen, A.J., and Beaston-Wimmer, P. (1986). Dendritic development in the rat superior cervical ganglion. Dev. Brain Res. 29:245-252.

Strack, A.M., Sawyer, W.B., Marubio, L.M., Loewy, A.D. (1988). Spinal origin of sympathetic preganglionic neurons in the rat. Brain Res. 455:187-191.

Swanson, L.M., Sawchenko, P.E., Berod, A., Hartman, B.K., Helle, K.B., Van Orden, D.E. (1981). An immunohistochemical study of the organization of catecholaminergic cells and terminal fields in the paraventricular and supraoptic nuclei of the hypothalamus. J. Comp. Neurol. 196:271-285.

Teitelman, G., Baker, H., Joh, T.H., Reis, D.J. (1979). Appearance of catecholamine synthesizing enzymes during development of rat embryo sympathetic nervous system: Possible role of tissue environment. Proc. Natl. Acad. Sci. U.S.A. 76:509-513.

Tucker, D.C., Saper, C.B., Ruggiero, D.A., Reis, D.J. (1987). Organization of central adrenergic pathways: I. Relationship of ventrolateral medullary projections to the hypothalamus and spinal cord. J. Comp. Neurol. 259:591-603.

Vaughn, J.E., Henrikson, C.K., Grieshaber, J.A. (1974). A quantitative study of synapses on motor neuron dendritic growth cones in developing mouse spinal cord. J. Cell Biol. 60:664-672.

Vaughn, J.E., Sims, T., Nakashima, M. (1977). A comparison of the early developmental axodendritic and axosomatic synapses upon embryonic mouse spinal motor neurons. J. Comp. Neurol. 175:79-100.

Verhofstad, A.A., Hökfelt, T., Goldstein, M., Steinbusch, H.W.M., Jooster, H.W.J. (1979). Appearance of tyrosine hydroxylase, aromatic amino acid decarboxylase, dopamine B-hydroxylase and phenylethanolamine N-methyltransferase during the ontogenesis of the adrenal medulla. Cell Tissue Res. 200:1-13.

EARLY EXPERIMENTAL INFLUENCES ON SEROTONIN PATHWAYS DURING BRAIN DEVELOPMENT

Jorge Hernández

Lab. Neurontogeny, Dept. Physiology, Biophysic and Neuroscience

Centro de Investigación, I.P.N., México

INTRODUCTION

In an attempt to gain information about the early influence of nutritional changes on the development of a specific neuronal system, we took advantage of the fact that an essential nutrient, namely L-Tryptophan (L-Trp), is the precursor molecule in the synthesis of a brain neurotransmitter, serotonin (5-hydroxytryptamine, 5-HT). Early experimental changes in the availability of such nutrient like protein-calorie malnutrition represented a useful experimental approach. It was also known, that L-Trp administration to adult rats induced an increase in brain 5-HT (1,2). The effect of L-Trp supplementation during the gestational period on the 5-HT biosynthetic pathway in the fetal brain was also tested. Both approaches, a defect of nutrients like early malnutrition and an excess of the precursor molecule, gave us interesting information about their early effects on the development of the brain serotonin system. Another experimental approach we used was the early administration of specific neurotoxics for serotonin neurons. So, we examined the effect of neonatal administration of 5,7-dihydroxytryptamine (5,7-DHT) and the developmental pattern of serotonin molecular markers during postnatal development. This approach gave us valuable information on the role of this neurotransmitter in determining the development of specific (^{3}H)5-HT binding sites.

In the present paper a brief summary of the results from those experimental approaches is presented together with some correlative comments.

Early L-Tryptophan supplementation

The free fraction of plasma L-Trp has been suggested to cross the blood-brain barrier and stimulate the synthesis of serotonin in the brain (1,3,4). One interesting observation concerning the role of a precursor molecule like L-Trp in the synthesis of brain neurotransmitters, has been an increase in the content of brain 5-HT in developmentally malnourished rats (5,6). It has been postulated that an increase in the free fraction of plasma L-Trp, could be the cause of the serotonin elevation (7). Bearing these antecedents in mind we proposed that L-Trp supplementation to normal rats during gestation could induce also important changes in the serotonin synthesis rate in the fetal brain and later in postnatal development. In order to test if the L-Trp administration was an effective

Molecular Aspects of Development and Aging of the Nervous System
Edited by J.M. Lauder
Plenum Press, New York, 1990

stimulus on the 5-HT system of the fetal brain, we studied the effect of loads of L-Trp, i.p.(100mg/kg) to rats on day 15th of gestation, on the activity of tryptophan-5-hydroxylase(T5-H, EC 1.14.16.4) in the fetal brain. Our results indicated that the precursor molecule readily passed through the feto-placental barrier and that it was effective in activating the fetal T5-H activity significantly as compared to the control values (Table I). Therefore, we proceeded to administer loads of the precursor molecule to pregnant rats twice a week during the whole gestational period. The activity of the rate limiting enzyme, T5-H, and the content of 5-HT were measured in the offspring brain from birth up to day 30th. The results showed a significant increase of T5-H activity and an elevation of 5-HT levels at all ages studied, Figures 1 and 2. Other results suggested also an elevation of the brain 5-HT metabolite, 5-hydroxyindole acetic acid (5-HIAA) in similarly treated animals (8).

Table I

Activity of T5-H

Brain	L-Trp	Saline
Mother	0.354$\pm$0.024	0.253$\pm$0.039**
Fetus	0.257$\pm$0.038	0.182$\pm$0.028*

T5-H activity in the brains of pregnant rats and fetuses 30 min after the acute-administration of L-Trp, 100mg/kg, i.p., and saline controls on day 15th of gestation. Activity is expressed as nmol indoles/mg protein/h. Numbers are mean values from 3 determinations $\pm$ S.D. (for fetuses a pool of about 30 brains was made, enzyme activity was measured in triplicate samples). *$p<0.05$; **$p<0.01$; Student's t-test.

These results give evidence of the relevance that a single nutrient molecule, in this case the essential amino acid L-Trp precursor of the synthesis of brain 5-HT, may have for the development of its own metabolic paths. L-Trp appears to regulate the biosynthesis of its final functional derivative, the neurotransmitter 5-HT, in the developing brain. This conclusion is supported by a parallel increase in the brain 5-HT content and in the activity of the key enzyme T5-H, as an effect of early stimulation by L-Trp. Its acute administration on day 15th of gestation showed to be an effective stimulus on its activity in the embryo's brain. Recently we have shown that in the fetal brain subjected to materno-placental insufficiency there is also an increase, since gestational day 17th up to postnatal day 10th, in the activity of T5-H, the content of 5-HT and L-Trp (9). So, the members of the whole biosynthetical path are elevated, strongly suggesting an increase in the rate of neurotransmitter synthesis in both conditions: early prenatal malnutrition and L-Trp supplementation.

It is possible that changes in the 5-HT biosynthetic machinery are provoked by the early L-Trp modifications. It is known that in the adult, brain T5-H has a low affinity for its substrate, L-Trp, resulting in a limitation of the neurotransmitter synthesis. Early L-Trp administration could account for a shift to a higher affinity of the enzyme for its substrate, resulting in a higher rate of 5-HT synthesis thereafter. We are currently studying the enzyme kinetics in the fetal brain of normal and malnourished rats to clarify this question.

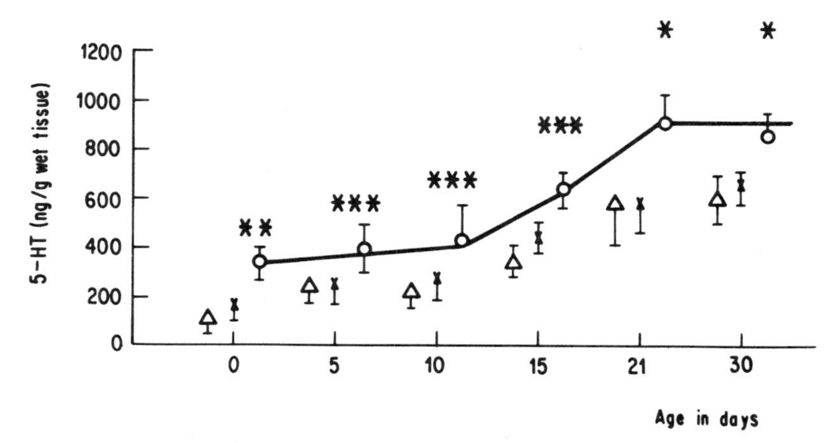

Figure 1. Serotonin content in the cerebral cortex of rats born to mothers treated during gestation with L-Trp 50mg/kg, i.p., 6X. △, Basal controls (animals minimally handled, mother not injected); X vehicle-injected controls., O L-Trp-treated animals.Each point represents mean values from 3 to 11 different experiments ± S.D. and pups were always taken from at least 3 different litters. Determinations were made in duplicate samples (* $p<0.05$; ** $p<0.01$; *** $p<0.001$; Student's t test, as related to controls).

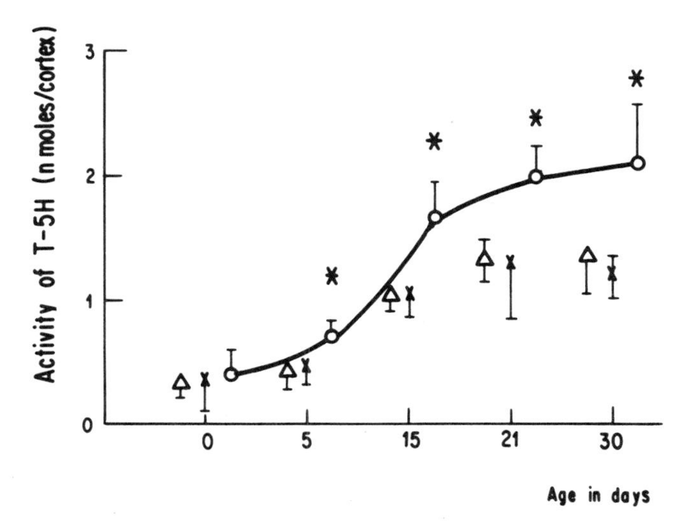

Figure 2. Postnatal development of T5-H activity in the cerebral cortex of rats born to mothers treated during gestation with L-Trp 50mg/kg, i.p., 6X. △ Basal controls (animals minimally handled, mothers not injected; X , vehicle-injected controls: O , L-Trp treated. Points represent mean values from 3 different experiments ± S.D. For each experiment in young animals several cortices were pooled and pups were always taken from at least 3 different litters. Determinations of enzyme activity were made in triplicate samples. ($p<0.05$; Student's t test, as related to controls).

Zeisel et al (10) have shown that fasting, in young rats, induces higher concentrations of serum L-Trp and a higher L-Trp/large neutral amino acid ratio. Miller et al (11), observed the same in chronically malnuorished. This together with an increase in brain L-Trp and 5-HIAA, suggest a regulatory role of L-Trp, on 5-HT synthesis since very early stages of brain development. Furthermore this regulation seems to be also on other important molecular systems that are very possibly functionally associated to 5-HT in the brain, such as (Na+,K+)ATPase (EC 3.6.1.3). Indeed we have observed that developmental changes on the 5-HT system are associated to an activation of (Na+,K+)ATPase (12). In conclusion, early stimulation of the serotonin system by its precursor molecule L-Trp, could determine important changes in its regulatory role during early neurogenesis like axogenesis or specific receptor formation, (13,14,15) and be the cause of permanent consequences for the whole brain maturation and functioning.

Early prevention of serotonin input and the ontogeny of specific receptors

Since general lines of evidence point to the existence of serotonin synthetizing cells in early stages of neurogenesis (16,17,8), questions about its possible role were also raised in our laboratory. We undertook together with the group of Neuropharmacology at Pasteur Institut (Prof. G. Fillion), the task of caracterizing developmental curves of (^{3}H)5-HT sites in the rat brain and spinal cord. Our results showed that the postnatal developmental pattern of these sites were parallel to general synaptogenesis (18). Presynaptic markers as T5-H and 5-HT content (19) also followed the same pattern suggesting the formation of specific 5-HT synapses. They are present in low percentages (3-4%) at birth and then increase explosively from about day 8 up to day 16, reaching a number (Bmax) similar to that in the adult, thereafter. This developmental pattern of (^{3}H)5-HT sites was very similar for the brain and the spinal cord, from birth up to day 30th (15,19), (Figs. 3,4). It was also observed that the coupling of the (^{3}H)5-HT receptor site to the transduction system, adenylate cyclase, develops asynchronically. Basal specific adenylate cyclase activity develops with a similar pattern than that of (^{3}H)5-HT binding sites with adult values about day 15th, but adenylate cyclase activity stimulated by 5-HT reaches adult values up to day 24th. This developmental lag suggests that the functional unit, the receptor coupled to the adenylate cyclase, requires a period of maturation of several days to become fully functional (20). Since in the rat the process of axogenesis and synaptogenesis is still underway during the postnatal period and coincides with receptor formation in the target cells, the possibility of an influence of 5-HT liberated by serotonergic neuroblasts, on the receptor expression was considered.

To test the early influence of serotonin on the formation of specific receptors on target cells, the specific neurotoxin 5,7-DHT was used. The neurotoxin was administered intracerebroventricularly to newborn rats, at a single dose of 3μg, preceded by the administration of an inhibitor of the uptake by catecholaminergic cells. Controls received the inhibitor but not 5,7-DHT. This treatment destroys the majority of 5-HT terminals in the brain and therefore would prevent most of 5-HT input on target developing neurons (20).

It was observed that in the brain cortex of vehicle-control rats, the developmental pattern of the (^{3}H)5-HT binding sites was not different from that observed in normal controls. On the contrary, in the cortex lacking 5-HT terminals, the number of binding sites (Bmax) was modified, from day 20. Indeed, a significant increase in the final number of (^{3}H)5-HT sites, 27%, (Student's t-test $p<0.01$) was observed in rats that received the intracerebral injection of 5,7-DHT on day 9th. Affinity constants

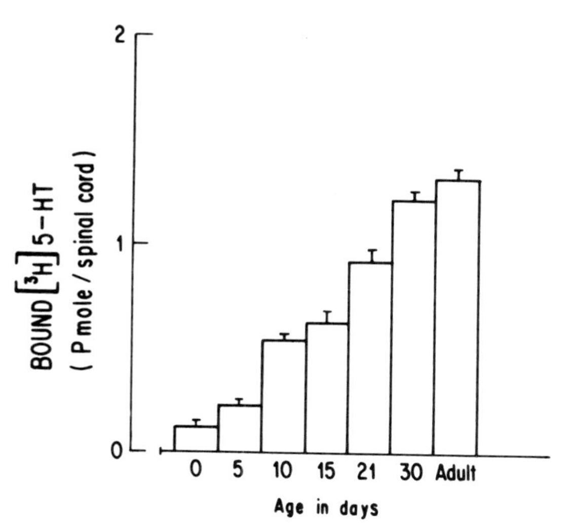

Figure 3. Developmental pattern of (^{3}H)5-HT receptors per spinal cord. Each bar corresponds to three independent experiment ± S.D. (triplicate samples). Bound (^{3}H)5-HT express Bmax calculated from Scatchard plots at the various ages. Binding of (^{3}H)5-HT was determined using 10-12 concentrations ranging from 2 to 30 nM.

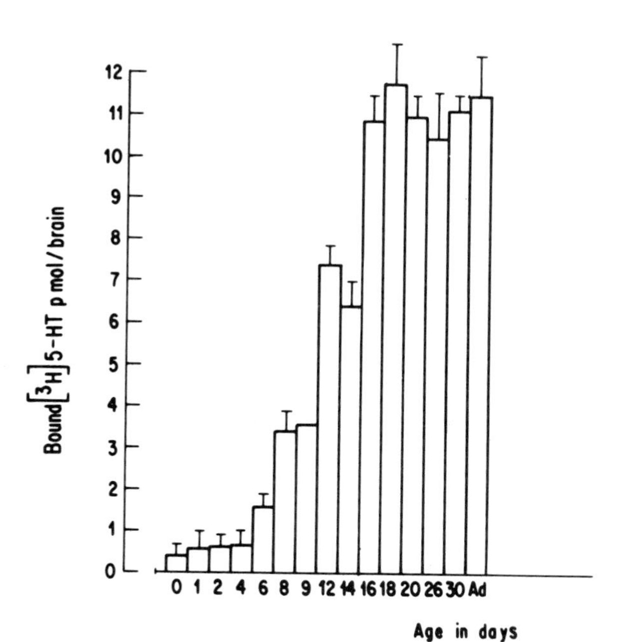

Figure 4. Postnatal development of the ^{3}H-5-HT binding sites in rat brain. Bound ^{3}H-5-HT express the number of specific, reversible binding sites, determined using the Scatchard plot of the saturation curve (ISD). The binding of ^{3}H-5-HT was measured at 6-9 concentrations ranging from 2 to 25 nM, in triplicate samples.

(KD) were not changed (KD, 12 days=17 nM; KD for 30 days=15nM). The persistence of (^{3}H)5-HT binding sites in the adult brain cortex, after destruction of most of 5-HT terminals by 5,7-DHT, indicated that they are located postsynaptically. Degeneration of 5-HT nerves from day 9th led to a significant increase in the number of (^{3}H)5-HT binding sites. These results indicate that the presynaptic growing terminals play an important role in the formation and development of postsynaptic receptors, suggesting that serotonin itself could play a trophic role for maturation and formation of serotonergic synapsis as suggested by Changeaux and Danchin (21). But other factors that may be crucial in this type of regulation are not excluded. Other authors have also showed in early periods of brain development a possible important role of 5-HT in cell and axon differentiation (11,13,22).

In conclusion our studies show that postsynaptic 5-HT receptors develop early after birth, mainly on the second week, corresponding to the period of specific 5-HT synaptogenesis, and that they became fully functional several days later. This might be a crucial period for brain development. The influence of 5-HT nerve terminals appears very important in the regulation of the final number of postsynaptic receptor sites.

REFERENCES

1. A. Tagliamonte, G. Biggio, L. Vargin and G. L. Gessa, Free Tryptophan in serum controls brain tryptophan level and serotonin synthesis, Life Sci., 12:277-287 (1973).
2. G. L. Gessa, G. Biggio, F. Fadda, G. U. Corsini and A. Tagliamonte, Effect of the oral administration of tryptophan-free amino acid mixtures on serum tryptophan, brain tryptophan and serotonin metabolism, J. Neurochem, 22:869-870 (1974).
3. G. L. Gessa and A. Tagliamonte, Serum free tryptophan control of brain concentrations of tryptophan and of synthesis of 5-hydroxytryptamine, in: Aromatic Amino Acids in the Brain, G.E.W Wolstenholme and D. W. Fitzsimons , eds., Elsevier, Amsterdam (1974).
4. A. Yuwiler, E. H. Oldendorf, E. Geller and L. Braun, Effects of albumin binding and amino acid competition on tryptophan up take into brain, J. Neurochem., 28:1015-1023 (1977).
5. R. J. Hernández, Effect of malnutrition and 6-hydroxydopamine on the early postnatal development of noradrenaline and serotonin content in the rat brain, Biol. Neonate, 30:181-186 (1976).
6. C. Stern, M. Miller, W. B. Forbes, P. J. Morgane and O. Resnick, Ontogeny of the levels of biogenic amines in various parts of the brain and in peripheral tissues in normal and protein malnourished rats, Exp. Neurol., 49:314-326 (1975).
7. M. Miller, J. P. Leahy, F. McConville, P. J. Morgane and O. Resnick, Effects of developmental protein malnutrition on tryptophan utilization in brain and peripheral tissues, Brain Res. Bull., 2:347-353 (1977).
8. R. J. Hernández and G. Chagoya, Brain serotonin synthesis and Na+,K+-ATPase activity are increased postnatally after prenatal administration of L-Tryptophan, Devl. Brain Res., 25:221-226 (1986).
9. G. Manjarrez, G. Chagoya and J. Hernández, Perinatal brain serotonin metabolism in rats malnourished in utero, Biol. Neonate, 54:232-240 (1988).
10. H. S. Zeisel, CH. Mauron, C. J. Watkins and R. J. Wurtman, Developmental changes in brain indoles, serum tryptophan and other serum neu tral amino acids in the rat, Devl. Brain Res., 1:551-564 (1981).
11. M. Miller, J. P. Leahy, W. C. Stern, P. J. Morgane and O. Resnick, Tryptophan availability: Relation to elevated brain serotonin in developmentally protein malnourished rats, Exp. Neurol, 57:142-157 (1977).

12. P. G. Haydon, D. P. McCobb and S. B. Kater, Serotonin selectively inhibits its growth cone motility and synaptogenesis of specific identified neurons, Science, 226:561-564 (1984).
13. J. M. Lauder, J. A. Wallace, H. Krebs and P. Petrusz, Serotonin as a timing mechanism in neuroembryogenesis, in: Progress in Psychoneuroendocrinology, F. Brambilla, G. Racagni and D. Wied, eds., Elsevier, North Holland (1980).
14. M. P. Fillion, R. J. Hernãndez, C. Baugen and G. Fillion, Postnatal development of high affinity neuronal recognition sites for (^{3}H)5-HT in rat brain, Dev. Neurosci., 5:484-491 (1982).
15. J. M. Lauder, J. A. Wallace, H. Krebs, F. Petrusz and K. McCarthy, In vivo and in vitro development of serotonergic neurons, Brain Res. Bull., 9:605-625 (1982).
16. H. Takahashi, S. Nakashima, E. Ohama, S. Takeda and F. Ikuta, Distribution of serotonin-containing cell bodies in the brainstem of the human fetus determined with immunohistochemistry using antiserotonin serum, Brain and Development, 8:355-365 (1986).
17. B. G. Cragg, The development of cortical synapses during starvation in the rat, Brain, 95:143-150 (1972).
18. R. J. Hernãndez, D. Martĩnez-Fong, G. Chagoya, M. P. Fillion and G. Fillion, Existence of (H)serotonin binding sites in the rat spinal cord: A developmental study, Int. J. Devl. Neurosci., 2:33-41 (1984).
19. G. R. Breese, R. A. Vagel and R. A. Mueller, Biochemical and behavioral alterations in developing rats treated with 5,7-dihydroxytryptamine. J. Pharmacol. Exp. Ther. 205:587-595 (1978).
20. E. Zifa, J. Hernãndez, C. Fayolle and G. Fillion, Postnatal development of 5-HT$_1$ receptors: (^{3}H)5-HT binding sites and 5-HT induced adenylate cyclase activations in rat brain cortex, Devl. Brain Res., 44: 133-140 (1988).
21. J. P. Changeaux and A. Danchin, Selectively stabilization of developing synapses as a mechanism for the specification of neuronal networks, Nature (London), 264:705-712 (1976).
22. G. Ahmad and S. Zamenhof, Serotonin as a growth factor for chick embryo brain, Life Sci., 22:963-970 (1978).

ONTOGENY OF THE GLUCOCORTICOID RECEPTOR IN THE RAT BRAIN

M.N. Alexis*, E. Kitraki+, K. Spanou*, F. Stylianopoulou+ and C.E. Sekeris*

*The National Hellenic Research Foundation
48, Vas. Constantinou Avenue, Athens 11635
+School Health Sciences, University of Athens
Box 14224, Athens, Greece

Glucocorticoids regulate the stress response *via* a negative feedback control on the brain-pituitary-adrenal axis[1,2]. In addition, in the brain glucocorticoids modulate the action of several monoamine, peptide and aminoacid neurotransmitters[2]. Glucocorticoid effects on the brain are likely to occur *via* receptor-mediated regulation of the rate of transcription of a set of genes[2,3,4]. In this light, it is interesting to study in parallel the ontogeny of glucocorticoid receptor and glucocorticoid regulation of glycerol-3-phosphate dehydrogenase (GPDH) activity[3,4] in order to follow their coupling during brain development.

Glucocorticoid receptors were initially detected in the brain of adrenalectomized rats following injection of tracer doses of tritiated corticosterone, the naturally occuring glucocorticoid in the rat, and dexamethasone, its synthetic analogue[5,6]. It was shown that corticosterone is retained primarily by the hippocampus, septum and amygdala, whereas dexamethasone labels preferentially the pituitary and uniformily, but to a lesser extent, the brain. The differences in the uptake of the two tritiated steroids is thought to reflect the differential distribution in the brain of two distinct types of steroid receptors, namely the mineralocorticoid receptor (MR) which prefers corticosterone and aldosterone to dexamethasone and the glucocorticoid receptor (GR) which preferentially binds dexamethasone[2]. For the experiments described in this study we used dexamethasone in order to lebel the glucocorticoid receptor *in vitro* and *in vivo*.

The concentration of the glucocorticoid receptor in the cytosol was determined by Scatchard plot analysis as previously described[7,8]. Studies were performed with extracts from the brain of intact as well as adrenalectomized animals. In the adrenalectomized rat, receptor concentration reached adult values during the second postnatal week (Fig. 1). Blot-hybridization analysis of total brain mRNA[9] demonstrated that a parallel increase in the concentration of glucocorticoid receptor mRNA (~ 6.5Kb) by a factor of ~ 5 occurs during the second week of life (data not shown). This increase in receptor mRNA and protein concentration most likely originates from differentiating oligodendrocytes, which in the rat exhibit a peak of proliferative activity at the end of the first week of life[10].

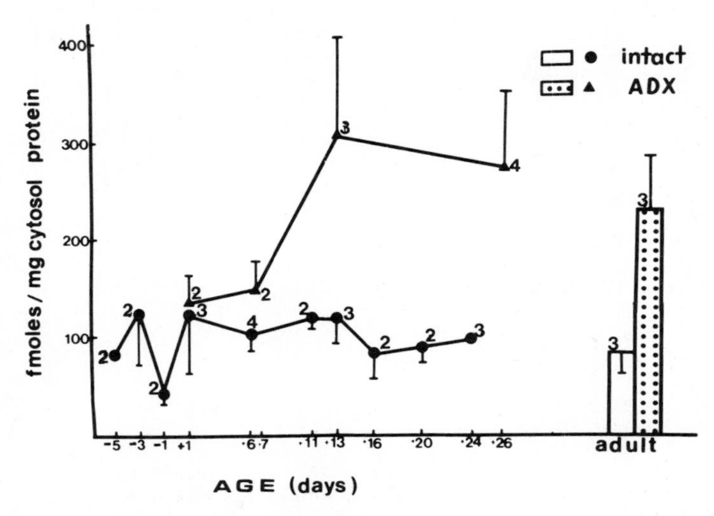

Fig. 1. Concentration of the glucocorticoid receptor in developing brain extracts from intact (●) and adrenalectomized (▲) rats, determined by Scatchard - plot analysis using tritiated dexamethasone as previously described[7,8]. Day 0 is the day of birth. Results are mean± S.D. of a number of separate estimations shown at each point.

When using non-operated animals, GR-like dexamethasone binding could be detected as early as the 17th day of gestation. (Day 0 of gestation was the day following that in which mating was allowed to occur). Dexamethasone binding drops just before birth and returns to prenatal values by day 1 of life (Fig. 1). The depletion of dexamethasone binding around birth may be due to receptor down-regulation by an excess of corticosterone[11], secreted during labour and birth[12,13].

By comparing the developmental patterns for intact and adrenalectomized rats, an estimate of percent receptor tranlocation from the cytoplasm to the nucleus can be made. Estimated percent translocation values range from 8% on postnatal day 1, to 26% on day 7, 60% on day 13 and 64% on day 26 (Fig. 1). Interestingly, reported non-stress plasma corticosterone concentration is 6.66μg per 100ml of plasma on postnatal day 1, 2.51μg/100ml on day 7 and about 4.2μg/100ml on day 14 and 21[13]. Thus, the estimated receptor translocation in the brain of intact rats during the first week of life, seems to be inversely proportional to the concentration of circulating corticosterone. Determination of hormone nuclear uptake following injection of saturating levels of tritiated dexamethasone ([^{3}H] D) in adrenalectomized developing rats corroborated this finding (Table 1). The limited nuclear uptake of the hormone on day 1 as compared to day 7 of life is not paralleled by an analogous difference in receptor concentration (Fig. 1), most likely due to an impairment of receptor translocation. Relevantly, a deficiency of the pituitary glucocorticoid receptor to translocate to the nucleus has also been reported for the newborn rat[14]. Since the magnitude of the biological response to the hormone is

Table 1 Nuclear uptake of (^{3}H)D *in vivo* and Kd values for GRC *in vitro* at different stages of brain development

Age[a] (days)	-5	+1	+7	+13	+26
Uptake[b] (fmoles D/mg DNA)	-	145±48(4)	360±170(4)	760±300(4)	420±170(7)
Kd[b] ($M \times 10^9$)	22.5±5(2)	20.8±5(3)	10.4±3(4)	6.0±3(3)	4.5±0.7(3)

[a]Day 0 is the day of birth
[b]Values are mean± S.D. of (N) separate estimations

proportional to the concentration of the nuclear glucocorticoid receptor complex[15], inhibition of receptor translocation might protect the developing brain from the high concentration of glucocorticoids present in the fetal and maternal circulation around the time of birth[12,13]. The inhibition is abolished by day 7 of life, at which time a large increase in hormone nuclear uptake is observed (Table 1). At this stage of brain development, a large increase in the amount of endogenous corticosterone bound to the particulate brain fraction has been reported[13]. This could reflect a period of openess to the influence of glucocorticoids, during which intensive "hormonal imprinting" in the brain[16] might take place.

Unexpectedly, the dissociation constant (Kd) of the glucocorticoid receptor complex (GRC) with tritiated dexamethasone, determined by Scatchard plot analysis as previously described[7,8] is 5-to 3-fold higher in prenatal and early postnatal brain cytosol as compared to the adult value (3.4×10^{-9}M)[7] (Table 1). This drop in binding affinity might be the result of receptor proteolysis during the lengthy equilibrium binding experiment. Relevantly, tryptic digestion of rat liver glucocorticoid receptor is reported to decrease receptor affinity by 3-fold[17]. In addition, calcium ions can cause rapid proteolysis of the receptor even at 0°C and in the presence of sodium molybdate, mediated by the calcium-activated protease(s) calpain[18,19]. Accordingly, as a result of receptor proteolysis, fragments could be identified in fetal and newborn rat brain extracts covalently labelled by tritiated dexamethasone-21-mesylate ([^{3}H] DM)[20] and analysed by sodium dodecylsulfate polyacrylamide gel electrophoresis (SDS-PAGE) and fluorography (Fig. 2). In the fluorograph, glucocorticoid receptor can be identified with the protein of M_r=94000 (94 KDa) (migrating close to phosphorylase b) which is not labelled in the presence of a 200-fold excess of radioinert dexamethasone competing for the occupancy of the steroid binding site. The truncated receptor of M_r= 48 KDa, is readily detectable in extracts of fetal and neonatal brain but not in 7- and 14- day old rat brain extracts (arrow). In addition to the specifically labelled bands, several other abundant proteins e.g. albumin of M_r=67 KDa, are also labelled by (^{3}H)DM, though not in a specific manner[20] (Fig. 2).

Cloning of the genes for the human, mouse and rat glucocorticoid receptor led to a domain structure for the receptor in which the DNA-binding domain is flanked by the steroid-binding domain at the C-terminal end and by an immunogenic domain comprising the N-terminal half of the protein[21]. Two major (^{3}H)DM- labelled receptor proteolytic fragments were immunoprecipitated from rat brain extracts with a polyclonal antibody to rat glucocorticoid receptor (Fig. 3, lanes 1,2). These can be identified as chymotryptic-like fragments, with M_r=39-48 KDa, comprising both the DNA- and steroid-binding sites and tryptic-like fragment, with M_r=27-29 KDa, comprising only the steroid binding site[22]. The 48 KDa form is that which is primarily detected in fetal extracts even in the presence of calcium-

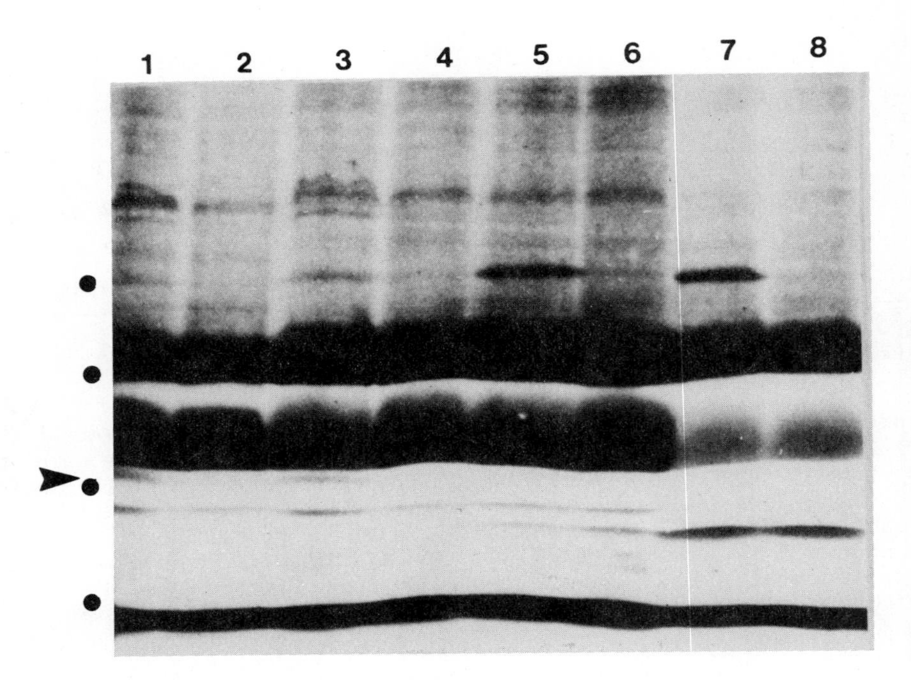

Fig. 2. Detection of (^{3}H)DM-labelled glucocorticoid receptor in brain cytosol from fetuses (lanes 1,2), 1-day (lanes 3,4), 7-day (lanes 5,6) and 14-day old rats (lanes 7,8). Cytosol was prepared in 3mM $MgCl_2$, 1mM EDTA, 1mM β-mercaptoethanol, 10mM Na_2MoO_4, 10% glycerol, 20mM sodium phosphate (pH 7.8) and labelled for 4hr at 0°C with 2.5×10^{-7}M (^{3}H)DM in the absence (1,3,5,7) and presence (2,4,6,8) of 20μM radioinert dexamethasone. Samples were analysed by SDS-PAGE and fluorography as previously described[22]. Dots correspond to the position of ^{14}C-labelled phosphorylase b (97 KDa), bovine serum albumin (67 KDa), ovalbumin (44 KDa) and carbonic anhydrase (30 KDa).

sequestering ethylene-diamine-tetraacetic acid (EDTA) (1mM) (Fig. 2). Calpain I, the form of the protease sensitive to calcium ions at the μmolar range[19], is most likely acting on the receptor in developing rat brain extracts: Addition of excess EDTA (5mM) and the calpain(s) inhibitor leupeptin (1mM) partly restored fetal receptor affinity for dexamethasone to near adult values and diminished the 48 KDa species (Fig. 3, lanes 3-10). The remaining truncated receptor forms may be the product of action of other proteases on different proteolytic hinges and/or the result of naturally occuring receptor turnover.

An adaptation of the endocrine system in order to serve morphogenesis and homeostasis during ontogeny, especially in the perinatal period, is thought to depend partly on the plasticity of the receptor system[2,16]. As far as the glucocorticoid receptor is concerned, our data suggest that its capacity for nuclear translocation and possibly its affinity for the hormone is affected during the perinatal period. The drop in affinity observed with fetal and newborn rat brain extracts is partly due to the extreme lability of the receptor in vitro. It is not clear whether the reduced capacity for nuclear translocation is the property of a genetically encoded distinct receptor form with a morphogenetic role during prenatal

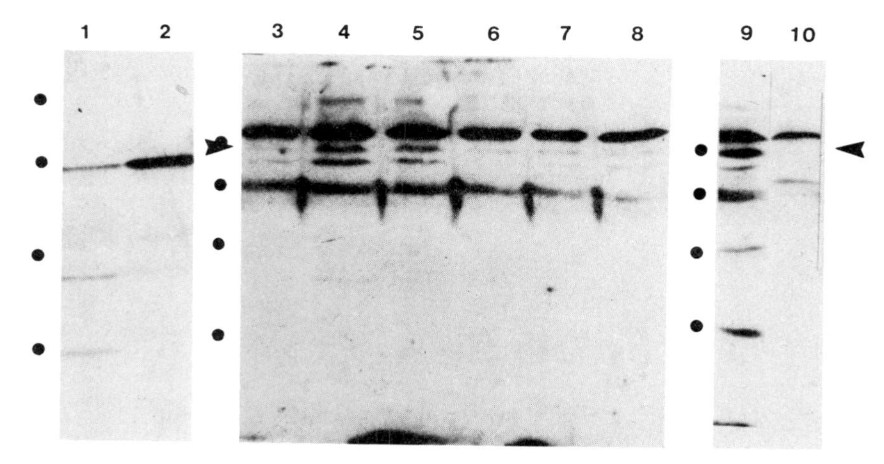

Fig. 3. Lanes 1,2: Rat brain cytosol was labelled by (^{3}H)DM in the absence (lane 1) and presence (lane 2) of excess dexamethasone as described in the legend to Fig. 2, incubated with anti-glucocorticoid receptor antiserum for 4hr at 0°C, the receptor-antibody complex precipitated with *Staph. aureus* fixed cell suspension and the precipitate analysed by SDS-PAGE and fluorography as previously described[22]. Non-specifically labelled bovine serum albumin is detected in both lanes.
Lanes 3-10: Cytosol from developing brain hemispheres (lanes 3-5) and stem (lanes 6-8) at gestation day 14 (3,6), 15 (4,7), 16 (5,8) and at postnatal day 15 (9,10) was analysed by SDS-PAGE and immunoblotting with 1:100 dilution of anti-glucocorticoid receptor antiserum (lanes 3-9) or preimmune serum (lane 10), followed by incubation of the nitrocellulose sheet with 0.5 µCi of ^{125}I-labelled protein A[27]. Dots correspond to the position of marker proteins described in the legend to Fig. 2. Arrow points to the position of the glucocorticoid receptor with M_r=94 KDa.

and early postnatal life, or a post-translational change of the receptor. However, on the basis of its proteolytic susceptibility in the cytosol (Figs. 2 and 3), its molecular weight as determined by SDS-PAGE (Figs. 2 and 3) and its immunochemical properties (Fig. 3), fetal and early neonatal receptor is indistinguishable from the adult. Furthermore, binding of fetal receptor to the glucocorticoid-regulated enhancer element of mouse mammary tumor virus (MMTV) genome *in vitro*[23] (Fig. 4), suggests that fetal receptor might mediate a glucocorticoid effect on the fetal brain *in vivo*[23,24]. Immunoblot analysis of brain extracts from the hemispheres and the brain stem (including pons and medulla) showed that the receptor is present in substantial quantities at gestational day 15 and 16 respectively (Fig. 3, lanes 3-8). These findings were corroborated by (^{3}H)DM-labelling of cytosol samples, followed by SDS-PAGE and fluorography of the labelled extracts (data not shown).

Analysis of developing brain extracts by immunoblotting (Fig. 3), (^{3}H)DM-labelling followed by SDS-PAGE and fluorography (Fig. 2), and equilibrium binding of tritiated dexamethasone (Fig. 1) as well as Northern blotting of total brain mRNA, demonstrated that an increase of receptor concentration by ~ 5-fold occurs between the first and second week of life. This increase is most likely due to a genetically-encoded, differentiation-associated event in oligodendrocytes, which show their proliferative peak at this time[10]. In these cells, glucocorticoids induce glycerol-3-phosphate dehydrogenase (GPDH), a soluble enzyme implicated in the process of myelination[3,4,10]. In the cerebrum of adult rats, GPDH activity decreases following adrenalectomy and enzyme activity is restored by administration of corticosterone[25]. In accordance with the model of steroid hormone action, glucocorticoid-regulation of GPDH has been shown

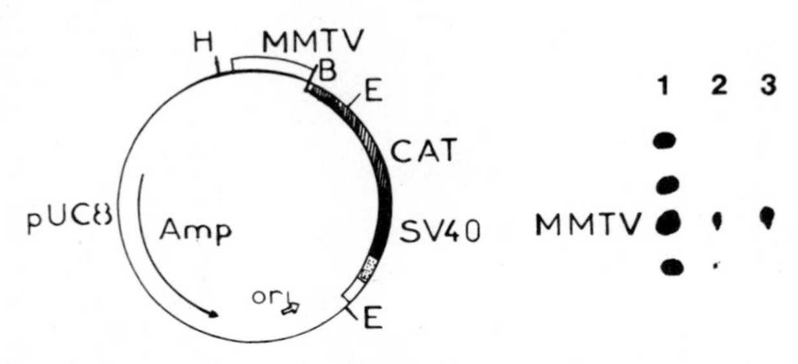

Fig. 4 Left: Plasmid MMTV* was digested with restriction endonucleases Hind III (site H), Bgl II (site B) and Eco RI (sites E) and the restriction fragments end-labelled with γ^{32}P ATP[24].
Right: Total γ^{32}P-labelled pure DNA (lane 1) and DNA incubated with fetal (lane 2) or adult (lane 3) rat brain cytosolic glucocorticoid receptor antiserum as described in the legend to Fig. 3, was analysed by agarose gel electrophoresis and autoradiography as described by Rusconi and Yamamoto[24]. Methods: Brain cytosol was prepared in 60mM NaCl, 3mM dithiothreitol, 5mM EDTA, 20mM Tris (pH 7.8), labelled for 1hr at 0°C with (^{3}H)D and heated for 30 min at 25°C to transform the receptor to the DNA-binding form. Following addition of $MgCl_2$ (10mM) and Na_2MoO_4 (10mM), the cytosol was incubated at 0°C for 30 min with a mixture of 50ng end-labelled DNA and 1.25 μg λ DNA prior to immune-precipitation of the glucocorticoid receptor-DNA complexes.

to occur *via* modulation of the rate of transcription of the gene and to depend primarily on the concentration of the receptor[3,4]. In Table 2, glucocorticoid induction of GPDH activity in the developing limbic system of the rat is compared to that in the adult. Implantation for two weeks of a corticosterone tablet in adrenalectomized adult animals[25] causes a 2.5-fold increase in GPDH activity in the limbic system. Interestingly, glucocorticoid regulation of GPDH activity is not detected in the developing limbic system earlier than postnatal day 25, in spite of the adult-like concentration and normal capacity for nuclear translocation of the receptor

*In the case of genes positively regulated by glucocorticoids, the receptor has been shown to bind to specific DNA sequences *in vitro* that represent glucocorticoid-responsive DNA elements that enhance transcription of these genes *in vivo*[9,23,24].

Table 2 The development of GPDH activity in the limbic system[a]

Age (days)	14	28	40	Adult
GPDH[b](intact)	6.7±0.6(6)	15.3±0.6(4)	23.9±2.7(4)	26.4±9.9(6)
GPDH[b](ADX)	-	13.6±3.5(5)	14.3±2.8(4)	20.0±7.8(8)
GPDH[b](ADX+CS)	-	-	33.0±4.8(7)	35.8±7.5(5)

[a]GPDH activity was determined by Breen and de Vellis[26] in the cytosol from the brain of intact, adrenalectomized (ADX) and ADX animals implanted at the time of adrenalectomy with a tablet of 100 mg of corticosterone (CS). GPDH was assayed 14 days after the operation.
[b]GPDH activity values ($M \times 10^9$ NADH/min/mg cytosol protein) are mean ± S.D. of (N) separate estimations.

already by day 13 of life (Fig. 1, Table 1). This latency in the appearance of glucocorticoid-regulated GPDH activity is likely to signify a miscoupling of a normal receptor to an immature oligodendrocyte acceptor system. The lack of glucocorticoid-regulation of GPDH activity coincides with the overall process of adaptation lasting for the first month of life[2,16]. Glucocorticoid-regulation is clearly established shortly after the adaptation response to stress has been developed in the rat.

In conclusion, the maturation of glucocorticoid responsivity of the brain can be considered to proceed through the following stages: a first stage, in which the fetal receptor displays reduced sensitivity to the hormone, followed by a stage of openess to the hormonal signal whose coupling to some of the adult brain functions is however still impaired.

Acknowledgements: We are grateful to Dr. K. Yamamoto for a gift of the glucocorticoid receptor cDNA and to Ms. A. Hatzistili for typing the manuscript.

REFERENCES

1. M.T. Jones, E.W. Hillhouse, and J.L. Burden, Dynamics and mechanics of corticosteroid feed-back at the hypothalamus and anterior pituitary gland, J. Endocr. 73:405 (1977).
2. B.S. McEwen, E.R. deKloet, and W. Rostene, Adrenal steroid receptors and actions in the nervous system, Physiol. Rev. 66:1166 (1986).
3. J.F. McGinnis, and J. de Vellis, Cell surface modulation of gene expression in brain cells by down regulation of glucocorticoid receptors, Proc. Natl. Acad. Sci. (USA) 78:1288 (1981).
4. S. Kumar, D.P. Weingarten, J.W. Callahan, K. Sachar, and J. de Vellis, Regulation of mRNAs for three enzymes in the glial cell model C_6 cell line, J. Neurochem. 43:1455 (1984).
5. J.L. Gerlach, and B.S. McEwen, Rat brain binds adrenal steroid hormone: Radioautography of hippocampus with corticosterone, Science 175:1133 (1972).
6. W.E. Stumpf, and M. Sar, The differential distribution of estrogen, progestin, androgen and glucocorticoid, in the rat brain. J. Steroid Biochem. 7:1170 (1976).
7. M.N. Alexis, F. Stylianopoulou, E. Kitraki, and C.E. Sekeris, The distribution and properties of the glucocorticoid receptor from rat brain and pituitary, J. Biol. Chem. 258:4710 (1983).
8. E. Kitraki, M.N. Alexis, and F. Stylianopoulou, Glucocorticoid receptors in developing rat brain and liver, J. Steroid Biochem. 20: 263 (1984).

9. R. Miesfeld, S. Rusconi, P.J. Godowski, B.A. Maler, S. Okret, A.C. Wikstrom, J.A. Gustafsson, and K.R. Yamamoto, Genetic complementation of a glucocorticoid receptor deficiency by expression of cloned receptor cDNA, Cell 46:389 (1986).
10. R.G. Wiggins, Myelin development and nutritional insufficiency, Brain Res. Rev. 4:151 (1982).
11. A. Sarrieau, M. Vial, B. McEwen, Y. Broer, M. Dussaillant, D. Philibert, M. Moguilewsky, and W. Rostene, Corticosterone receptors in rat hippocampal sections: Effect of adrenalectomy and corticosterone replacement, J. Steroid Biochem. 24:721 (1986).
12. D. Pechinot, and A. Cohen, The determination of maternal and foetal rat plasma corticosterone concentration in late pregnancy by competitive protein binding analysis, J. Steroid Biochem. 18:601 (1983).
13. J.C. Butte, R. Kakihana, M.L. Farnham, and E.P. Noble, The relationship between brain and plasma corticosterone stress response in developing rats, Endocrinology 92:1775 (1973).
14. M. Sakly, and B. Koch, Ontogenesis of glucocorticoid receptors in anterior pituitary gland: Transient dissociation among cytoplasmic receptor density, nuclear uptake and regulation by corticotropic activity, Endocrinology 108:591 (1981).
15. E. Bloom, D.T. Matulich, N.C. Lan, S.J. Higgins, S. Simons, and J. Baxter, Nuclear binding of glucocorticoid receptors. Relations between cytosol binding, activation and the biological response, J. Steroid Biochem. 12:175 (1980).
16. G. Csaba, Receptor ontogeny and hormonal imprinting, Experientia 42: 750 (1986).
17. O. Wrange, and J.A. Gustafsson, Separation of the hormone and DNA-binding sites of the hepatic glucocorticoid receptor by means of proteolysis, J. Biol. Chem. 253:856 (1978).
18. D.B. Mendel, N.J. Holbrook, and J.E. Bodwell, Degradation without apparent change in size of molybdate-stabilized non-activated glucocorticoid-receptor complexes in rat thymus cytosol, J. Biol. Chem. 260:8736 (1985).
19. T. Murachi, Calpain and calpastatin, Trends Biochem. Sci. 8:167 (1983).
20. S.S. Simons, and P.A. Miller, Affinity-labelling steroids as biologically active probes of antiglucocorticoid hormone action, J. Steroid Biochem. 24:25 (1986).
21. S. Green, and P. Chambon, A superfamily of potentially oncogenic hormone receptors, Nature 324:615 (1986).
22. M.N. Alexis, L. Baki, C. Elefteriou, and C.E. Sekeris, Glucocorticoid receptor structure as probed by endogenous proteases, J. Steroid Biochem. 29:407 (1988).
23. A.C.B. Cato, R. Miksicek, G. Schütz, J. Armemann, and M. Beato, The hormone regulatory element of mouse mammary tumour virus mediates progesterone induction, EMBO J. 5:2237 (1986).
24. S. Rusconi, and K.R. Yamamoto, Functional dissection of the hormone- and DNA-binding activities of the glucocorticoid receptor, EMBO J. 6: 1309 (1987).
25. J.S. Meyer, D.J. Micco, B.S. Stephenson, L.C. Krey, and B.S. McEwen, Subcutaneous implantation method for chronic glucocorticoid replacement therapy, Physiol. Behav. 22:867 (1979).
26. G.A.M. Breen, and J. de Vellis, Regulation of glycerol phosphate dehydrogenase by hydrocortisone in dissociated rat cerebral cell cultures, Develop. Biol. 41:255 (1974).
27. W.N. Burnette, "Western Blotting": Electrophoretic transfer of proteins from sodium dodecyl sulfate-polyacrylamide gels to unmodified nitrocellulose and radiographic detection with antibody and radioiodinated protein A., Anal. Biochem. 112:195 (1981).

MOLECULAR GENETIC APPROACHES IN THE THERAPY OF ALZHEIMER'S DISEASE

Ezio Giacobini

Department of Pharmacology
Southern Illinois University
School of Medicine
Springfield, IL 62794-9230

INTRODUCTION

Definitive treatment of Alzheimer's disease (AD) will depend upon understanding the etiology and pathogenesis of this degenerative process. Techniques derived from studies of molecular genetics will identify individuals at high risk [1]. These studies may also characterize the molecular defect(s) derived from the genetic expression of the disease, but we view them as future developments without clinical application for 1 to 2 decades. Alternatively, a shorter-term progress in the understanding of basic alterations of neurotransmitters, neuromodulators and growth factors may lead to the development of ameliorative or palliative therapies that attenuate the expression of symptoms and possibly even slow the disease process [2].

Fundamental aspects of the basic pathology of Alzheimer's disease are not yet fully understood. There is still a basic lack of understanding of the relationships among clinical pathological changes such as plaques, tangles and amyloid formation and the cause and course of the disease itself [3,4]. The most widely held formulation of the disease pathogenesis has associated plaques, tangles and amyloid accumulation with the progression of the disease [5]. Recently this view has been challenged, and the importance of neuronal loss has been stressed [6].

The accumulation of the pathological stigma (plaques and tangles) is not in synchrony with the disease progression and there is not sufficient accumulation of amyloid to produce the characteristic symptoms [3]. Two points can be argued: dementia can occur without (e.g. Pick disease) or with few plaques and tangles, and there is not a close enough relationship between pathology and dementia. There is a subgroup of AD patients with preserved mental status and numerous neocortical plaques [4]. If there is no direct causal sequence between amyloid formation and dementia, then the former could be only a secondary effect. Therefore, a therapy directed to prevent or decrease accumulation of amyloid may not alter the course of the dementia. A second possibility is that excessive amyloid formation or accumulation is a quantitative expression of aging which is present in the normal individual as well as in the AD patient.

Molecular Aspects of Development and Aging of the Nervous System
Edited by J.M. Lauder
Plenum Press, New York, 1990

Based on our present understanding of the process, there are at least three potentially therapeutic strategies to address Alzheimer's disease. First, the accumulation of amyloid could be arrested or reversed in the individuals at risk. To date, there is no means to identify these persons. Second, the rate of progression could be slowed in individuals already diagnosed. Third, symptomatic expression could be suppressed or reduced without attentuation in the pathology. If neuronal death is indeed the primary process and amyloid is a secondary event, then the first task will be particularly difficult to accomplish. Thus, the second and the third approaches may be more feasible and realistic. In this review we will address the first approach (Table I). The other strategies have been discussed in detail by Giacobini and Becker (1988) [2].

MOLECULAR GENETIC FINDINGS IN ALZHEIMER DISEASE

Within the last three years, evidence has accumulated that a genetic abnormality is present in some form of AD. It has been suggested that the primary defect in the pathogenesis of one form of AD is a regulatory disturbance involving the amyloid gene represented on chromosome 21. Recent studies have shown that duplication of the amyloid gene in Down's syndrome (trisomy 21) can indeed be correlated to the excessive accumulation of amyloid in the brain. This has not been found in AD [7]. In the absence of a demonstration of gene duplication, several possibilities are still open to explain the excessive deposition: a defect in amyloid gene regulation, a defect in the β-amyloid protein cleavage, specific amyloid protein modifications and changes in amyloid protein turnover regulation. Some of these mechanisms may be amenable to pharmacological intervention. Using hybridization techniques, Lewis et al. [8] found that mRNA encoding amyloid-β-protein precursor was present in neocortical neurons from the brain of both control subjects and patients with AD. This suggests that this protein precursor is a natural constituent of certain cortical cells. The same authors found that a subpopulation of neurons that show a decrease in precursor amyloid-β-protein mRNA in healthy brain developed NFT (neurofibrillary tangles) but not neuritic plaques. However, it appears that the expression of precursor amyloid-β-protein mRNA is necessary but not a sufficient prerequisite for NFT formation. Other factors such as: 1) long axonal projections; 2) distinctive pattern of cortico-cortical connections; 3) high levels of nonphosphorylated neurofilament protein may all contribute to make a neuron more vulnerable to deposition of amyloid-β-protein [8]. It is interesting to note that the amyloid protein precursor mRNA is specifically increased in nucleus basalis and locus coeruleus, suggesting that increased production of the precursor in these nuclei may play an important role in the deposition of cerebral amyloid in AD [9]. Amyloid deposition represents a biochemical and pathological process that is not restricted to AD or Down's syndrome but is a rather common feature not only of the human aging brain but also of other mammalian species. Isolation and cloning of the amyloid gene will make it possible to transmit the defective gene to laboratory animals producing models of excessive amyloid expression. This step will certainly help the search for drugs that alter the production and accumulation of the amyloid protein.

THE PROTEASE-INHIBITOR APPROACH

Which could be a feasible strategy at the molecular-genetic level? The recent studies of the genetic aspects of AD have demonstrated that gene coding and expression of the β-amyloid precursor is not increased in AD patients, only the accumulation of amyloid in the CNS is increased.

The studies of Tanzi et al. [10] and Kitaguchi et al. [11] suggest the presence of a proteoglycan receptor-type site related to chromosome with a possible function of activation of a protease inhibitor specific for the β-amyloid molecule cleavage. The aminoacid sequence encoded by HL124i is homologous to Kunitz-type protease inhibitors which act specifically on proteases such as trypsin, chymotrypsin, elastase, plasmin and cathepsin G. Depending on its specificity this domain could either promote amyloidogenesis by inhibiting proteases which degrade amyloid or suppress amyloidogenesis by inhibiting proteases involved in the generation of amyloid from the precursor. Several cellular roles for the glycosilation of the amyloid peptide precursor have been discussed, including cell adhesion, ion transport and regulation of an endogenous protease inhibitor. It is possible that this protease inhibitor which is encoded by a DNA sequence designated HL124i, may influence the metabolism of the amyloid protein precursor (APP), protecting the molecule from degradation by certain not well identified proteases [10,11]. Inhibition of this protease could prevent amyloidogenesis by blocking upregulation of the gene. A therapy could be developed to alter the function of the protease inhibitor. On the other hand, the protease inhibitor may be beneficial in decreasing amyloidosis because it prevents the cleavage of the β-protein from the membrane. Therefore, inactivation of this inhibitor could achieve the opposite effect and actually increase amyloidosis instead of decreasing it. With regard to the type of proteases involved in this process, it has been demonstrated that the APP molecule blocks trypsin activity _in vitro_ and could therefore be similar to a chimotrypsin-G [11]. Additional proteases of α-chimotrypsin type have been postulated for the amyloidotic process. Any therapy directed to influence such a protease inhibitor needs to be targeted specifically to brain cells at concentrations high enough to act selectively on the specific cells carrying abnormally high levels of the inhibitor. Besides expression in nerve cells, cerebral vessels and meninges, the amyloid gene is also expressed in cells of the kidney, heart, and other organs. It can be assumed that adverse effects may follow complete suppression of amyloid formation or excessive breakdown. Because amyloid is ubiquitously expressed in cell tissues of the body as a universal protein, high regional selectivity might be required for therapeutic effect. Fetal brain shows a uniform high expression of the amyloid gene [9]. In the adult brain cortex, particularly in the temporal cortex, it seems to be more represented [10]. Recent findings by Tanzi et al. [10,12] indicate the presence of two amyloid messages (APP_1 and APP_2), the levels of which are affected differently by AD. APP_2 mRNA increases specifically in the basal forebrain and in the entorinal cortex. If this is confirmed, it may allow higher specificity for potential drugs, inhibiting selectively the protease inhibitor (APP_1) which is expressed by the AD brain. One question is: If available, how would such an inhibitor be administered? The intraventricular route would be particularly suitable to concentrate and amplify the effect of the drug in brain. The delivery of other macromolecules (peptides, proteins) which might also be considered in future therapy, such as neuropeptides or growth factors, may also be more effective if performed i.c.v. or may require this route because of the blood brain barrier or other similar factors. Before effectively using this route of administration, we need to obtain detailed information with regard to ventricular penetration, intracerebral diffusion and transport of these potential drugs and macromolecules (Table I).

THE NERVE GROWTH FACTOR APPROACH

A second molecular genetic approach to AD treatment is based on the fact that NGF (nerve growth factor) has been implicated in the survival of basal forebrain cholinergic neurons. These neurons are selectively and progressively eliminated in the course of AD. A continuous intraventricular infusion of NGF or the intracerebral transplant of NGF-producing cells may stop the degeneration of these cells. Two recent approaches in these directions have been reported at the Neuroscience Meeting (Toronto, 1988). Human NGF (hNGF) expression vectors were constructed by Bruce and Heinrich (1988) [13] to allow for the production of recombinant hNGF for infusion and also for the potential to transform cells for grafting. Three expression vectors were constructed containing the entire coding region for mature HNGF derived from the 3' exon (exon 4) of the hNGF gene. These vectors are being transiently expressed in COS, CHO and PC12 cells. Olson et al (1988) [14] reported at the same meeting the establishment and use of stable cell lines that overexpress a transfected β-NGF gene. These cell lines have been assayed by injecting them into the anterior chamber of the eye of mice or intracerebrally. Similar cell lines were established using cells of the submandibular gland of the mouse. These experiments show that these cell lines may function as a source of hNGF that may stimulate injured NGF-receptive neurons in the host brain and may support cografted NGF-dependent neurons or chromaffine cells. The obvious route for this experimental treatment is the intraventricular one.

TABLE I

FUTURE DEVELOPMENTS IN TREATMENT OF ALZHEIMER'S DISEASE

Mechanism of Action	Class of Drugs	Preferrable Route of Administration
Classical Neurotransmission	new cholinominimetics (inhibitors or agonists)	oral or i.c.v.
Modulation of Neurotransmission	neuropeptides (somatostatin, NPY)	i.c.v.
Trophic effects or neuronal rescue	NGF or other trophic factors	i.c.v.
Inactivation of protease inhibition (β-amyloid cleavage)	specific enzyme inhibitors or reactivators	oral or i.c.v.

REFERENCES

1. R.N. Rosenberg. Neurogenetics, Principles, and Practice. Raven Press, NY, Alzheimer's disease, pp. 95-99 (1986).
2. E. Giacobini and R. Becker. 1988. Advances in therapy of Alzheimer's disease. In: Familial Alzheimer's Disease: Molecular Genetics, Clinical Perspectives, α Promising New Research. eds.G. Miner, J. Blass, R. Richter & J. Valentine. Marcel Dekker, Inc. New York (In press)
3. R.D. Terry, L.A. Hansen, R. DeTeresa et al. Senile dementia of the Alzheimer type without neocortical neurofibrillary tangles. J. Neurophathol. Exp. Neurol. 46, 262-268 (1987).
4. R. Katzman, R. Terry, R. DeTeresa, T. Brown, P. Davies, P. Fuld, X. Renbing and A. Peck. Clinical, pathological and neurochemical changes in dementia: A subgroup with preserved mental status and numerous neocortical plaques. Ann. Neurol. 23, 138-144 (1988).

5. H.M. Wisniewski, G.S. Merz and R.I. Carp. Current hypothesis of the etiology and pathogenesis of senile dementia of the Alzheimer type. Interdiscipl. Topics Geront. 19, 45-53 (1985).
6. R.D. Terry. What, where and why are plaques. Alzheimer Disease and Associated Disorders 2:223 (1988).
7. R.N. Rosenberg. Molecular genetics of Alzheimer's disease; recent advances. Proceedings of the International Symposium on Alzheimer's disease Abst., Kuopio, Finland (June, 1988).
8. D.A. Lewis, G.A. Higgins, W.G. Young, D. Goldgaber, D.C. Gajdusek, M.C. Wilson and J.H. Morrison. Distribution of precursor amyloid-β-protein messenger RNA in human cerebral cortex: Relationship to neurofibrillary tangles and neuritic plaques. Proc. Natl. Acad. Sci 85, 1691-1695 (1988).
9. M.R. Palmert, T.E. Golde, M.L. Cohen, D.M. Kovacs, R.E. Tanzi, J.F. Gusella, M.F. Usiak, L.H. Youngkin and S.G. Youngkin. Amyloid protein precursor messenger RNAs; differential expression in Alzheimer's disease. Science, Vol. 241, pp. 1080-1084 (1988).
10. R.E. Tanzi, A.I. McClatchey, E.D. Lamperti, L. Villa-Komaroff, J.F. Gusella, R.L. Neve. Protease inhibitor domain encoded by an amyloid protein precursor mRNA associated with Alzheimer's disease. Nature, 528-530 (1988).
11. N. Kitaguchi, Y. Takahashi, Y. Tokushima, S. Shiojiri and H. Ito. Novel precursor of Alzheimer's disease amyloid protein shows protease inhibitory activity. Nature 331, 530-532 (1988).
12. R.E. Tanzi, J.F. Gusella, P.C. Watkins, G.A. Bruns, P. St. George-Hyslop, M.L. VanKeuren, D. Patterson, S. Pagan, D.M. Kurnit and R.L. Neve. Amyloid β protein gene: cDNA, mRNA distribution, and genetic linkage near the Alzheimer ocus. Science 235, 880-884 (1987).
13. G. Bruce and G. Heinrich. The construction of human NGF expression vectors: Their potential role in the treatment of Alzheimer disease. 18th Annual Meeting Society for Neuroscience. Abst 331.17 pp 827. (1988).
14. L. Olson, P. Enfors, T. Ebendal, P. Mouton, I. Stromberg and H. Persson. The establishment and use of stable cell lines that overexpress a transfected β-NGF gene with studies in vitro, in oculo and intracranially. 18th Annual Meeting Society for Neuroscience Abst 276.6 pp 684. (1988).

TAU PROTEIN: ITS PRESENCE AND METABOLISM IN HUMAN NEUROBLASTOMA CELLS

H. Sternberg, G. Mesco, G. Cole, P.S. Timiras

University of California
Department of Physiology-Anatomy
Berkeley, California

INTRODUCTION

The structural basis of the cellular disturbance in Alzheimer's disease (AD) may involve the cytoskeleton. One of the major constituents of the cytoskeleton is the microtubule network. This is composed primarily of tubulin which has a molecular weight of 55 kd and assembles, under certain conditions, to form the microtubules (Cleveland et al., 1977). Other microtubular components include the microtubule associated proteins (MAPS), MAP I and MAP II, and Tau proteins, which co-purify with MAPs. "Tau" represents a class of several proteins which will be referred to collectively as Tau protein.

Below is a review of Tau protein, its metabolism, biochemistry, and relation to the pathogenesis of AD. Also included is a discussion of (1) whether Tau protein is present in human neuroblastoma cells and how this presence is affected by the degree of cell differentiation, and (2) whether Tau protein is involved in the induction of neurofibrillary tangles that might be induced by various neurotoxins (e.g. doxorubicin) known to promote degenerative lesions (tangles and amyloid) in cultured human cells (Cole and Timiras, 1987a,b).

RELATION TO ALZHEIMER'S DISEASE

In the adult brain, Tau protein appears to be primarily restricted to the axons (Binder et al.,1986). It is a major antigenic component of the paired helical filaments (PHF), which contribute to neurofibrillary tangles. They have been implicated in the pathology of various neurologic diseases (e.g. sclerosing panencephalitis, postencephalitic Parkinsonism, Parkinson' dementia, dementia pugilistica) but are rare in normal aging (Cole and Timiras, 1987a). PHFs accumulate in large 'neurofibrillary' tangles in the cell bodies of selected brain nuclei in AD.

PHF formation is currently interpreted as a type of human-specific cell-damage response, that is accompanied by structural, biochemical and molecular alterations of the neuron, and involving Tau protein in some

Molecular Aspects of Development and Aging of the Nervous System
Edited by J.M. Lauder
Plenum Press, New York, 1990

capacity. Tau, a normal constituent of axonal microtubules, becomes part of a high molecular weight polymer which eventually accumulates in tangles in the cell body (Grundke-Iqbal et al., 1986). This damage response is presumably reflected in the abnormal "sprouting response" (that is, de novo neuronal process generation) which occurs in AD neuritic plaques and also in isolated neurites (not associated with plaques) throughout the neuropil. This damage response may be visualized by staining Tau protein (Kowall and Kosik, 1987). Tau protein may also be identified by a monoclonal antibody, Alz 50, which recognizes an antigenic determinant greatly elevated in AD supernatant fractions (Wolozin et al., 1986). Northern blots with cDNA probes do not show any evidence for altered expression of Tau (Goedert et al., 1988) although *in situ* hybridization may reveal some different Tau forms. Yields of Tau protein from AD brain are generally low and Tau protein fractions isolated from AD brain appear to be less capable of promoting microtubule assembly (Iqbal et al., 1986) suggesting a post-translational defect in the available Tau. Several reports have established an abnormal phosphorylation of Tau in AD (Binder et al., 1986), but its origin and functional significance remain unknown.

"NEURITE SPROUTING" AND TAU PROTEIN

Growing evidence supports the idea that neuritic plaques involve an abnormal "sprouting response" (Probst et al., 1983; Geddes et al., 1985). Because sprouting in the injured adult brain appears to involve a re-expression of numerous fetal antigens associated with process generation in the fetal brain, it is not surprising that fetal antigens may be expressed by AD neurons, particularly those involved in plaque and tangle production. For example, fetal gangliosides and neuron-specific enolase are expressed in AD brain (Chapman et al., 1988; Lamour et al., 1987). Anti-PHF sera display a greater reactivity with fetal than adult forms of Tau (Kosik et al.,1986) and the Alz-50 antibody does not stain formalin fixed adult neurons but does stain neuronal cell bodies in fetal brain sections. Alz 50 is also known to stain PHF (Love et al., 1988); this suggests that an antigen, possibly Tau, present in fetal and sprouting neurites is abnormally elevated in AD brain and contributes to PHF. Because PHF, with immunological and ultrastructural characteristics similar to in-vivo states, form in cultured human neuroblastoma cells in reaction to damage, it is reasonable, as our studies propose, to investigate Tau protein metabolism in sprouting and damaged human neuroblastoma cells in-vitro.

TAU PROTEIN METABOLISM

Tau proteins are the low molecular weight (42-68kd) microtubule associated proteins of brain. They have similar amino acid content and are immunologically related (Kosik et al., 1986). Tau protein exists in varying degrees of phosphorylation (Lindwall and Cole, 1984). In the dephosphorylated state four bands can be distinguished by sodium dodecyl sulfate-polyacrylamide gel electrophoresis (SDS-PAGE). Phosphorylation of Tau alters the electrophoretic migration and four to eight bands are resolved by SDS-PAGE. Two dimensional gel electrophoresis reveals further microheterogeneity. Due to differential phosphorylation, as many as 30 species can be detected on 2-D gels (Butler and Shelanski,1986). Virtually nothing is known concerning the functional significance of this modification of Tau proteins. Phosphorylation may inhibit the interaction of tubulin with actin (Selden and Pollard, 1983) and dephosphorylation may promote microtubule assembly (Lindwall and Cole, 1984). In developing animals the predominant dephosphorylated form of

Tau may promote microtubule assembly in growing neurites. The view that Tau protein plays a role in neurite extension is supported by the following evidence: Tau proteins are not apparent in fetal and migrating neuroblasts but develop simultaneously with sprouting of processes, consistent with a role in the assembly of microtubules in developing neurites. In undifferentiated PC12 cells (human pheochromocytoma cells) which lack neuronal processes, Tau proteins are virtually undetectable; when process formation is induced by NGF or cAMP, Tau becomes more prevalent (Drubin et al., 1985; Drubin and Kirschner, 1986). The gene for mouse Tau protein has been cloned and although there seems to be only one gene, there are several related mRNAs apparently produced by alternate splicing (Lee et al.,1988; Goedert et al., 1988). The remaining heterogenenity is apparently due to post-translational modifications. The cloned gene sequences indicate a number of serine and threonine phosphorylation sites and one potential tyrosine site.

TAU PROTEIN PHOSPHORYLATION

Tau protein preparations can be phosphorylated by protein kinases A, B, and C. Protein kinase C phosphorylation of serine residues reduces the ability of Tau protein to promote tubulin and actin assembly (Hoshi et al.,1987). However, it is difficult to reconcile this with the observation of increased PK-C activity induced by neurite extending agents and the reduced PK-C found in Alzheimer brain (Cole et al., in press). Calcium/calmodulin dependent kinase II phosphorylates purified Tau and unlike kinase C is reported to induce a gel mobility shift of Tau proteins similar to that reported in AD (Baudier and Cole, 1987). This is particularly interesting in view of the known autonomous calcium-independent kinase activity of the autophosphorylated form of this enzyme. However, phosphorylation of bovine Tau by this kinase does not alter immunoreactivity with Tau-1 antibody which recognizes a site reported to be abnormally phosphorylated in AD and PHF (Baudier and Cole, unpublished observations). Tau protein is also a good substrate for several tyrosine kinases (Miyata et al.,1986). The kinase responsible for the abnormal Tau phosphorylation in AD brain is presently unknown, as is the functional significance of this phosphorylation. Limited epitope mapping of the Tau-1 site using cloned cDNA digests has identified the Tau-1 site to be in a region including serine, threonine and tyrosine residues, thus placing no restrictions on possible kinases responsible for Tau-1 phosphorylation.

In our studies, Tau protein purified from bovine brain was dephosphorylated with alkaline phosphatase and subsequently phosphorylated using protein kinase C, Ca/calmodulin protein kinase II, and cAMP dependent protein kinase. The phosphorylated proteins were then electrophoretically separated on an SDS PAGE. They were blotted onto nitrocellulose and stained using an antibody to the Tau-1 site. Presumably, the antibody to the Tau-1 site does not recognize the site when it is abnormally phosphorylated. However, the antibody reacted with similar strength to Tau protein which was phosphorylated to each of the kinases mentioned. This indicates that none of the kinases, tested under the conditions used, phosphorylates the Tau-1 site which is abnormally phosphorylated in AD brains (Sternberg et al., 1988b).

Postranslational Tau protein modifications may play a role in the accumulation of Tau in PHF. Tau protein may belong to the family of short-lived regulatory proteins which normally turn over by an ATP-dependent mechanism involving ubiquitination (a covalent binding of the polypeptide ubiquitin which marks a protein for degradation) (Haas and Bright, 1985). This is particularly interesting because of the presence of ubiquitin conjugates in AD lesions (Cole and Timiras, 1987b). That is, PHF may contain ubiquitin because abnormal proteins or short-lived regulatory proteins are targets for ubiquitination. In PC12 cells, Tau proteins turned over with a 2 hour half-life while the bulk protein turnover was 12 hours (M. Kirschner, personal communication) which is consistent with a normal role for ubiquitin in Tau turnover. This is also consistent with the observation that ubiquitin is normally associated with microtubules in cultured cells (Murti et al.,1988).

DETECTION OF TAU PROTEIN IN HUMAN NEUROBLASTOMA CELLS

Using culture cells we have been able to induce neurofibrillary-like and amyloid-like lesions by treatment of neuroblastoma and teratocarcinoma cells with agents capable of disrupting cytoskeletal elements (Cole et al., 1985; Cole and Timiras, 1987a,b). Moreover, we have been using these cells to study the biochemistry and metabolism of Tau.

Tau protein was detected in both normally dividing and differentiated LAN-5 cells (Sternberg et al., 1988a,c; Mesco et al., 1988). Exposure of growing cells to retinoic acid (vitamin A) causes cessation of cell division and induces developmental changes, which are representative of adult neuronal morphology by several criteria (Beidler et al., 1973; Andrews, 1984). The differentiated neural networks become increasingly complex with time. The differentiated cells persist in apparently excellent health for weeks but not longer than several months. After performing a Western Blot of the electrophoretically separated proteins of LAN-5 cells (using a rabbit antisera against Tau protein, a gift of Dr. Marc Kirschner) we were able to clearly visualize the presence of 4-5 protein bands corresponding to Tau protein (Sternberg et al., 1988a). Interestingly, there is a difference in the relative abundance of the various Tau forms in differentiated versus undifferentiated cells. There is an additional lower band corresponding to a molecular weight of approximately 45 kd present only in the undifferentiated cells (Argasinski et al., in press).

DOXORUBICIN EFFECTS TAU PROTEIN

LAN-5 cells were treated with 4×10^{-8}M doxorubicin (Cho et al., 1980) for 48 hours. This dose of doxorubicin (a heterocyclic compound which interferes with DNA synthesis and generates free radicals) was previously found, in our laboratory, to induce neurofibrillary tangles in cultured human teratocarcinoma cells (Cole et al., 1985; Cole, 1986). In a preliminary experiment we observed the accumulation of a 45kd species of Tau in undifferentiated cells (Argasinski et al., 1988). The accumulation of this Tau species could only be induced with doxorubicin in undifferentiated cells and not differentiated cells. It is possible that in the differentiated cells the lower molecular weight forms of Tau protein remain associated with microtubules within the neurites. Thus, the Tau that associates to microtubules is not readily extractable and detectable.

CONCLUDING REMARKS

Interest in Tau protein has grown since it has been found to be a major antigenic component of PHF (a principal lesion of AD brains). While considerable progress has been made, still little is known about the various Tau protein forms including (1) the mechanism involved in regulating their presence, (2) the enzymes involved in their generation, or (3) their different cellular roles and functions. In our laboratory we have recently developed an in-vitro approach using human neuroblasoma cells to address some of these questions. Tau protein appears to be influenced by cell differentiation, the presence of doxorubicin (a chemotherapeutic drug that can also induce AD-like lesions), and subtle changes in culture conditions as well. Our studies suggest that subtle neurochemical changes may alter Tau metabolism and possibly result in specific brain lesions. We are optimistic that this approach along with studies performed both in-vivo and in cell-free systems will greatly improve our current knowledge regarding Tau.

ACKNOWLEDGEMENT

The original experiments quoted in this review were supported by a grant from the State of California, Alzheimer's Disease Program.

REFERENCES

Andrews, P.W. Retinoic acid induces neuronal differentiation of a cloned human embryonal carcinoma cell line in vitro. Dev. Biol. 103, 285-293, (1984).

Argasinski, A., Sternberg, H., Fingado, B., Huynh, H., Timiras, P.S. Doxorubicin Effects Tau Protein Metabolism In Human Neuroblastoma Cells. Neurochemical Research, in press).

Argasinski, A., Fingado, B., Huynh, H., Sternberg, H., and Timiras, P.S. "Tau Protein in Alzheimer's Disease: Doxorubicin Effects in Cultured Human Cells" (abstract) American Physiological Soc. 1988.

Baudier, J., and Cole, R.D., Phosphorylation of tau proteins to a state like that in Alzheimer brain is catalyzed by a calcium/calmodulin-dependent kinase and modulated by phospholipids, J. Biol. Chem. 1987, 262:17584-17590.

Biedler, J.L., Helson, L., and Spengler, B.A. Morphology and growth, tumorigenicity, and cytogenetics of human neuroblastoma cells in continuous culture. Cancer Res., 33, 2643-2652 (1973).

Binder, L.I., Frankfurter, A., and Rebhun, L.I. Differential localization of MAP-2 and tau in mammalian neurons in situ" Ann. N.Y. Acad. Sci. 1986 466:145-66.

Butler, M., and Shelanski, M.L., Microheterogeneity of microtubule-associated tau proteins in living cells" J. Neurochem. 47,1517-22 (1986).

Chapman, J., Sela, B.A., Wertman, E., Michaelson, D., "Antibodies to Ganglioside GM1 in Patients with Alzheimer's Disease" Neurosci. Lett. 1988 Mar 31 86(2):235-240.

Cho, E.S., Spencer, P.S., and Jortner, B.S. Doxorubicin, in Experimental and Clinical Neurotoxicology. P.S. Spencer and H.H. Schaumberg, eds. Wilkins and Wilkins, Baltimore, pp. 430-439 (1980).

Cleveland, D.W., Hwo, S.Y., Kirschner, M.W. "Physical and Chemical Properties of Purified Tau Factor and the Role of Tau in Microtubule Assembly" J. Mol. Biol. 1977 116:227-247.

Cole, G.M. An in vitro Model for Alzheimer's Disease Pathology. Ph.D. Dissertation, University of California, Berkeley, (1986).

Cole, G.M. and Timiras, P.S. Aging-related pathology in human neuroblastoma and teratocarcinoma cell lines. In Model Systems of Development and Aging of the Nervous System, A. Vernadakis et al., eds., Martinus Nijhoff Publ., Boston, pp. 453-473 (1987a).

Cole, G.M. and Timiras, P.S. Ubiquitin-protein conjugates in Alzheimer's lesions. Neurosci. Let. 79, 207-212, (1987b).

Cole, G.M., Wu, K., and Timiras, P.S. A culture model for age-related human neurofibrillary pathology. Int. J. Dev. Neurosci. 3, 23-32 (1985).

Cole, GM. Dobkins, K.R., Hansen, L.A., Terry, R.D. and Saitoh, T. (1988) Brain Res., in press.

Drubin, D.G., Kirschner, M.W. "Tau Protein Function in Living Cells" J. Cell. Biol. 1986 103 (No.6 Pt. 2) p. 2739-2746.

Drubin, D.G., Feinstein, S.C., Shooter, E.M., and Kirschner, M.W. Nerve Growth Factor Induced Neurite Outgrowth in PC12 Cells Involves the Coordinated Induction of Microtubule Assembly and Assembly Promoting Factors. J. Cell. Biol. 101, 1799-1807 (1985).

Geddes, J.W., Monaghan, D.T., Cotman, C.W., Cott, I.T., Kim, R.C., and Chui, H.C. Plasticity of hippocampal circuitry in Alzheimer's disease. Science, 230(4730) 1179-81 (1985).

Goedert, M., Wischik, C.M., Crowther, R.A., Walker, J.E., and Klug, A. Cloning and sequencing of the cDNA encoding a core protein of the paired helical filaments of Alzheimer's disease: identification as the microtubule-associated protein Tau. Proc. Natl. Acad. Sci. USA, 85, 11, 4051-4055 (1988).

Grundke-Iqbal, I., Iqbal, K., Tung, Y.C., Quinlan, M., Wisniewski, H.M., and Binder, L.I. Abnormal phosphorylation of microtubule-associated protein (Tau) in Alzheimer cytoskeletal pathology. Proc. Natl. Acad. Sci. USA, 83, 4913-4917 (1986).

Haas, A.L., and Bright, P. M. The immunochemical detection and quantitation of intracellular ubiquitin-protein conjugates. J. Cell Biol., 260, 23, 12464-12473 (1985).

Hoshi, M., Nishida, E., Miyata, Y., Sakai, H., Miyoshi, T., Ogawara, H., and Ayikama, T. "Protein kinase C phosphorylates Tau and induces its functional alterations" FEBS Letters 1987, June 15, 217(2):237-41.

Iqbal, K., Grundke-Iqbal, I., Zaidi, T., Merz, P.A., Wen, G.Y., Shaikh, S.S., Wisniewski, H.M. "Defective brain microtubule assembly in Alzheimer's Disease" Lancet, 1986 Aug. 23, 2(8504):421-426.

Kosik, K.S. Joachim, C.L. and Selkoe, D.J. Microtubule-associated protein tau is a major antigenic component of paired helical filaments in Alzheimer disease. Proc. Natl. Acad. Sci. USA, 83, 4044-4048 (1986).

Kowall, N.W., Kosik, K.S. "The Cytoskeletal Pathology of Alzheimer's Disease is Characterized by Aberrant Tau Distribution" Ann. Neurol. 22:639-43, 1987.

Lamour, Y., Scarna, H., Roudier, M., Safer, S., and Davous, P. Serum Neuron-specific enolase in senile dementia of the Alzheimer's type, Neurology 1987 37:768-772.

Lee, G., Cowan, N., and Kirschner, M.W. "The Primary Structure and Heterogeneity Tau Protein from Mouse Brain" Science 239:285-89, 1988.

Lindwall, G., and Cole, D. The purification of Tau protein and the occurrence of two phosphorylation states of Tau in brain. J. Biol. Chem, 259, 19, 12241-12245 (1984).

Love, S., Saitoh, T., Quiada, S., Cole, G.M., and Terry, R.D. "Alz-50, ubiquitin, and Tau immunoreactivity of neurofibrillary tangles, Pick's bodies, and Lewy bodies" J. Neuropath. Exp. Neurol. 1988 Jul. 47(4):393-405.

Mesco, E.R., Sternberg, H., Timiras, P.S. Immunological Identification of Tau Protein in Neuroblastoma Cells. American Aging Association, San Francisco, CA (1988).

Miyata, Y., Hoshi, M., Mishida, E., Minami, Y., and Sakai, H. "Binding of MAP-2 and Tau to the intermediate filament reassembled from the neurofilament 70kDa subunit protein; Its regulation by Calmodulin" J. Biol. Chem. 1986, 261(28): 13026-13030.

Murti, K.G., Smith, H.,T., and Fried, V.A. "Ubiquitin is a Component of the Microtubule Network" Proc. Nat. Acad. Sci. U.S.A. 85(9):13019-23, 1988.

Probst, A., Basler, V., Bron, B., and Ulrich, J. Neuritic plaques in senile dementia of Alzheimer's type: a golgi analysis in the hippocampal region. Brain Res. 268(2), 249-254 (1983).

Selden, S.C. and Pollard, T.D. Phosphorylation of microtubule associated protein regulates their interation with filaments. J. Biol. Chem. 258, 7064-7071 (1983).

Sternberg, H., Mesco, G., Argasinski, A.B., Sanchez, I., and Timiras, P.S. Tau protein in LAN-5 cells. American Society for Neurochemistry (abstracts), New Orleans, LA, (1988a).

Sternberg, H., Baudier, J., Akizuki, K., Cole, G., Martin, W.H., Creutz, C.E., and Timiras, P.S., Similarities and differences between Tau protein and chromobindin A. Neurochem. Inter., (accepted Jan. 1988b).

Sternberg, H., Mesco, G., Fingado, B.H., Petrie, R., Dao, Q., Cole, G.M., and Timiras, P. S. Differentiation of a human neuroblastoma cell line influences Tau protein. Int. Soc. Develop. Neurosci. (abstracts), Jerusalem, Israel, (1988c).

Wolozin, B.L., Pruchnicki, A., Dickson, D.W., and Davies, P. A neuronal antigen in the brains of Alzheimer patients. Science, 232, 4750, 648-50 (1986).

GENETICS OF THE ALZHEIMER AMYLOID PROTEIN PRECURSOR

Rachael L. Neve

Division of Genetics
The Children's Hospital
300 Longwood Avenue
Boston, MA 02115

Alzheimer's disease (AD) is a progressive neurodegenerative disorder characterized by gradual loss of memory, reasoning, orientation, and judgment (Katzman, 1983). AD generally occurs as a sporadic disorder of unknown cause. However, a proportion of cases, which have been termed familial Alzheimer's disease (FAD), are caused by a genetic defect that is transmitted in an autosomal dominant fashion (Heston et al., 1981), and that has been mapped to chromosome 21 (St. George-Hyslop, 1987a). One of the hallmarks of AD is the presence of numerous neuritic plaques in postmortem brain tissue that are revealed by neuropathological examination (Glenner, 1983). The degree of intellectual impairment in patients appears to be correlated with the frequency of neuritic plaques in the cortex (Roch et al., 1966). The mature neuritic plaque consists of degenerating axons and neurites surrounding an amyloid plaque core (APC) composed of 5- to 10-nm filaments (Mertz et al., 1983) that stain with Congo red (Terry et al., 1981). Similar filaments are also found outside of plaques as independent bundles in the cortical neuropil and in the walls of meningeal and intracortical blood vessels [termed cerebrovascular amyloid (CVA)] (Glenner, 1983). A 4.2-kD polypeptide, called the β protein (due to its partial β-pleated sheet structure), has been isolated from both CA and APC. An identical 28-amino acid sequence (with the exception of a Glu to Gln substitution) was obtained from the CA and APC β proteins, suggesting a possible common origin for both types of amyloid (Glenner and Wong, 1984a, 1984b; Masters et al., 1985). The brains of aged individuals with Down syndrome (DS, trisomy 21) also have both APC and CA that contain a β protein with the same stretch of 28 amino acids (Glenner and Wong, 1984b; Masters et al., 1985); this similarity indicates that a common mechanism may underlie the formation of amyloid in AD and DS.

Isolation and Characterization of Alzheimer Amyloid Protein Precursor cDNA Clones

Several groups used oligonucleotides based on the 4.2 kD amyloid polypeptide sequence to isolate brain amyloid protein precursor (APP) cDNAs that were shown to hybridize with a 3.4-3.6 kb mRNA doublet expressed normally in the brain and other tissues (Tanzi et al., 1987a; Kang et al., 1987; Goldgaber et al., 1987; Robakis et al., 1987). Thus, the APP gene encodes a protein considerably larger than the amyloid

Molecular Aspects of Development and Aging of the Nervous System
Edited by J.M. Lauder
Plenum Press, New York, 1990

polypeptide found deposited in neuritic plaques in AD and DS. These findings suggested the possibility that amyloid deposits may result from either an abnormal expression or a posttranslational modification of a normal molecular constituent. This possibility was strengthened by the finding that the APP gene is found on chromosome 21 (Tanzi et al., 1987a; Kang et al., 1987; Goldgaber et al., 1987; Robakis et al., 1987) and hence may be the cause of the amyloid deposits common to Down syndrome and Alzheimer's disease. Additional studies, however, revealed that the APP gene is distinct from the genetic defect on chromosome 21 responsible for familial Alzheimer's disease (Tanzi et al., 1987b; Van Broeckhoven et al., 1987) and that the APP gene is not duplicated in sporadic or familial Alzheimer's disease (Tanzi et al., 1987c; St. George-Hyslop et al., 1987b; Podlisney et al., 1987). Furthermore, no significant allelic association was detected between sporadic AD and the APP locus (St. George-Hyslop et al., 1987b), suggesting that a defect in the APP gene is not the primary cause of sporadic AD.

The APP gene was shown to be expressed in brain and in most other tissues examined (Tanzi et al., 1987a), but the specific biochemical events leading to deposition of amyloid are not known. We screened complementary cDNA libraries constructed from non-neural tissues to determine whether the messenger RNA encoding APP in these tissues is identical to that expressed in brain, and we identified a second APP mRNA (Tanzi et al., 1988; also reported by Kitaguchi et al., 1988, Ponte et al., 1988) that encodes an additional internal domain with a sequence characteristic of a Kunitz-type serine protease inhibitor (Fig. 1). Kitaguchi et al. (1988) showed that this new domain functions as an active trypsin inhibitor. They also found a third APP cDNA encoding not only the protease inhibitor domain but also a further contiguous sequence of 19 amino acids of unknown homology (1988). Genomic analysis suggests that the protease inhibitor domain and the third, functionally unidentified, domain arise from two extra exons within the APP gene, which are alternatively spliced to generate three mRNA species. These three mRNA species will hereafter be referred to as APP-695 and APP-751/770 (containing the protease inhibitor domain and/or the additional 19 residues).

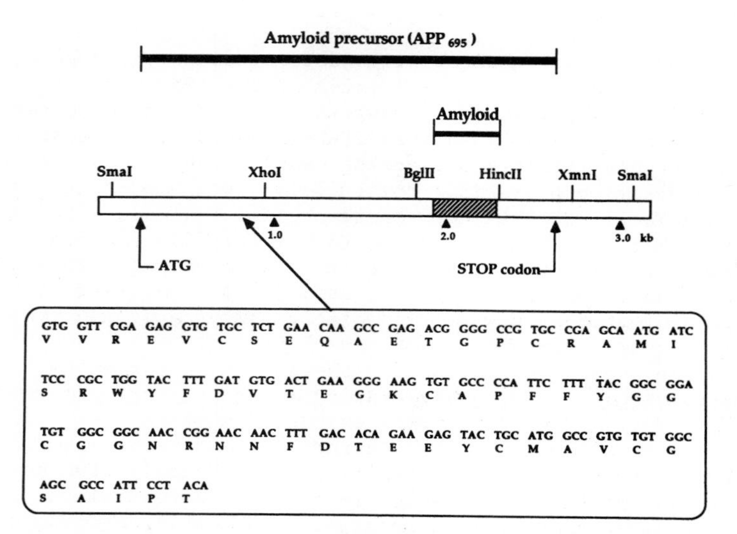

Figure 1. Sequence and position of the Kunitz-type serine protease inhibitor domain-encoding segment of an alternative APP mRNA.

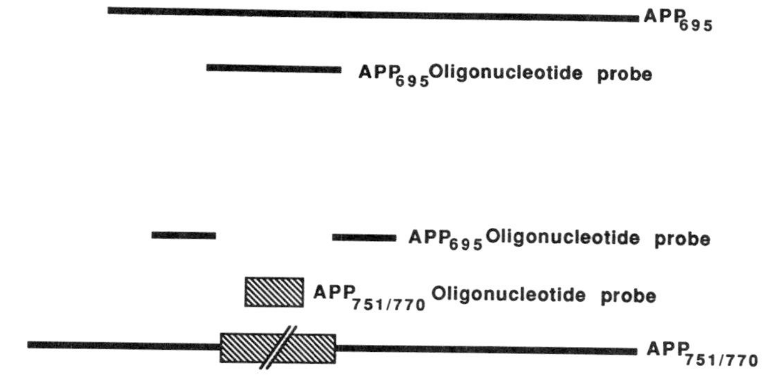

Figure 2. Schematic diagram showing the design of APP RNA-specific oligonucleotide probes. The sequence of the APP-695 specific oligonucleotide is 5'CTGGCTGCTGTTGTAGGAACTCGAACCACCTTTCCACAGA3'. The sequence of the APP-751/770 specific oligonucleotide is 5'CTTCCCTTCAGTCACATCAAAGTACCAGCGGGAGATCAT3'.

Expression of Amyloid Protein Precursor Messenger RNAs in the Brain

We utilized oligonucleotide probes specific for APP-695 and APP-751/770 to characterize their expression in human brain (Neve et al., 1988). RNA (Northern) blots were used to examine tissue distribution and overall regional distribution of these APP RNAs within the brain; in situ hybridization studies allowed us to identify specific cellular populations within the brain that express the mRNAs. The design of the oligonucleotide probes is shown in Fig. 2. The 40 base oligonucleotide specific for APP-695 encompasses 20 bases on either side of the potential splice junction sites in the APP-751/770 RNA, and therefore represents 40 contiguous bases in RNAs lacking the protease inhibitor domain. The 40 base oligonucleotide specific for APP-751/770 mRNAs is homologous to a relatively non-conserved portion of the nucleotide sequence encoding the protease inhibitor domain.

RNA blot analysis using these oligonucleotides revealed several differences in the patterns of expression of the APP RNAs. A survey of twelve fetal tissues showed that, while APP-751/770 is expressed in all tissues examined, APP-695 is selectively expressed in nervous tissue (Neve et al., 1988). Hybridization of an APP cDNA, which detects all APP mRNAs characterized thus far, to human fetal and adult human brain region RNAs (Fig. 3A,B) revealed striking developmental and regional pattern of expression. The APP cDNA hybridized intensely to RNA from most regions surveyed in a 19-week fetal brain. In the adult human brain, on the other hand, expression of APP RNAs showed a striking regional variation, exhibiting greatest abundance in associative regions of the neocortex, specifically in Brodmann areas A10, A20/21, A40 and A44. Use of oligonucleotide probes to distinguish the APP mRNAs showed that this regional variation was due to the expression of APP-695. The ubiquitously expressed APP-751/770 transcript was expressed homogeneously across brain regions (Tanzi et al., 1988). In contrast, APP-695 expression was highest in the association cortex (Neve et al., 1988). Notably, the contrast between APP-695 and APP0751/770 mRNA levels is greatest in the hippocampus, where the APP-695 transcript is almost undetectable by Northern blot analysis. To quantify our observations, we carried out a slot blot analysis in which both oligonucleotides were radiolabeled to the same specific activity and were hybridized to a slot blot containing RNAs

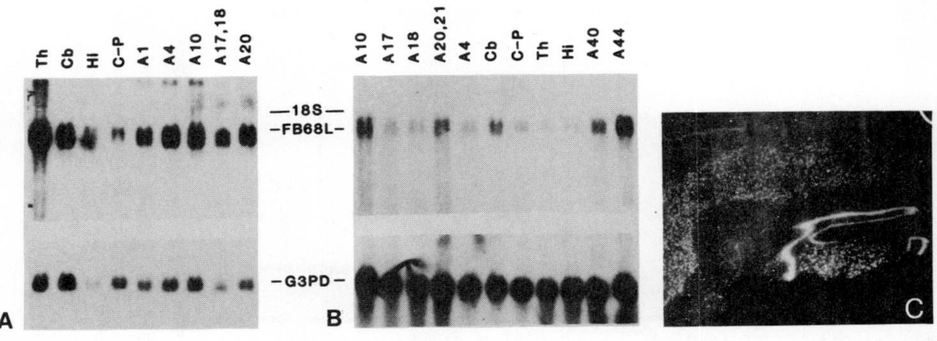

Figure 3. A and B. RNA blot analysis of APP expression in human fetal and adult brain subregions. RNA was isolated, electrophoresed on agarose-formaldehyde gels, transferred to Biotrans membrane, and hybridized with labeled probe as described (Neve et al., 1986). A. Hybridization of the APP cDNA FB68L (Tanzi et al., 1987a) to 19-week fetal brain subregion RNAs. Th, thalamus, Cb, cerebellum; Hi, hippocampus; C-P, caudate-putamen; A1, primary somatosensory area; A4, motor cortex; A10, frontal pole; A17/18, striate and extrastriate cortex; A20, temporal associative cortex. B. Hybridization of FB68L to adult brain subregion RNAs. A10, frontal pole, A17, striate cortex; A18, extrastriate cortex; A20/21, temporal associative cortex; A4, motor cortex; Cb, cerebellar cortex; C-P, caudoputamen; Th, thalamus-VP: nucleus; Hi, hippocampus; A40, posterior perisylvian cortex; A44, anterior perisylvian cortex. Each Northern blot was also hybridized with a cDNA for glyceraldehyde-3-phosphate dehydrogenase as a control (shown below FB68L hybridization). C. In situ hybridization of an APP riboprobe to adult human hippocampus.

from hippocampus, frontal cortex (A10), and inferior temporal cortex (A20) from four different adult brains (Neve et al., 1988). Densitometric analysis revealed that the relative abundance of APP-695 RNA relative to APP-751/770 RNA in each region was 0.6, 1.4 and 1.2, respectively. These data confirm the qualitative impression derived from the Northern blot analyses. The preferential increase in APP-695 mRNA in associative neocortex relative to primary sensory cortex is intriguing in that this regional variation parallels the regional distribution of neurofibrillary tangles found in visual cortices of Alzheimer's disease brains (Lewis et al., 1987). Moreover, earlier studies have postulated a greater number of neuritic plaques and cerebrovascular amyloid deposits in association than in primary sensory areas (Mlandybar, 1975; Morimatsu et al., 1975; Vinters and Gilbert, 1983).

In situ hybridization of the oligonucleotides to sections of adult human brain revealed a quantitative rather than a qualitative difference in their patterns of expression (Neve et al., 1988). Within the cortex, the overall pattern of hybridization for each probe reflected the laminar distribution of pyramidal neurons in the cortex, which varies among regions according to their cytoarchitecture. This pattern of distribution roughly parallels that of neocortical senile plaques (Pearson et al., 1985). App-695 RNA levels were higher than those of APP-751/770, particularly in associative regions of the cortex, in agreement with the results of the RNA blot analyses. These relative levels of APP-695 and APP-751/770 RNAs shifted in subcortical areas: in situ hybridization demonstrated higher levels of APP-751/770 RNA than of APP-695 in the adult caudate-putamen and in the hippocampus (Neve et al., 1988), although both RNAs revealed an identical pattern of distribution in the hippocampus (Fig. 3C): RNA levels were high in the pyramidal cell layer of Ammon's horn, and modest in the densely packed granule cells of the dentate gyrus.

Amyloid Protein Precursor Expression in Down Syndrome and Alzheimer's Disease

RNA blot analysis in which APP oligonucleotides were hybridized to RNAs from brain tissue of individuals with Down syndrome and Alzheimer's disease (Figs. 4 and 5) showed significant differences. A comparison of APP-695 and APP-751/770 expression in normal and Down syndrome fetal brain (Fig. 4A) revealed that, while the APP-695 transcript is expressed normally at higher levels than is APP-751/770 in 19-week fetal brain (as confirmed with in situ hybridization, Fig. 4B), both RNAs are increased several-fold relative to normal in 19-week Down syndrome fetal brain. By densitometric analysis, the average abundance of APP-695 RNA in DS fetal brain compared with normal was 4.6; that of APP-751/770 in Ds relative to normal was 3.8. This result is particularly striking in light of the fact that APP expression is not increased more than 1.5-fold over normal in non-neuronal DS tissues that were examined (Fig. 5A).

Hybridizations of the APP-specific oligonucleotides to RNA from normal and AD cerebellum (Fig. 5B; see Neve et al., 1988), a region relatively spared in Alzheimer's disease, show that both transcripts are present at approximately normal levels in AD cerebellum. However, in AD frontal

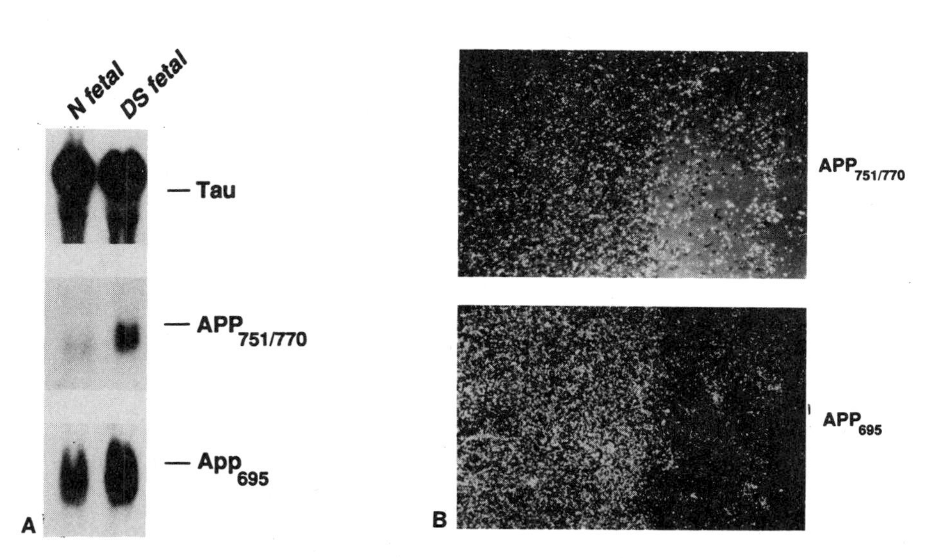

Figure 4. Expression of APP RNAs in normal and Down syndrome fetal brain. A. RNA blot hybridization analysis. B. In situ hybridization of oligonucleotide probes specific for APP-695 and APP-751/770 to human 19- to 20-week temporal cortex. Details of the hybridization are described by Neve et al. (1988).

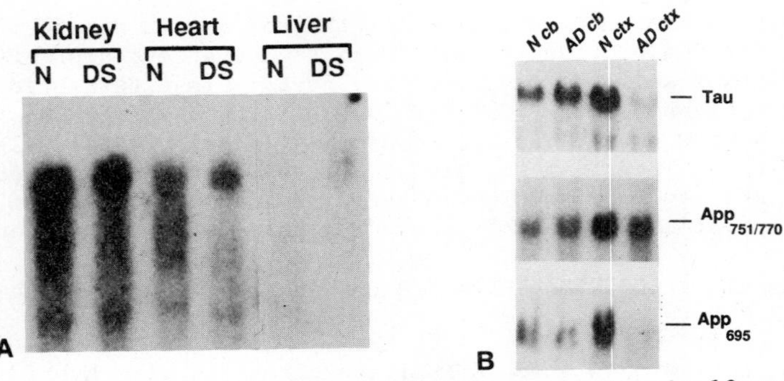

Figure 5. A. RNA blot analysis of APP RNA expression in 19-week DS non-neuronal tissues. Only APP-750/751 is expressed in these tissues. B. RNA blot analysis of APP RNA expression in AD cerebellum and cortex. Hybridization with a cDNA for the neuron-specific microtubule associated protein tau (Neve et al., 1986) is shown for comparison.

cortex the level of APP-751/770 RNA is near normal, while the APP-695 transcript appears to be selectively lost. This loss may be due to death of the neurons that normally express the APP-695 transcript, or to decreased expression of this transcript in affected regions of the AD brain. The level of APP-751/770 could appear relatively unaffected in AD if the cells expressing this transcript are spared, or if the transcript is overexpressed to the extent that, despite loss of cells expressing APP-751/770 RNA, a significant level of the message remains. We again quantified our results using slot blot analysis; the average ratio of APP-695 RNA levels between normal and AD cortex was 3.5; for APP-751/770 it was 1.5.

Surveys of regional APP expression in two aged DS brains (ages 37 and 57 years) have been described (Neve et al., 1988), and reveal differential patterns of relative expression of the two APP RNAs in DS brain compared to normal. Two major shifts in expression were noted: (1) While APP-695 RNA is normally present at relatively high levels in associative neocortex, its expression in these areas is depressed in Down syndrome brains compared to normal. This depression is seen as early as 6 months postnatally (unpublished data), even though APP-695 is clearly overexpressed in DS fetal cortex compared to normal. (2) The normal contrast in levels of expression between APP-695 and APP-751/770 in the hippocampus and in the caudate-putamen becomes more marked in aged Down

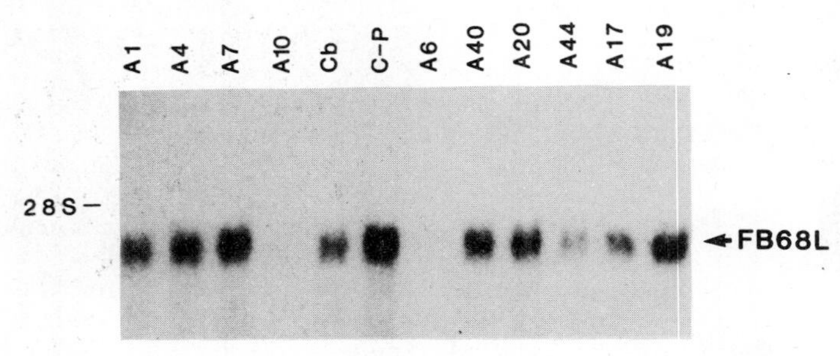

Figure 6. Hybridization of the APP cDNA FB68L (Tanzi et al., 1987a) to RNA from brain regions of a 57-year DS individual. Abbreviations for brain regions are as in Fig. 3, with the addition of A7, parietal cortex.

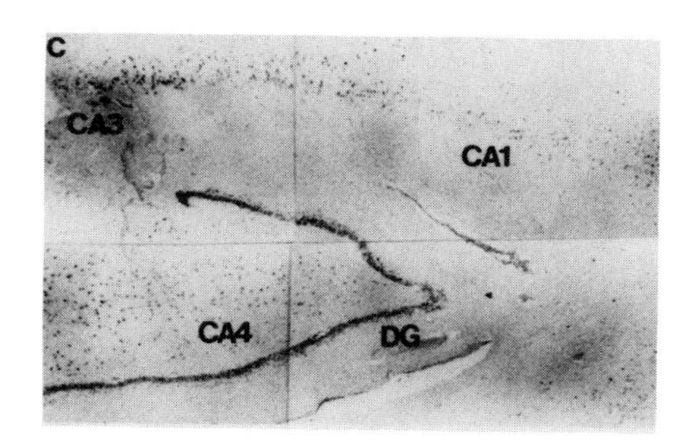

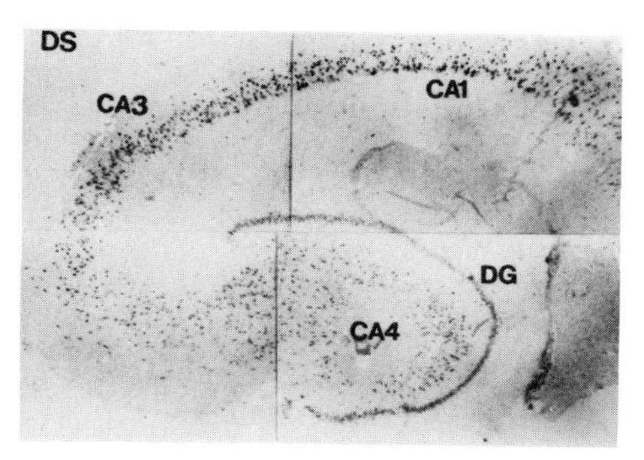

Figure 7. In situ hybridization of a riboprobe synthesized from the APP cDNA FB68L (Tanzi et al., 1987a) to hippocampus of a normal adult (C) and a 57-year DS individual (DS). Details of hybridization are described by Neve et al., (1988).

syndrome brains. This first became evident when we observed heightened APP RNA expression in the caudate-putamen of a 57-year DS individual when we hybridized with APP cDNA (Fig. 6), and is probably due to unusually high levels of APP-751/770 RNA. In situ hybridization of the oligonucleotide probes to normal and Down syndrome adult hippocampus suggests that APP-751/770 mRNA is more abundant than APP-695 mRNA pyramidal cells of Ammon's horn (Neve et al., 1988; Higgins et al., 1988b). The in situ hybridization results shown in Fig. 7, in which normal and aged DS hippocampi were probed with an APP riboprobe that detects both APP RNAs, show that the increased levels of APP mRNA seen in DS hippocampus on Northern blots are partly due to heightened levels of expression specifically in CA1 pyramidal cells of aged DS brains.

This result parallels those we have obtained immunohistochemically, which reveal that in the brains of subjects diagnosed as having AD, the density of anti-APP antibody staining in pyramidal cells of Sommer's sector (CA1) was markedly increased, and was marked by atrophy of many of these cells (Benowitz et al., unpublished). It is significant that these neurons are known to be at risk in AD, and raises the possibility that the dense concentration of APP in these neurons and in other affected areas may account for the accumulation of amyloid polypeptide in neuritic plaques. A shift in the level of one of the APP mRNAs, or of the ratio of the different forms, may play a role in the disease process. Two reports utilizing in situ hybridization with APP probes recognizing all three APP transcripts add additional evidence that differential APP mRNA expression does occur in AD hippocampus (Higgins et al., 1988a) and in AD nucleus basalis (Cohen et al., 1988) compared to normal. Higgins et al. (1988a) demonstrated that whereas neurons of the dentate gyrus and cornu Ammonis (CA) fields contain a 2.5-times-greater APP hybridization signal than neurons of the subiculum and entorhinal cortex in normal brain, the levels of APP RNA in CA3 and parasubiculum are equivalent in AD brain. This suggests that neurons of the subicular complex and entorhinal cortex, which are sensitive to the pathological consequences of AD (Ball, 1978; Burger, 1983; Gibson, 1983; Hyman et al., 1986), may express elevated levels of APP in the disease. Cohen et al. (1988) reported that nucleus basalis neurons from AD patients consistently hybridized with more APP probe than those from controls, but that this increase was not seen in cerebral cortex neurons of AD patients. This increase in APP expression was proposed to be correlated with the greater severity of neuronal degeneration in nucleus basalis than in cerebral cortex of AD brains (Whitehouse et al., 1982). The same laboratory later reported that the increase in APP expression in AD nucleus basalis was due exclusively to the APP-695 transcript (Palmert et al., 1988). Careful in situ hybridization analysis of the level and pattern of expression of the APP transcripts in the associative cortex and subcortical regions of normal and affected brains may shed some light on the participation of the APP gene products in the process of neuronal degeneration in Alzheimer's disease and Down syndrome.

References

Ball MJ. Acta Neoropathol 1978; 42:73-80.

Burger PC. In: Katzman R, ed. Banbury Report 15: Biological Aspects of Alzheimer's Disease. Cold Spring Harbor, NY: Cold Spring Harbor Laboratory, 1983; pp. 37-44.

Cohen ML, Golde TE, Usiak MF, Younkin LH, Younkin SG. Proc Natl Acad Sci USA 1988; 85:1227-1231.

Gibson PH. Neuropathol Appl Neurobiol 1983; 9:379-389.

Glenner G. Arch Pathol Lab Med 1983; 107:281-282.

Glenner GG, Wong CW. Biochem Biophys Res Commun 1984; 120:885-890.

Glenner GG, Wong CW. Biochem Biophys Res Commun 1984; 122:1131-1135.

Goedert M. The EMBO J 1987; 6:3627-3632.

Goldgaber D, Lerman MI, McBride W, Saffiotti U, Gajdusek DC. Science 1987; 235:877-880.

Heston LL, Mastri AR, Anderson VE, White J. Arch Gen Psychiatry 1981; 38:1085.

Higgins GA, Dawes LR, Neve RL. Soc Neurosci Abst 1988; 14:637.

Higgins GA, Lewis DA, Bahmanyar S, Goldgaber D, Gajdusek DC, Young WG, Morrison JH, Wilson MC. Proc Natl Acad Sci USA 1988; 85:1297-1301.

Hyman BT, Van Hoesen GW, Kormer LS, Damasio AR. Ann Neurol 1986; 20:472-479.

Kang J, Lemaire H-G, Unterbeck A, Salbaum MJ, Masters CL, Grzeschik KH, Multhaup G, Beyreuther K, Muller-Hill B. Nature 1987; 325:733-736.

Katzman R, (ed.) Banbury Report 15: Biological Aspects of Alzheimer's Disease (Cold Spring Harbor Laboratory, Cold Spring Harbor, NY, 1983).

Kitaguchi N, Takahashi Y, Tokushima Y, Shiojiri S, Ito H. Nature 1988; 331:530-532.

Lewis DA, Campbell MJ, Terry RD, Morrison JH. J Neurosci 1987; 7:1799.

Masters CL, Simms G, Weinman NA, Multhaup G, McDonald BL, Beyreuther K. Proc Natl Acad Sci USA 1985; 82:4245-4249.

Mlandybar TI. Neurology 1975; 25:120-126.

Morimatsu M et al. J Am Geriatr Soc 1975; 23:390-406.

Neve RL, Kosik KS, Kurnit DM, Donlon TA. Mol Brain Res 1986; 1:271-280.

Neve RL, Finch EA, Dawes LR. Neuron 1988; in press.

Palmert MR, Golde TE, Cohen ML, Kovacs DM, Tanzi RE, Gusella JF, Usiak MF, Younkin LH, Younkin SG. Science 1988; 241:1080-1084.

Pearson RCA, Esiri MM, Hiorns RW, Wilcock GK, Powell TPS. Proc Natl Acad Sci USA 1985; 82:4531-4534.

Podlisney MB, Lee G, Selkoe DJ. Science 1987; 238:669-671.

Ponte P, Gonzalez-DeWhitt P, Schilling J, Miller J, Hsu D, Greenberg B, Davis K, Wallace W, Lieberburg I, Fuller F, Cordell B. Nature 1988; 331:525-527.

Robakis NK, Ramakrinshna N, Wolfe G, Wisniewski HM. Proc Natl Acad Sci USA 1987; 84:4190-4194.

Roch M, Tomlinson BE, Blessed G. Nature (London) 1966; 209:109-110.

St. George-Hyslop PH, Tanzi RE, Polinsky RJ, Haines JL, Nee L, Watkins PC, Myers RH, Feldman RG, Pollen D, Drachman D, Growdon J, Bruni A, Foncin

JF, Salmon D, Frommelt P, Amaducci L, Sorbi S, Piacentini S, Stewart GD, Hobbs WJ, Conneally PM, Gusella JF. Science 1987a; 235:885-890.
St. George-Hyslop PH, Tanzi RE, Polinsky RJ, Neve RL, Pollen D, Drachman D, Growdon J, Cupples LA, Nee L, Myers RH, O'Sullivan D, Watkins PC, Amos JA, Deutsch CK, Bodfish JW, Kinsbourne M, Feldman RG, Bruni A, Amaducci L, Foncin J-F, Gusella JF. Science 1987b; 238:664-666.
Tanzi RE, Gusella JF, Watkins PC, Bruns GAP, St. George-Hyslop P, Van Keuren ML, Patterson D, Pagan S, Kurnit DM, Neve RL. Science 1987a; 329:156-157.
Tanzi RE, St. George-Hyslop PH, Haines JL, Polinsky RJ, Nee L, Foncin JF, Neve RL, McClatchey AI, Conneally PM, Gusella JF. Nature 1987b; 329:156-157.
Tanzi RE, Bird ED, Latt SA, Neve RL. Science 1987c; 238:666-669.
Tanzi RE, McClatchey AI, Lamberti ED, Villa-Komaroff L, Gusella JF, Neve RL. Nature 1988; 331:528-530.
Terry RD, Peck A, DeTeresa R, Schechter R, Horoupian DS. Ann Neurol 1981; 10:184-192.
Van Broeckhoven C, Genthe AM, Vandenberghe A, Horsthemke B, Backhoven H, Raeymaekers P, Van Hul W, Wehnert A, Gheuens J, Cras P, Bruyland M, Martin JJ, Salbaum M. Nature 1987; 329:153-155.
Vinters HV, Gilbert JJ. Stroke 1983; 14:924-928.
Whitehouse PJ, Price DL, Struble RG, Clark AW, Coyle JT, DeLong MR. Science 1982; 215:1237-1239.

ABERRANT PROTEIN KINASE C CASCADES IN ALZHEIMER'S DISEASE

Tsunao Saitoh, Gregory Cole, and Tuan V. Huynh

University of California, San Diego, Department of Neurosciences, M-024 and the Center for Molecular Genetics, La Jolla, CA 92093, U.S.A

INTRODUCTION

Alzheimer's disease (AD) is characterized by the dysfunction and eventual death of selected sets of central nervous system cortical neurons. This deterioration of neurons is responsible for the cognitive impairment of patients. The reason for this neuronal death is not known. In 1981, Appel proposed that the lack of a neurotrophic factor might be responsible for neuronal loss in AD. Two years later, Hefti suggested that NGF may be the neurotrophic factor missing in AD. However, normal levels of NGF mRNA have been found in AD brain (Goedert et al., 1986), indicating that a new hypothesis might be necessary to explain neuronal dysfunction and loss in AD. Indeed, we now know that many classes of neurons that are not responsive to NGF are also lost in AD. Furthermore, there is direct evidence for increased neurotrophic activity in AD brain (Uchida et al., 1988), suggesting that rather than a generalized deficit of trophic factors, a defect exists in the responsive machinery within the target cells.

In addition to providing an explanation for the neuronal loss, proposed mechanisms underlying this cell loss phenomenon should also contribute to an explanation of other AD pathology, for example the formation of neuritic plaques and neurofibrillary tangles.

In this paper, an hypothesis about the biochemical mechanisms underlying neuronal degeneration in AD is proposed which can integrate many aspects of the disease. This hypothesis is based on the following three findings made in our laboratory. First, amyloid ß-protein precursor (ABPP) may be a trophic factor that is abnormally processed in AD, producing amyloid. Second, protein kinase C (PK-C), an inevitable target of trophic factors, is aberrant in AD. Third, several biochemical abnormalities found in AD may be induced by the down-regulation of PK-C. Thus, we postulate that aberrant ABPP regulation (synthesis or processing) is primary in the disease. This causes the amyloid formation on the one hand

Molecular Aspects of Development and Aging of the Nervous System
Edited by J.M. Lauder
Plenum Press, New York, 1990

and the aberrant PK-C on the other. The aberrant PK-C, in turn, triggers many biochemical reactions causing neurofibrillary tangle formation and neuronal death.

PROTEIN KINASE C IS DIMINISHED IN THE BRAIN TISSUE OF ALZHEIMER'S DISEASE PATIENTS

PK-C has been proposed and proved to be a key regulatory enzyme involved in the control of cell physiology (Nishizuka, 1986). The function of many growth factors involves the activation of PK-C. Should the PK-C cascade reactions be deficient, even with a sufficient concentration of growth factors, cells may not survive. In brain tissues of AD patients the concentration of NGF, a potent growth factor for cholinergic neurons, is comparable to that in healthy individuals. As demonstrated by Guroff and his colleagues (Hama et al., 1986), the function of NGF seems to be through the activation of PK-C. Thus, it is possible that in AD patients the PK-C cascade reactions are not properly activated by NGF or other trophic factors, resulting in the death of cholinergic and other neurons.

1. PK-C Activity Is Reduced in Alzheimer Samples. The assay of PK-C activity was performed using histone III S as the substrate, PMA/PS as the activator, and frontal cortex homogenate as the enzyme source. Postmortem time and age of AD cases were matched with those of controls in this study. The PK-C activity on histone III S found in the brain homogenate was markedly reduced in the samples prepared from AD patients (3.4 pmol/mg/min in AD versus 7.2 pmol/mg/min in controls; see details in Cole et al., 1988).

2. PK-C Level Is Lower in Alzheimer's Disease Samples. One way to quantify the number of PK-C molecules is to use a radioactive activator of PK-C such as ^{3}H-phorbol 12, 13 dibutyrate (PDB). Because the interaction is stoichiometric, the quantity of PK-C can be determined by counting the radioactivity bound to the enzyme. The comparison of ^{3}H-PDB binding in AD samples to age- and postmortem time-matched controls showed that ^{3}H-PDB binding activity in the AD brain homogenate was about one-half of that in the control. Thus, we can assume that AD tissue contains fewer PK-C molecules, and therefore, less PK-C activity. We may then ask, are there any particular PK-C substrates which are affected by this reduced PK-C activity?

3. The Phosphorylation of a M_r 86,000 Protein Is Reduced in the Alzheimer's Disease Samples. Autoradiograms of phosphoproteins, labeled *in vitro* with [^{32}P]ATP under a condition where PK-C is activated, showed a reduced phosphorylation of certain proteins in the AD sample compared with the control. Among the reduced phosphoprotein bands were those of 150,000, 120,000, 100,000, 86,000, and 67,000, with the major reduction in the M_r 86,000 band (P86). The comparison of P86 phosphorylation, under PK-C activating conditions, between AD and control samples showed that its reduction in AD patients was highly significant. The specificity of the reduced P86 phosphorylation was then studied comparing AD, Pick's disease, and multi-infarct dementia cases, all exhibiting neuronal degeneration in the frontal cortex. We detected reduced P86 phosphorylation only in the AD cases, indicating its possible involvement in the pathogenesis of AD.

We do not know the molecular mechanisms for the reduced P86 phosphorylation in the AD samples. The reduced P86 phosphorylation mediated by PK-C was observed in the AD cytosol fractions. However, the PK-C activity measured in the AD cytosol fraction, using exogenous histone III S, was not reduced as compared to controls. Furthermore, the direct measurement of PK-C concentration with ^{3}H-PDB did not demonstrate reduced PK-C in the AD cytosol. Also, mixing normal and AD samples gave a degree of P86 phosphorylation neither more nor less than the average of the two, a result which does not support the possibility of inhibitory molecules, such as phosphoprotein phosphatases or kinase inhibitors in the AD sample. Nevertheless, the reduced P86 phosphorylation in the AD samples should be a good measure of altered PK-C-mediated reactions.

PROTEIN KINASE C IS DIMINISHED IN CULTURED FIBROBLASTS DERIVED FROM ALZHEIMER'S DISEASE PATIENTS

Interpretation of the reduced levels of PK-C and the diminished *in vitro* P86 phosphorylation in AD brain tissue is not an easy task. It is possible that this abnormality is primarily involved in the pathogenesis of AD or merely a consequence of the dysfunction or death of neurons. To gain more insight into the nature of aberrant PK-C in AD, it is desirable to study tissues which are spared in the disease. Because of mounting evidence that AD is systemic (Blass and Zemcov, 1984), and because we found the reduced P86 phosphorylation dependent on PK-C in the AD brain, we studied the *in vitro* phosphorylation of fibroblast extracts and found the reduced phosphorylation of a M_r 79,000 protein (P79), a fibroblast homologue of the brain P86 (Saitoh et al., 1988a). Details will be published elsewhere (Huynh et al., 1989). We then asked if this reduced phosphorylation was due to reduced PK-C activity or to reduced substrate. We chose to use an immunological technique to measure the PK-C concentration in the homogenate. To do so, we generated three 11-amino-acid peptides which are unique to PK-C and conjugated these peptides to hemocyanin and raised antibodies against PK-C. Using these antisera, we found reduced levels of PK-C in the AD fibroblast.

SOME BIOCHEMICAL ABNORMALITIES FOUND IN ALZHEIMER'S DISEASE CAN BE INDUCED BY DOWN-REGULATING PROTEIN KINASE C

If reduced PK-C and the reduced P79 phosphorylation mediated by PK-C are found in AD fibroblasts, it is likely that the abnormal PK-C biochemistry is not merely the consequence of neuronal loss in the AD brain. Because PK-C plays an important role in regulating many cellular functions through the phosphorylation of many proteins, it may be asked if the reduced PK-C levels in AD are causing other biochemical abnormalities found in

AD. We approached this question by down-regulating PK-C in a human neuroblastoma cell line, LAN-5, and analyzing two biochemical markers of AD brain, A68 (Wolozin et al., 1986) and P60 phosphorylation (Saitoh and Dobkins, 1986; Saitoh et al., 1988b). Cells were cultured for 36 hr with various concentrations of phorbol ester to down-regulate PK-C (Stabel et al., 1987; Matthies et al., 1987). Cells were harvested, homogenized, and fractionated into particulate and soluble fractions. A68 was analyzed by immunostaining the dot blot with a monoclonal antibody, Alz-50, and P60 phosphorylation was analyzed by autoradiography of SDS–polyacrylamide gels which separates *in vitro* phosphorylated P60. As shown in Fig. 1, the Alz-50 immunoreactivity found in the soluble fraction was induced by the down-regulation of PK-C. And the *in vitro* phosphorylation of proteins showed an enhanced level of P60 phosphorylation in the homogenate prepared from PK-C-depleted cells relative to control homogenate (Fig. 2). Thus, certain AD markers were expressed in cells depleted of PK-C. It should be interesting to study this system so that biochemical abnormalities in AD are classified into two categories; one independent of PK-C down-regulation and the other caused by PK-C down-regulation.

In this context, it may be interesting to consider other instances of neurodegeneration. As discussed above, this PK-C abnormality was relatively specific to AD, in that other chronic neurodegenerative disorders, such as Pick's disease and multi-infarct dementia, did not show aberrant PK-C biochemistry. However, reduced levels of PK-C activity were found in an acute spinal cord ischemia (Kochhar et al., 1987). In this system, *in vitro* phosphorylation levels of a M_r 63,000 protein were found to be elevated. Although more studies need to be done, an attractive working hypothesis is that the M_r 63,000 protein is a homologue of AD P60, the *in vitro* phosphorylation of which is induced by reducing the PK-C levels.

WHAT CAUSES THE REDUCED LEVELS OF PROTEIN KINASE C IN ALZHEIMER'S DISEASE?

As suggested above, it is possible that aberrant PK-C is positioned close to the basic defect in the cascade of reactions which lead to AD pathology. We then must ask which biochemical abnormality precedes and causes the PK-C abnormality in AD pathogenesis. Unfortunately, at present there are too few data available to answer this question. We may speculatively propose, nevertheless, one possible hypothesis based on three observations. First, neuritic plaques are the most specific pathological changes in AD, because the neurofibrillary tangles, the other prominent neuropathological hallmark in AD, are found in many other neurodegenerations (Wisniewski et al., 1979; Love et al., 1988) and because tangles are not found in the cortical neurons of elderly AD patients (Terry et al., 1987). Second, the precursor of the core component of neuritic plaques, ABPP, appears to be a trophic factor as described below. All previously described trophic factors regulate PK-C activity and eventually the concentration of PK-C in cells, thus we postulate this will also

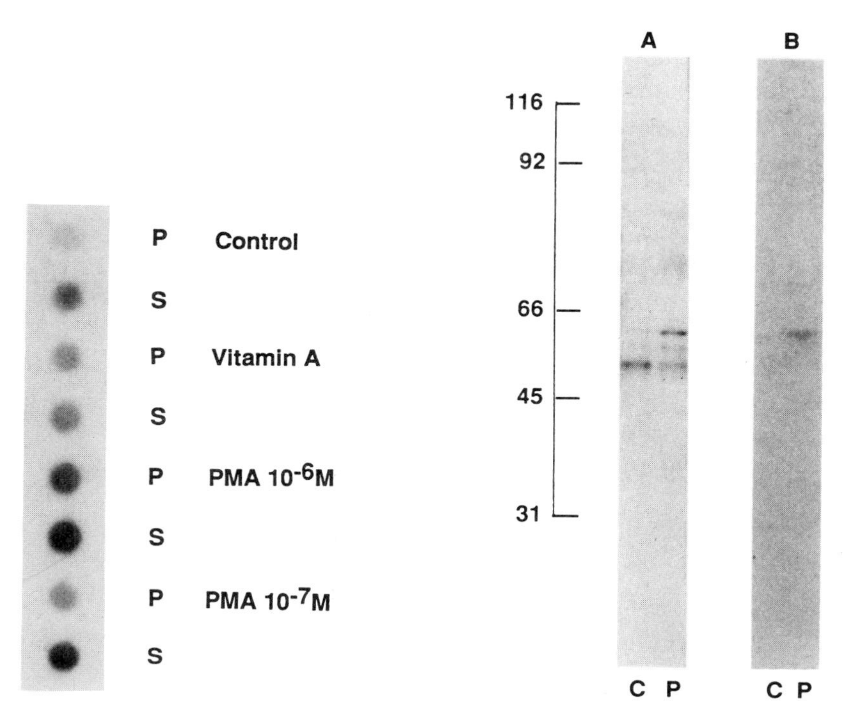

Fig. 1. *Left: Effect of PK-C down-regulation on Alz-50 immunoreactivity of cultured neuroblastoma.* Log phase LAN-5 human neuroblastoma cells in DMEM with 5% fetal calf serum were grown in 75-ml square flasks in triplicate and incubated with 100 nM PMA, 10 nM PMA or 1 part in 100,000 ethanol vehicle, or 10 μM retinoic acid. The cultures were treated with phorbol ester for 36 hr and with retinoic acid for 7 days. Cells were then washed with PBS, scraped, frozen, and thawed in homogenization buffer (Saitoh et al., 1988b) with a protease inhibitor cocktail, sonicated, and fractionated by centrifugation. The protein concentrations of the 100K pellet (P) and supernatant (S) fractions were measured and equal amounts of protein dot blotted to nitrocellulose paper which was then dried, blocked, and incubated with Alz-50 hybridoma supernatant (1/100) in 3% BSA PBS. After washing, the blot was incubated with rabbit anti-mouse IgG serum (1/4000), washed, and incubated further with ^{125}I-protein A. Autoradiographs depicted the increased Alz-50 immunoreactivity in the supernatant (S) fractions caused by treatment with increasing concentrations of PMA. This was not simply an effect of process extension or morphological differentiation because retinoic acid did not increase Alz-50 immunoreactivity, although it caused the differentiation of cells into neuron-like structures.

Fig. 2. *Right: Effect of down-regulation of PK-C on the in vitro phosphorylation of P60 in cultured neuronal cells.* PK-C was down-regulated in LAN-5 human neuroblastoma cells with 100 nM PMA and the cells were harvested, homogenized, fractionated, and the protein concentration determined as described for Fig. 1. Homogenates were then assayed for the 60K phosphoprotein previously found elevated in AD brain essentially as described in Saitoh and Dobkins (1986). Briefly, 6.3 μg of protein in 10 μl was added to 15 μl of reaction mixture to give a final concentration of 50 mM Tris–HCl, 10 mM $MgSO_4$, 100 μg/ml protein kinase A inhibitor, 5 mM 2-mercaptoethanol, 10 μM ATP, 0.2 mM EDTA, 0.5 M sucrose, 4 mM $MnCl_2$, and 1 μCi of [^{32}P]ATP. Reactions were run 12 min at 32°C, stopped with 5X Laemmli sample buffer, and electrophoresed on 6–16% gradient gels. Gels were then fixed, Coomassie stained, dried, and autoradiographed to indicate ^{32}P incorpora-tion into specific bands. Control cell supernatants (C) and supernatants from phorbol ester-treated cultures (P) were compared for 60K ^{32}P incorporation. In A, we see that PK-C down-regulation resulted in elevated 60K ^{32}P incorporation, similar to our finding in AD. We have also found that the 60K band elevated in AD is alkali resistant and therefore probably tyrosine phosphorylated. In B, the gel was fixed and treated 1 hr with 1 M KOH at 55°C prior to Coomassie staining, a procedure known to eliminate serine and threonine phosphate. Again, as shown in B, the alkali-stable 60K ^{32}P incorporation was elevated in PK-C down-regulated cells, reinforcing the similarity with AD brain.

prove true of ABPP. Third, ABPP synthesis and/or processing is aberrant in AD. It has been demonstrated by many investigators that ABPP concentration is altered in AD brain at the mRNA level (Neve et al., 1988; Cohen et al., 1988; Higgens et al., 1988). Furthermore, because AD has an amyloid deposit derived from ABPP, and because physiologically ABPP does not form amyloid, the processing of ABPP should be aberrant.

Among these three arguments, the second point may need further elaboration, as this is the most important and newest concept.

The amyloid protein of the AD neuritic plaque core is derived from a precursor protein. The biological function of this protein is, as yet, unknown. To shed light on the function of this amyloid ß-protein precursor (ABPP), an antisense ABPP cDNA conjugated to an inducible eukaryotic promoter has been constructed, and transfected into fibroblasts. The fibroblasts harboring this construct (pNCA; neuritic core antisense) grew poorly in the presence of 10% serum relative to non-transfected control cells. Normal growth was restored when either medium conditioned with the parent cells or purified ABPP was provided. The capacity of the conditioned medium (CM) to restore cell growth was entirely abolished by passage through an anti-ABPP immunoaffinity column; the activity was recovered in the bound fraction. The CM from parent cells contained a M_r 100,000 protein that was recognized by the antibody against ABPP (anti-GID). As expected, the relative concentration of this protein was reduced in the medium conditioned by pNCA-transfected cells. To demonstrate that this activity was synonymous with ABPP, CM from ABPP cDNA-transfected cells which expressed high levels of ABPP was tested. This CM was 10 times more potent than non-transfected parent cells in promoting the growth of pNCA-transfected cells. These results are consistent with the hypothesis that ABPP is released from cells into the medium and has an autocrine function suggesting a basic role in growth regulation.

Thus, ABPP is a candidate molecule, whose defective metabolism or synthesis may, after a cascade of reactions, produce abnormal PK-C which in turn can amplify this defect to result in the malfunction of many molecules and the devastation seen in the AD brain.

AN HYPOTHESIS FOR NEURONAL DEATH IN ALZHEIMER'S DISEASE

Integrating the above mentioned facts, and explaining neuronal death, neurofibrillary tangle formation, and neuritic plaque formation, we can hypothesize a primary deficiency in the vicinity of ABPP which is linked to the activation of the PK-C cascades and which leads to the neuronal death in the brain tissues of AD patients. This is schematized in Fig. 3. In this hypothesis, ABPP plays a dual role. First, ABPP is processed to the core of the neuritic plaque, which, by itself, is not involved in the sequence leading to neuronal death. Second, being aberrantly processed in the AD brain, ABPP does not appropriately activate its receptor, leading to altered PK-C cascade reactions. This altered PK-C cascade is postulated to be important for the formation of neurofibrillary tangles, which may or may not participate

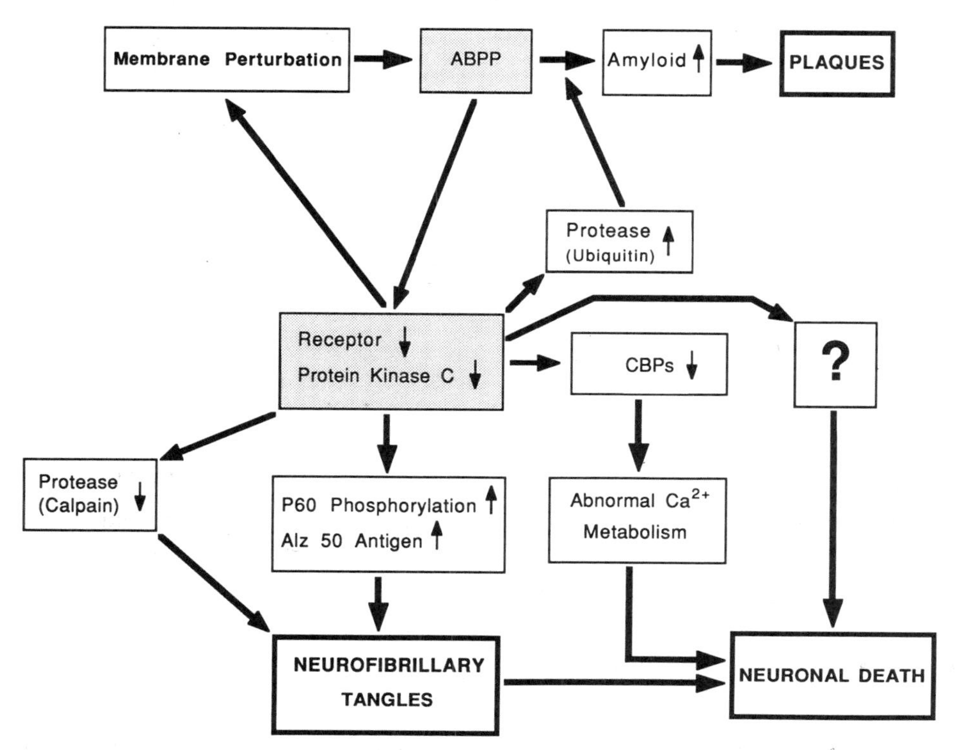

Fig. 3. *An hypothesis for the neuronal death in Alzheimer's disease.* The primary deficit in Alzheimer's disease is postulated to be in the deregulation of ABPP which is linked to protein kinase C (PK-C). The resulting aberrant PK-C reactions induce many abnormal biochemical reactions, including membrane perturbation, increased ATP-dependent proteolysis (which may be involved in the plaque formation), decreased expression of Ca^{2+}-binding proteins (CBPs, which leads to abnormal Ca^{2+} metabolism), increased P60 phosphorylation (which may be involved in tangle formation), and decreased Ca^{2+}-dependent proteolysis (which is involved in tangle formation). Neuronal death is postulated to be brought about by the combination of aberrant gene expression, abnormal Ca^{2+} metabolism, and altered cytoskeletons. The "?" in the figure denotes some proteins with altered expression. They may include IMCAL, calpain II, ubiquitin, A68, P86, GAP-43, and ChAT.

in the process of neuronal death. Altered gene expression caused by aberrant PK-C is hypothesized to be of central importance in the neuronal death.

FIBROBLASTS FROM FAMILIAL ALZHEIMER'S DISEASE PATIENTS EXPRESS LESS AMYLOID ß-PROTEIN PRECURSOR mRNA THAN DO CONTROL FIBROBLASTS

If our hypothesis is correct and ABPP regulates the concentration and activity of PK-C, AD fibroblasts may have altered levels of ABPP, because their PK-C levels are altered. Therefore, the ABPP mRNA of fibroblasts derived from AD patients was compared with that of control fibroblasts (Uéda et al., 1989). Northern blot analysis demonstrated that, in both AD-derived and control samples, the EcoR1 fragment of ABPP cDNA detected a major band at around 4.0 kb and a minor band at around 3.8 kb. This heterogeneous mRNA may reflect the multiple splicing sites in the ABPP gene (Tanzi et al., 1988; Ponte et al., 1988; Kitaguchi et al., 1988). Densitometric quantification of hybridization signals of the ABPP and actin probes in the AD sample revealed that the relative ABPP mRNA expression was reduced in fibroblasts derived from familial AD patients as compared to fibroblasts from controls but was not reduced in cells from sporadic AD patients. Here, an actin probe was used as an internal standard to reduce experimental errors. To exclude the possibility that actin mRNA was excessively expressed in familial AD fibroblasts, we also tested the expression of monoamine oxidase mRNA and compared it to that of ABPP mRNA expression. The reduced ABPP mRNA expression in familial AD fibroblasts was again apparent.

To confirm the result of Northern blot studies that showed reduced ABPP mRNA in AD fibroblasts, the concentration of proteins which reacts with anti-ABPP antibody (anti-GID) was studied in conditioned media with AD and control fibroblasts. Western blot analysis of proteins secreted from fibroblasts showed a M_r 110,000 band reactive to anti-GID antibody in several samples, as well as various lower molecular weight bands in others. These latter may well be degradation products of the M_r 110,000 protein. Quantification of all the anti-GID antibody-reactive protein demonstrated a reduction in AD fibroblasts as compared to control. When cell-associated ABPP was assayed instead of secreted proteins, essentially the same result was obtained (66.6% ABPP found in AD relative to control). This result negates the possibility that the AD cells secrete less ABPP into the culture medium but retain more in a cell-associated form, and reinforces our hypothesis of a primary defect in ABPP metabolism in AD.

CONCLUSION

We have begun investigations to determine the hierarchy of biochemical abnormalities in AD. Our preliminary results led us to postulate an hypothesis in which a PK-C abnormality plays a major role in AD pathogenesis. However, this PK-C abnormality is apparently not the primary deficit in the disease. It is possible that an ABPP abnormality precedes and causes the PK-C abnormality, because ABPP is a growth-regulating protein with altered processing in AD. We have demonstrated abnormal levels of ABPP and PK-C in AD fibroblasts consistent with the idea that they are intrinsic factors in AD and not the secondary effects of neuronal degeneration in the disease. Our working hypothesis and approach should prove helpful in discriminating secondary changes and primary changes in AD pathology and eventually help elucidate the etiology of the disease. We would like to stress, however, that our working hypothesis is merely a starting point which may be substantiated or discarded in the future.

ACKNOWLEDGMENTS

We thank Robert Davignon and Isabelle Hafner for preparing the manuscript. This work has been supported by grants from the McKnight Endowment Fund for Neuroscience, the Pew Charitable Trust, Sandoz Foundation for Gerontological Research, the Alzheimer's Disease and Related Disorders Association, the California State Department of Health Services, and the National Institutes of Health (AG05131 and AG08205).

REFERENCES

Appel, S. H., 1981, A unifying hypothesis for the cause of amyotrophic lateral sclerosis, parkinsonism, and Alzheimer disease, Ann. Neurol., 10:499.

Blass, J. P., Zemcov, A., 1984, Alzheimer's disease: a metabolic systems degeneration? Neurochem. Pathol., 2, 103.

Cohen, M. L., Golde, T. E., Usiak, M. F., Younkin, L. H., Younkin, S. G., 1988, In situ hybridization of nucleus basalis neurons shows increased beta-amyloid mRNA in Alzheimer disease, Proc. Natl. Acad. Sci. U. S. A., 85:1227.

Cole, G., Dobkins, K. R., Hansen, L. A., Terry, R. D., and Saitoh, T., 1988, Decreased levels of protein kinase C in Alzheimer brain, Brain Res., 452:165.

Goedert, M., Fine, A., Hunt, S. P., and Ullrich, A., 1986, Nerve growth factor mRNA in peripheral and central rat tissues and in the human central nervous system: lesion effects in the rat brain and levels in Alzheimer's disease, Mol. Brain Res., 1:85.

Hama, T., Huang, K.-P., and Guroff, G., 1986, Protein kinase C as a component of a nerve growth factor-sensitive phosphorylation system in PC12 cells, Proc. Natl. Acad. Sci. U. S. A., 83:2353.

Hefti, F., 1983, Is Alzheimer disease caused by lack of nerve growth factor? Ann. Neurol., 13:109.

Higgins, G. A., Lewis, D. A., Bahmanyar, S., Goldgaber, D., Gajdusek, D. C., Young, W. G., Morrison, J. H., Wilson, M. C., 1988, Differential regulation of amyloid-beta-protein mRNA expression within hippocampal neuronal subpopulations in Alzheimer disease, Proc. Natl. Acad. Sci. U. S. A., 85:1297.

Huynh, T. V., Cole, G., Katzman, R., Huang, K.-P., and Saitoh, T., 1989, Reduced M_r 79,000 protein phosphorylation and PK-C immunoreactivity in AD fibroblasts, submitted for publication.

Kitaguchi, N., Takahashi, Y., Tokushima, Y., Shiojiri, S., and Ito, H., 1988, Novel precursor of Alzheimer's disease amyloid protein shows protease inhibitory activity, Nature, 331:530.

Kochhar, A., Zivin, J. A., and Saitoh, T., 1987. The effects of ischemia on protein phosphorylation in rabbit spinal cord, Soc. Neurosci. Abstr., 13:1499.

Love, S., Saitoh, T., Quijada, S., Cole, G., and Terry, R. D., 1988, Alz-50, ubiquitin and tau immunoreactivity of neurofibrillary tangles, Pick bodies and Lewy bodies,

J. Neuropathol. Exp. Neurol., 47:393.

Matthies, H. G. J., Palfrey, H. C., Hirning, L. D., and Miller, R. J., 1987, Down regulation of protein kinase C in neuronal cells: effects on neurotransmitter release, J. Neurosci.7:1198.

Neve, R. L., Finch, E. A., and Dawes, L. R., 1988, Expression of the Alzheimer amyloid precursor gene transcripts in the human brain, Neuron, 1:669.

Nishizuka, Y., 1986, Studies and perspectives of protein kinase C, Science, 233:305.

Ponte, P., Gonzalez-DeWhitt, P., Schilling, J., Miller, J., Hsu, D., Greenberg, B., Davis, K., Wallace, W., Lieberburg, I., Fuller, F., and Cordell, B., 1988, A new A4 amyloid mRNA contains a domain homologous to serine proteinase inhibitors, Nature, 331:525.

Saitoh, T., and Dobkins, K. R., 1986, *In vitro* phosphorylation of a M_r 60,000 protein is elevated in Alzheimer brain, Proc. Natl. Acad. Sci. U. S. A., 83:9764.

Saitoh, T., Cole, G., Huynh, T., Katzman, R., and Sundsmo, M., 1988a, Abnormal protein kinase C in Alzheimer fibroblasts, Soc. Neurosci. Abstr., 14:154.

Saitoh, T., Hansen, L. A., Dobkins, K. R., and Terry, R. D., 1988b, Increased M_r 60,000 protein phosphorylation is correlated with neocortical neurofibrillary tangles in Alzheimer's disease, J. Neuropathol. Exp. Neurol., 47:1.

Stabel, S., Rodriguez-Pena, A., Young, A., Rozengurt, E., and Parker, P. J., 1987, Quantitation of protein kinase C by immunoblot — expression in different cell lines and response to phorbol esters, J. Cell. Physiol., 130:111.

Tanzi, R. E., McClatchey, A. I., Lamperti, E. D., Villa-Komaroff, L., Gusella, J. F., and Neve, R. L., 1988, Protease inhibitor domain encoded by an amyloid protein precursor mRNA associated with Alzheimer's disease, Nature, 331:528.

Terry, R. D., Hansen, L. A., DeTeresa, R., Davies, P., Tobias, H., and Katzman, R., 1987, Senile dementia of the Alzheimer type without neocortical neurofibrillary tangles, J. Neuropathol. Exp. Neurol., 46:262.

Uchida, Y., Ihara, Y., and Tomonaga, M., 1988, Alzheimer's disease brain extract stimulates the survival of cerebral cortical neurons from neonatal rats, Biochem. Biophys. Res. Commun. 150:1263.

Uéda, K., Cole, G., Sundsmo, M., Katzman, R., and Saitoh, T., 1989, Decreased adhesiveness of Alzheimer fibroblast: is ß-protein precursor involved? Ann. Neurol., in press.

Wisniewski, K., Jervis, G. A., Moretz, R.C., and Wisniewski, H. M., 1979, Alzheimer neurofibrillary tangles in diseases other than senile and presenile dementia, Ann. Neurol., 5:288.

Wolozin, B. L., Pruchnicki, A., Dickson, D. W., and Davies, P., 1986, A neuronal antigen in the brains of Alzheimer patients, Science, 232:648.

DEVELOPMENTAL EXPRESSION OF AMYLOID PRECURSOR PROTEIN IN NORMAL AND TRISOMY 16 MICE

Shannon Fisher and Mary Lou Oster-Granite

Developmental Genetics Laboratory of the Department of Physiology
Johns Hopkins University School of Medicine
Baltimore, Maryland

INTRODUCTION

Down Syndrome (DS) or Trisomy 21 (Ts21) is the most common genetic cause of mental retardation in human newborns [1]. While some DS individuals manifest cognitive decline as they age, those DS individuals who survive to become adults invariably have the neuropathologic stigmata of Alzheimer's disease (AD) at autopsy [2]. Overexpression of individual genes on human chromosome 21 (HSA 21) may contribute to the pathogenesis of these neuropathologic changes in both DS individuals and in AD patients [3]. To explore this possibility, we study a model system, the trisomy 16 (Ts16) mouse. Mouse chromosome 16 (MMU 16) and HSA 21 exhibit significant genetic homology for a cluster of genes whose overexpression as a result of triplication is thought to contribute to the phenotypic characteristics of DS [4].

One such shared gene, APP (Amyloid Precursor Protein), encodes a cell surface glycoprotein that is composed in part of the A4 peptide, a major component of the cerebrovascular amyloid plaques and extraneuronal senile plaques found in the brains of AD patients and aged DS individuals [5,6,7]. Based on sequencing the A4 peptide, the cDNA for the human amyloid precursor protein (APP) was isolated, and the gene mapped to HSA 21[8,9,10]. Subsequently, it was shown that the message for this gene in human beings exists in at least three distinct forms: the form isolated originally, APP_{695}, and two additional splicing forms of the message, APP_{751} and APP_{770} [11,12,13]. These two latter forms contain an insert with significant homology to the Kunitz family of serine protease inhibitors (KPI). These forms are distributed heterogeneously in various tissues and APP_{695}, the shortest form, is expressed in virtually no tissue in the adult other than the brain [14].

We have cloned a mouse APP cDNA, corresponding to APP_{695}, and have found a 97% homology with human APP_{695} at the amino acid level (Morgan et al., unpublished data). This sequence is virtually identical to that published for mouse APP_{695} by Yamada et al [15]. We have localized App to MMU16, and have used linkage analysis to show that it is located very close to other genes shared with HSA 21[16]. Using in situ hybridization methods, we have demonstrated that APP mRNA is expressed abundantly and heterogeneously in normal fetal and adult tissues, particularly in brain, and that it is overexpressed in the tissues of the head of trisomy 16 (Ts16) mice [17].

To understand the effects that such overexpression has on development, we wanted to determine when APP message is first observed, the levels of its mRNA expression in the pre- and postnatal period, and the temporal regulation of splicing to generate different messages in various tissues of normal and Ts16 mice relative to their normal littermates. We isolated RNA from brain and kidney at different developmental stages in C57BL/6J mice, Ts16 mice, and from their euploid littermate controls. We subjected those RNAs to Northern blot analysis, using probes recognizing all forms of APP, as well as probes designed to distinguish among the alternate splicing forms. To analyze whether alternate splicing forms of APP correspond at the protein level to significant differences in function or processing, we generated polyclonal antibodies to a synthetic peptide taken from the A4 portion of APP (peptide M) and to a synthetic peptide taken from the putative KPI sequence (peptide I). Using these antibodies on Western blots, we compared the levels of APP protein in various adult mouse tissues to the levels of mRNA we detected on northern blots. Specifically, we sought differences in the pattern of proteins detected with the two antibodies that could correspond to a processing event of the full-length precursor.

MATERIALS AND METHODS

Generation of Ts16 Mice

Ts16 conceptuses are among the progeny generated by mating normal C57BL/6J female mice with males doubly heterozygous for Robertsonian translocations with monobranchial homology for MMU 16[18]. The detection of a copulation plug was judged as day 0 of gestation (E0). The percentage of trisomic progeny in individual litters varied at different gestational ages. It was maximal (as much as 40%) at late embryonic and early fetal stages, and declined progressively thereafter. This decline was principally the result of fetal death [19]. Ts16 progeny were distinguished from the euploid littermates on the basis of karyotype analysis of fetal membranes (E10-12) and/or liver (E12-14) before E14 [20] or by phenotypic appearance thereafter [21].

Collection of Tissues

iEmbryos or fetuses were removed from the uterus and dissected, following cervical dislocation and hysterotomy of the pregnant mouse. Only live conceptuses (with active heartbeat and reflex movements) were selected for study. While various tissues were removed for study at gestational stages, at embryonic stages (E10-13) the conceptuses were decapitated and the entire head was used for study. At fetal stages (E14-17), brains were dissected from the cranial vault and used as whole brains for study. Adult C57BL/6J mice were also killed by cervical dislocation before removal of tissues. All tissues for RNA isolations were dissected rapidly and frozen on dry ice, then stored at -70°C. Tissues prepared for membrane isolations were immediately chilled in homogenization buffer at 4°C after their removal.

Northern Blots

RNA isolations were performed by the guanidine thiocyanate technique [21]. Ten to 20 μg total RNA per sample loaded into slots onto a 1.1% agarose-formaldehyde gel, and subjected to electrophoresis. Following staining with ethidium bromide, the gels were photographed, and then the RNA was transferred to nylon filters and fixed by UV cross-linking.

A 1.4 kb EcoR1 fragment from the 3' end of mouse App was subcloned into Bluescript, and anti-sense RNA was transcribed in vitro from the T7 promoter and labeled with α-[^{32}P]-CTP. Dr. Rachael Neve generously provided the clone for the KPI domain of human APP, subcloned into pGEM3. We used this clone to transcribe anti-sense RNA in vitro. A 30 base synthetic oligonucleotide, specific for APP_{695}, was synthesized that consisted of the 15 bases on either side of the insert site of the KPI domain (CTGCTGTCGTGGGAACTCGGACCAC-CTCCT). This probe was end-labeled with T4 polynucleotide kinase and the labeled DNA was separated from the unincorporated nucleotide on a Sepharose G-50 column.

All hybridizations were performed in 5X SSC, 50% formamide, 50 mM sodium phosphate (pH 6.8), 1 mM EDTA, 1% SDS, .05% Ficoll and PVP, at 65°C for the 1.4 kb App probe and at 55°C for the KPI domain probe; the probe was added at 10^6 cpm/μl. At the appropriate hybridization temperatures, the blots were then washed in 1X SSC and 0.5% SDS for 5 minutes, followed by 3 washes for 20 minutes in 0.1X SSC and 0.5% SDS.

The hybridizations with the oligonucleotide probe were performed at 36°C, the calculated T_m-15° for the 30 base duplex, but still above the T_m for the duplexes formed by either half of the probe. Washes were 5 minutes in 1X SSC and 0.1% SDS at 36°C, followed by 20 minutes in the same buffer at room temperature. After washing, all blots were exposed to X-ray film at -70°C for 1-3 days. Between probings, the blots were stripped in 2 mM Tris (pH 8.0), 1 mM EDTA, and 0.1% SDS at 65°C for 45 minutes.

Generation of Antibodies

The peptides M (DAEFGHDSGFEVRHQ) and I (SRWYFDVTEG) were used to generate antibodies M4 and I2. New Zealand white rabbits were immunized by injections at multiple subcutaneous sites with a total of 350 μg of peptide emulsified in Complete Freund's Adjuvant. This immunization was followed by 2 injections of 100 μg peptide in Incomplete Freund's Adjuvant. The rabbits were bled by arteriopuncture. Antibody M4 was purified by passing the serum over an affinity column prepared by linking the M peptide to CNBr-activated Sepharose (Pharmacia). Preabsorption of this antiserum with the M peptide eliminated completely staining on Western blots. The antiserum I2 was used without purification or concentration by either affinity column chromatography or by ammonium sulfate precipitation.

Western Blots

Tissues were homogenized in 10 mM Tris (pH 8.0) containing 1 mM EDTA, and were centrifuged at 500x g for 10 minutes. After the initial pellets were discarded, the supernatants were centrifuged at 14,000x g for 30 minutes to pellet the membrane associated proteins. These membrane pellets were resuspended in homogenization buffer with 0.5% SDS, and protein concentration determined spectrophotometrically by a nomogram using the fluorescence at 260 and 280 nm [22].

RESULTS

Developmental expression of total App mRNA in normal mouse brain increased steadily throughout development when actin was used as a control (data not shown), but fluctuated among individual animals during the increase when 28S RNA was used as the control (Figure 1). When regulation of slicing during the same period was examined, using probes specific for total APP mRNA (Figure 2A) and for the mRNA lacking the KPI domain (Figure 2B), levels of APP_{695} rise during the prenatal period (Figure 2C).

To determine differential distribution of the alternate splicing forms, APP levels were compared by northern blot analysis in kidney and brain at E15 and E17, in both Ts16 and euploid littermate control animals (Figure 3). The same blot was probed sequentially for total APP (Figure 3A), APP containing the KPI domain (Figure 3B), and APP lacking the KPI domain (Figure 3C). Although in adult mouse, APP_{695} is brain specific, as it is in human beings (data not shown), this form is detectable also in fetal mouse kidney.

When scanning densitometry was used to compare the levels of APP mRNA between brains of individual Ts16 mice and euploid littermates at three different gestational ages, the degree of overexpression in the Ts16 mice ranged from 1.3 fold to to 3.2 fold (Figure 4).

Western blot analysis of membrane associated proteins from a variety of adult mouse tissues, performed using the M4 antibody, detected two bands, at 80 and 135 kDal, respectively (Figure 5A). There was, however, poor correspondence between the amount of APP protein and the mRNA levels in various tissues, suggesting that there may be a difference in stability or processing of the protein in different tissues. When the same tissues were analyzed with the I2 antibody, similar levels of protein were detected (Figure 5B). In addition to the bands at 135 and 80 kDal recognized by M4, I2 antibody detected an additional band at 50 kDal in all tissues examined (Figure 5B). Such an additional band may correspond to a processing product of the full-length precursor.

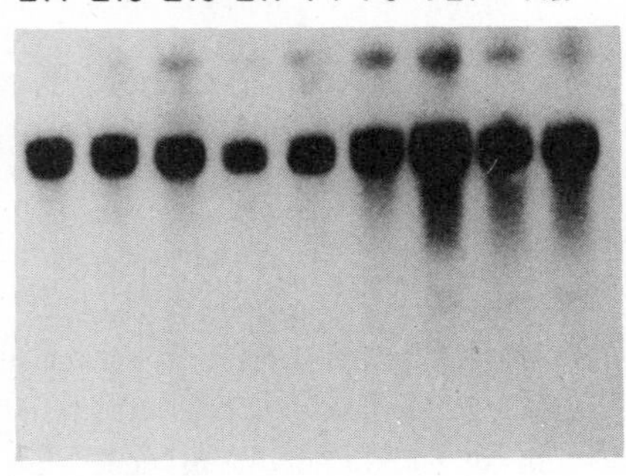

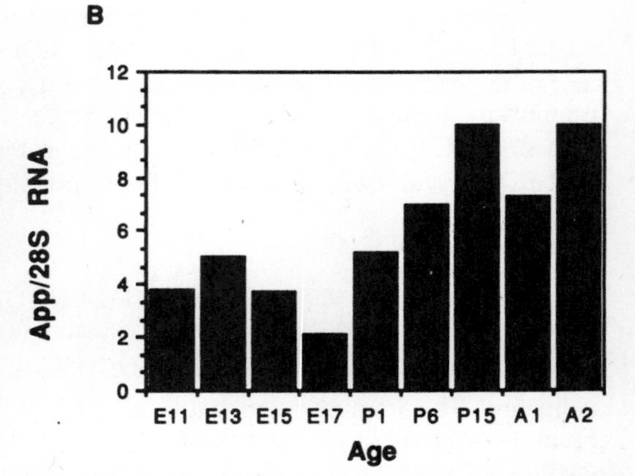

Figure 1. RNA was prepared as described from heads at gestational days 11.5 and 13.5, and from total brains at all other timepoints. The gel was run and blotted as described, with approximately 20 μg total RNA per lane. The samples were E11, E13, E15, E17, postnatal day 1 (P1), P6, P21, and adult, 3.5 months (2 animals). In **A**, the blot was hybridized with App RNA probe; **C** is a graph of the expression of App, normalized to total RNA, at the ages shown. The autoradiograms of the blots and pictures of the ethidium bromide stained gels were analyzed by scanning densitometry (data not shown). The values for App hybridization for individual samples were normalized to total RNA as judged by the 28S ribosomal bands; the data are expressed as a unitless ratio.

DISCUSSION

Multiple splicing forms of APP mRNA occur in mice, just as they do in human beings. In normal mouse brain, we found that total APP mRNA increased throughout pre- and postnatal development. When we examined the levels of alternate splicing forms during prenatal development in the brain, we found that the relative amount of APP_{695} increased steadily. If the alternate splicing corresponds to important functional differences at the protein level, this developmental regulation would lead to alterations in APP function during prenatal development. The fact that APP_{695} is found exclusively in the central nervous system in adults [14], and increases steadily in amount during prenatal development, may indicate that this form is associated with the maturation of neurons in the central nervous system. To test whether this tissue specificity was also present during development, we examined both brain and kidney at different gestational ages for evidence of the expression of the alternate splicing forms. Although we did not detect APP_{695} in adult mouse kidney, we did find appreciable levels in fetal kidney, showing that this form is not associated exclusively with cells in the adult central nervous system. As yet, we have not determined the precise cellular localization and distribution of the various APP messages in either the fetal kidney or brain.

To determine the effect of gene triplication on APP mRNA levels, we compared Ts16 mice with euploid littermates at different gestational ages. At some ages (E14, E17) the increase in total APP expression exceeded the expected 1.5 fold that would be the result of a simple gene dosage effect. While such an exaggerated overexpression has been observed associated with expression of other genes located in the same shared region of these chromosomes [18], such overexpression of APP could have important implications in DS individuals. For example, APP overexpression might tend to unbalance a normal processing pathway of the protein, leading to accumulation of fragments of the precursor that prove to be amyloidogenic or perhaps, even toxic. The degree of APP overexpression could then be an important factor in the development of AD pathology in DS brains.

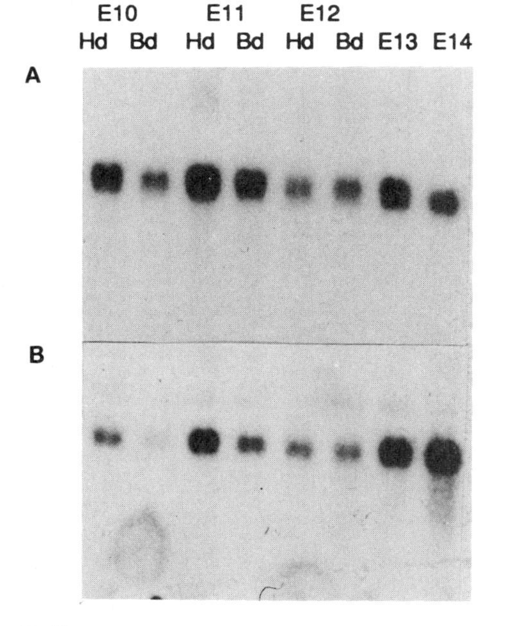

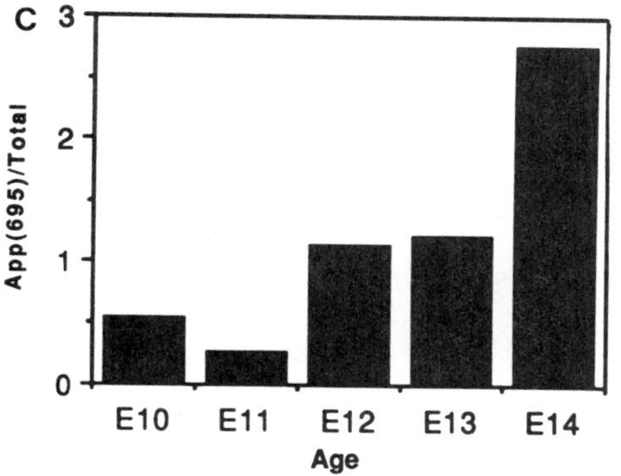

Figure 2. RNA was prepared as described from heads and bodies to E13.5, and from whole brain at E14.5. The gel was run and blotted as described, with approximately 12 μg total RNA per lane. The samples were E10 head, E 10 body, E11 head, E11 body, E12 head, E12 body (-liver), E13 head, and E14 brain. In **A**, the blot was hybridized with App RNA probe, and in **B**, the same blot was stripped and rehybridized with the 30 base oligonucleotide probe. **C** is a graph of the ratio of hybridization intensities with the oligonucleotide probe to those with the RNA probe for total App. The ratios were calculated after analyzing both blots by scanning densitometry (data not shown).

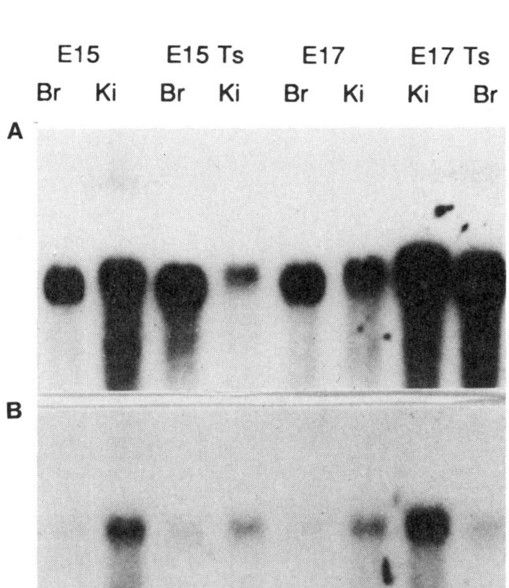

Figure 3. A Northern blot of RNA from brain and kidney of euploid and trisomic mice was probed for total App and specifically for the forms with and without the KPI domain. In each lane 10-15 μg total RNA was loaded, and the samples were brain and kidney, E15; brain and kidney, E15 trisomic; brain and kidney, E17; and kidney and brain, E17 trisomic. In **A**, the blot was hybridized with the 1.4 kb App probe. In **B**, the blot was stripped and reprobed for the KPI domain. In **C**, the blot was reprobed with the oligonucleotide.

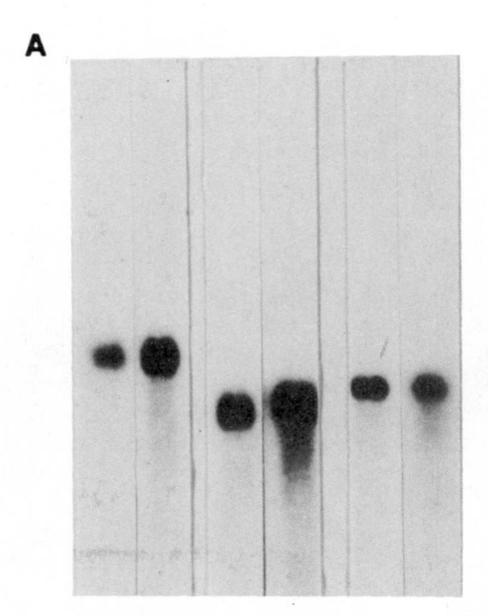

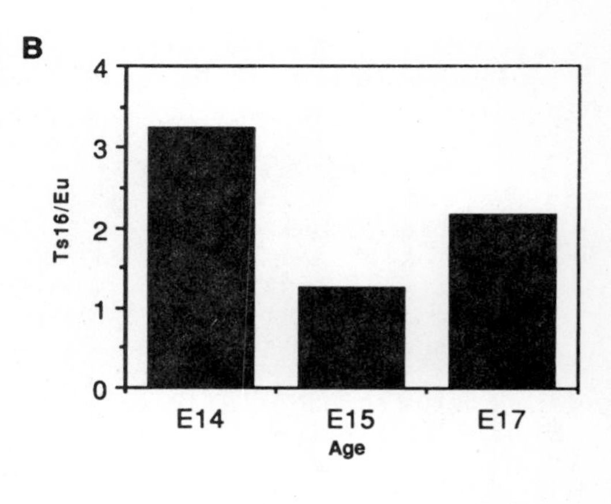

Figure 4. RNA was prepared from the brains of Ts16 mice and normal littermates as described. The gels were run as described, with approximately 12 μg total RNA per lane. The samples were E14 brain, euploid and trisomic, E15 brain, euploid and trisomic, and E17 brain, euploid and trisomic (2 gels). **A** shows the lanes hybridized with the App RNA probe. **C** shows a graph of the App levels in the Ts16 mice relative to normal at the different stages; the E17 bar is an average of the values obtained from the two different gels. The ratios were obtained by first normalizing each value to 28S rRNA as described in Figure 1, then taking the ratio of the normalized values for Ts16 and euploid samples.

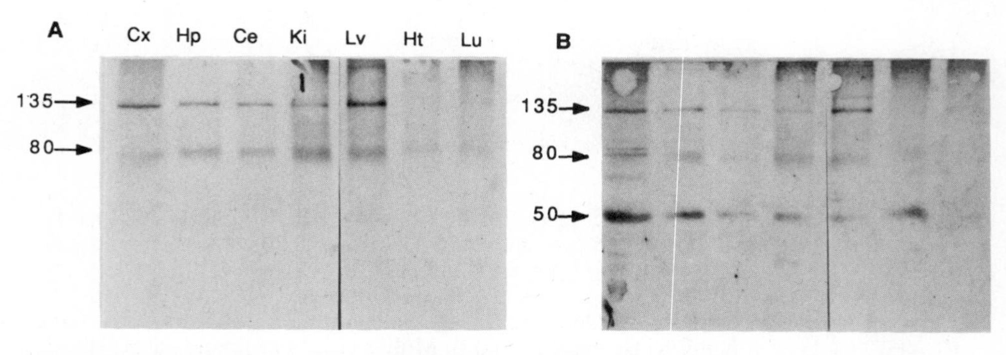

Figure 5. Membrane associated protein was prepared from several mouse tissues and two identical Western blots prepared and run in parallel as described. Ten μg total protein was run in each lane; the samples are cortex, hippocampus, cerebellum, kidney, liver, heart, and lung. In **A**, the blot was reacted with antibody M4, and in **B**, the blot was reacted with serum I2. The major bands and their molecular weights are indicated by the arrows on the left.

Interestingly, we found that antibody M4, which should detect all forms of the protein arising from alternate splicing, detected levels of protein that varied significantly from the total APP mRNA levels seen by northern blot analysis in the same adult tissues (data not shown). Specifically, we found much less protein in the brain than would have been predicted. In contrast, antibody I2, which should have detected only the two forms of the protein containing the KPI domain, detected similar relative levels of protein in the same tissues. Thus, even antibody M4 may be detecting mainly the forms of APP with the KPI domain. One possible explanation for this discrepancy might be that APP_{695}, the predominant form in the adult brain, has a shorter half-life than the other forms of the protein. Therefore, it may exist in a lower steady-state level, producing a lower than predicted level of protein. The detection of an additional band at 50 kDal when antibody I2 was used leads us to suggest that in the normal processing of APP, the epitopes detected by these two antibodies are separated by a proteolytic cleavage.

ACKNOWLEDGEMENTS

The authors acknowledge gratefully the secretarial assistance of Cindy Lapinsky and the tireless efforts of Karen Smith-Connor, who maintains the breeding colonies of the various stocks used in this study and who performs the karyotype analyses expertly. Helpful comments by and useful discussions with B. O'Hara, R. Morgan, R. Reeves, and J. Gearhart were appreciated. This research was funded in part by a grant from the Alzheimer's Disease and Related Diseases Association (PRG 87-043 to MLOG) and by NIH grants PO1 HD 19920 (Joseph T. Coyle, MLOG) and RO1 HD 19932 (MLOG). SF is the recipient of a Medical Scientist Training Program (M.D.-Ph.D.) Fellowship at the Johns Hopkins University School of Medicine.

REFERENCES

1. J. T. Coyle, M. L. Oster-Granite, and J. D. Gearhart. The neurobiologic consequences of Down Syndrome, Brain Res. Bull. 16: 773 (1986).

2. K. E. Wisniewski, H. M. Wisniewski, and G. Y. Wen. Occurrence of neuropathologic changes and dementia of Alzheimer's disease in Down's syndrome, Ann. Neurol. 17: 278 (1985).

3. J. T. Coyle, M. L. Oster-Granite, R. H. Reeves, and J. D. Gearhart. Down syndrome, Alzheimer's disease, and the trisomy 16 mouse. TINS 11: 390 (1988).

4. R. H. Reeves, J. D. Gearhart, and J. W. Littlefield. Genetic basis of a mouse model of Down Syndrome, Brain Res. Bull. 16: 803 (1986).

5. G. G. Glenner and C. W. Wong. Alzheimer's disease: initial report of the purification and characterization of a novel cerebrovascular amyloid protein, Biochem. Biophys. Res. Comm. 120: 885 (1984).

6. G. G. Glenner and C. W. Wong. Alzheimer's disease and Down syndrome: sharing a unique cerebrovascular amyloid fibril protein, Biochem. Biophys. Res. Comm. 122: 1131 (1984).

7. C. L. Masters, G. Simms, N. A. Weinman, G. Multhaup, B. L. McDonald, and K. Beyreuther. Amyloid core protein in Alzheimer's disease and Down syndrome, Proc. Natl. Acad. Sci. USA 82: 4245 (1985).

8. J. Kang, H-G. Lemaire, A. Unterbeck, J. M. Salbaum, C. L. Masters, K-H. Grzeschik, G. Malthaup, K. Beyreuther, and B. Muller-Hill. The precursor of Alzheimer's disease amyloid A4 protein resembles a cell surface receptor, Nature 325: 733 (1987).

9. R. E. Tanzi, J. F. Gusella, P. C. Watkins, G. A. P. Bruns, P. St. George-Hyslop, M. L. van Keuren, D. Patterson, S. Pagan, D. M. Kurnit, and R. L. Neve. Amyloid β protein gene: cDNA, mRNA distribution, and genetic linkage near the Alzheimer locus, Science 235: 880 (1987).

10. D. Goldgaber, M. I. Lerman, O. W. McBride, U. Saffiotti, and D. C. Gajdusek. Characterization and chromosomal localization of a cDNA encoding brain amyloid of Alzheimer's disease, Science 235: 877 (1987).

11. P. Ponte, P. Gonzalez-DeWhite, J. Schilling, J. Miller, D. Hsu, B. Greenberg, K. Davis, W. Wallace, I. Lieberberg, F. Fuller, and B. Cordell. A new A4 amyloid mRNA contains a domain homologous to serine protease inhibitors, Nature 331: 525 (1988).

12. R. E. Tanzi, A. I. McClatchey, E. D. Lamperti, L. Villa-Komaroff, J. F. Gusella, and R. L. Neve. Protease inhibitor domain encoded by an amyloid protein precursor mRNA associated with Alzheimer's disease. Nature 331: 528 (1988).

13. N. Kitaguchi, Y. Takahashi, Y. Tokushima, S. Shiojiri, and H. Ito. Novel precursor of Alzheimer's disease amyloid protein shows protease inhibitor activity, Nature 331: 530 (1988).

14. R. L. Neve, E. A. Finch, and L. R. Dawes. Expression of the Alzheimer amyloid precursor gene transcripts in the human brain, Neuron, in press.

15. T. Yamada, H. Sasaki, H. Furuya, T. Miyata, I. Goto, and Y. Sakaki. Complementary DNA for the mouse homolog of the human amyloid beta protein precursor, Biochem. Biophys. Res. Comm. 149: 665 (1987).

16. R. H. Reeves, N. K. Robakis, M. L. Oster-Granite, H. M. Wisniewski, J. T. Coyle, and J. D. Gearhart. Genetic linkage in the mouse of genes involved in Down syndrome and Alzheimer's disease in man, Molec. Brain Res. 2: 215 (1988).

17. C. Bendotti, G. L. Forloni, R. A. Morgan, B. F. O'Hara, M. L. Oster-Granite, R. H. Reeves, J. D. Gearhart, and J. T. Coyle. Neuroanatomical localization and quantification of amyloid precursor protein mRNA by in situ hybridization in the brains of normal, aneuploid, and lesioned mice. Proc. Natl. Acad. Sci. USA 85: 3628 (1988).

18. J. D. Gearhart, M. T. Davisson, and M. L. Oster-Granite. Autosomal aneuploidy in mice: Generation and developmental consequences. Brain Res. Bull. 16: 789 (1986).

19. A. Gropp. Chromosomal animal model of human disease. Fetal trisomy and developmental failure, in: "Teratology, " C. L. Berry and D. E. Poswillo, eds., Springer-Verlag, New York (1975).

20. J. D. Gearhart, H. S. Singer, T. H. Moran, M. Tiemeyer, M. L. Oster-Granite, and J. T. Coyle. Mouse chimeras composed of trisomy 16 and normal (2N) cells: Preliminary studies, Brain Res. Bull. 16: 815 (1986).

21. S. Miyabara, A. Gropp, and H. Winking. Trisomy 16 in the mouse fetus associated with generalized edema and cardiovascular and urinary tract abnormalities. Teratology 25: 369 (1982).

22. J. W. Chirgwin, A. E. Przybyla, R. J. MacDonald, and W. J. Rutter. Isolation of biologically active ribonucleic acid from sources enriched in ribonuclease, Biochemistry 18: 5294 (1979).

23. V. F. Kalb, Jr. and R. W. Benlohr. A new spectrophotometric assay for protein in cell extracts, Anal. Biochem. 82: 362 (1977).

24. U. K. Laemmli. Cleavage of structural proteins during the assembly of the head of bacteriophage T4, Nature 227: 680 (1970).

25. W. N. Burnette. "Western blotting": Electrophoretic transfer of proteins from sodium dodecyl sulfate - polyacrylamide gels to unmodified nitrocellulose and radiographic detection with antibody and radioiodinated Protein A, Anal. Biochem. 112: 195 (1981).

INDEX